Electronic Constitution:
Social, Cultural, and Political Implications

Francesco Amoretti
University of Salerno, Italy

INFORMATION SCIENCE REFERENCE

Hershey · New York

Director of Editorial Content: Kristin Klinger
Senior Managing Editor: Jamie Snavely
Managing Editor: Jeff Ash
Assistant Managing Editor: Carole Coulson
Typesetter: Carole Coulson
Cover Design: Lisa Tosheff
Printed at: Yurchak Printing Inc.

Published in the United States of America by
      Information Science Reference (an imprint of IGI Global)
      701 E. Chocolate Avenue
      Hershey PA 17033
      Tel: 717-533-8845
      Fax: 717-533-8661
      E-mail: cust@igi-global.com
      Web site: http://www.igi-global.com/reference

and in the United Kingdom by
      Information Science Reference (an imprint of IGI Global)
      3 Henrietta Street
      Covent Garden
      London WC2E 8LU
      Tel: 44 20 7240 0856
      Fax: 44 20 7379 0609
      Web site: http://www.eurospanbookstore.com

Library of Congress Cataloging-in-Publication Data

Electronic constitution : social, cultural and political implications / Francesco Amoretti, editor.
   p. cm.
 Includes bibliographical references and index.
 Summary: "This book provides main political problems about digital information technology in world politics, relating them to the processes of transformation of the current historical system"--Provided by publisher.
  ISBN 978-1-60566-254-1 (hardcover) -- ISBN 978-1-60566-255-8 (ebook)
 1. Internet in public administration. 2. Public administration--Data processing. 3. Information technology--Political aspects. 4. Information society--Political aspects. I. Amoretti, Francesco.
  JF1525.A8E42 2009
  352.3'802854678--dc22
                    2008041544

British Cataloguing in Publication Data
A Cataloguing in Publication record for this book is available from the British Library.

All work contributed to this book is new, previously-unpublished material. The views expressed in this book are those of the authors, but not necessarily of the publisher.

Table of Contents

Section III
Themes and Issues

Detailed Table of Contents

Section I
The *Longue durée*

In "Electronic Constitution: A Braudelian Perspective", Francesco Amoretti presents a model for the analysis of time and space structures of digital networks based on the braudelian triad of times: structure, conjuncture and event. Taking the incipit by a short description of this tri-partition, he proposes a historical method to frame the products of innovation processes such as the Internet in their wider socio-economical context. He thus argues that the world wide Web (WWW), with its enormous quantity of easy-accessible and easy-produced information, is just the evenementielle of a historical process that could represent a new conjuncture, with its own dynamics and phases based on the structure of capitalist world system. The thesis at the base of this work is that Internet, rather than being a revolutionary technology that will subvert the current organization of social and economic production, is a technological instruments that gives to institutions and organizations a way to re-organize their assets and processes in order to start a new conjuncture of capitalistic structures. Most of the authors and scholars debating the transformative power of the Internet have up to day focused their attention on the WWW as the locus of a democratising and participative movement that take the technology in service of civil society. With this chapter Amoretti shed light on the character of continuity that the Internet has regarding such traditional categories of political economy as hierarchy and institutional enforcement.

This chapter offers a long-term perspective on citizenship, questioning one of the basic assumptions of most of the literature on this topic, that is, the nation-state as unit of analysis. Through the adoption of a world-systemic perspective, two basic aspects of the history of citizenship stand out. Firstly, the fundamentally exclusive nature of this category, as it emerged and developed over the history of the modern world-system, since at least the "long 16th Century". And, secondly, that well before the so-called "information revolution" of the last decades, "technology" has shaped the Western social imagination, acting, in various and changing historical forms, as an effective instrument of control and supremacy, producing asymmetric and inegalitarian effects, and providing a yardstick of the different "levels of development" of Western and non-Western peoples. In this view, the most recent phase of the history of citizenship, his e-form, seems to replicate, in new ways, the explanations of the gap existing both between and within countrie—now conceptualized as "digital divide"—and, at the same time, the illusory universalistic promise of an expansion of the citizenship and the rights associated to it.

Section II
The *Conjuncture*: The Geopolitics of Technological Innovations

Chapter III
Oreste Ventrone, University of Naples Federico II, Italy

Following the diffusion of e-government in the high income countries, international organizations, notably UN, OECD, World Bank, have promoted the implementation of e-government practices in developing countries. However, the few researches conducted in the field show that the overwhelming majority of e-government projects end up in total or partial failure. Despite the recognition of the need to take into account local specificities and to get the locals involved in the process, e-government in developing countries still appears essentially as a mere transfer operated by donor countries' firms with western technologies. Moreover, as these technologies are mostly proprietary, they prevent institutions and users from developing countries to modify and adapt the tools to their particular needs and lock them in a position of permanent technological dependency. The causality chain between e-government, good governance, and democracy, if at all plausible, looking at history should be probably read the other way around. In fact, some scholars consider the contribution of e-government to overall development irrelevant, if not negative, in the measure in which it diverts funds from higher priorities.

Chapter IV
Fortunato Musella, University of Naples Federico II, Italy

The chapter is dedicated at analyzing the strategic use of new technologies in the United States. An evident synergy has been noted between the digital policy projects and the neo-liberal ideology wave that has traced origin in the fiscal crisis of the State in the 1970s. About four decades have transformed some political directions in true imperatives: public sector downsizing, cost-cutting in public agencies, decision-making privatization, and the principle of efficiency as a measure of collective action. If new public management has been imposed as a dominant paradigm for administrative restructuring, ICTs programs sustain reform objectives by putting emphasis on the sure advantages of technological applications. In addition to this, administrative reforms seem to be in continuity with some American historical tradition, in reasserting a central role of private actor in public activities and realizing a significant "fusion of political and economic power". Digital era seems to have added a new chapter to the American corporate liberalism history, with the difference – and the aggravating circumstance – that private organizations have now more powerful instruments to control and regulate society. New technological instruments seem to be used essentially to produce a neo-liberal interpretation of government activities.

Chapter V

Francesco Amoretti, University of Salerno, Italy
Fortunato Musella, University of Naples Federico II, Italy

The challenge of convergence has become a core issue in the European agenda, as the existence of widely accepted administrative standards represents one of the most important preconditions to promote socio-political development and to reinforce the single Market. Indeed many initiatives have been launched by European institutions to ensure uniformity in terms of administrative action and structures, and several communications by the European Commission have considered the impact of new technologies in creating systems of integrated and interoperable administration in the Old Continent. In this chapter it will be investigated the role of communication and information technologies in the formation of an European administrative space, the process for which administrations become more similar and close to a common European model. The contribution will consider ICTs as a key element of Europe's economic competitiveness agenda as well as the interconnection between e-government programs and the social dimension of development. In addition to this, in the final part of the chapter it will be also analyzed the nature and implications of the process of uniformity produced by the new digital infrastructures, a peculiar mix of attractiveness and imposition.

Chapter VI

Clementina Casula, University of Cagliari, Italy

The rhetoric used worldwide by policymakers in promoting the uptake of Information and Communication Technologies (ICT) emphasizes the advantages deriving for all citizens from the advent of the Information Society (IS). Among the democratic features of the IS particularly praised are despatialisation processes, leading to a sort of "death of distance" mainly benefitting the inhabitants of territories traditionally located in peripheral and backward areas, as well as the enlarged global market. However, research shows that the uptake of ICT varies territorially, mainly following wealth distribution, among other variables. This consideration would corroborate the view of those reading the rhetoric over IS as a facade covering the restructuring of capitalist economy at the global level and arguing that the uptake of ICT, based on an unequal model of development, further strengthens rather than reduces the territorial and socio-economic divides between centres and peripheries. The chapter confronts those two readings of the main rationale behind policymaking for the development of an IS by looking at the case of the European Union (EU). The argument is that, although global economic competition in the ICT sector seems to be the mainspring that led the EU to promote policies for the IS, social concerns are emerging as the flagship of the policy, increasingly tuned with other policies within a wider European developmental strategy, which may start up a new field on which to compete for global leadership.

Chapter VII

Many of African States are focusing on ICTs and developing e-government infrastructures in order to fasten and improve their "formalisation strategy". This philosophy drives the South African State in its impressive efforts to deploy an efficient and pervasive e-government architecture for its citizens to enjoy accurate public services and for this young democracy to be "useful" to them. By focusing on the South African case, people will be able to understand the role of ICTs as tools to register, formalise and normalise, supporting the final objective of Weberian rationalisation. The author will consider the historical process of this strategy, across different political regimes (from Apartheid to democracy). He will see how it is deployed within a young democracy, aiming at producing a balance between two poles: a formal existence of citizens for them to enjoy a "delivery democracy" in which they are to be transparent; an informal existence of citizens for them to live freely in their private and intimate sphere. In this tension, South Africa, given its history, is paradigmatic and can shed light on many other countries, beyond Africa.

Chapter VIII

Several African countries have begun to introduce and implement Information and Communication Technology (ICT) policies. In the context of such developing countries, it is important to assess the nature of research focus on the ongoing ICT revolution and its potential to stimulate institutionalization of democracy in Africa. This chapter reviews and integrates literature by scholars focusing on ICT in Africa in general and more specifically on Ghana. The authors incorporate several key points in their discussion. First, they provide a summary of ICT trends and policies in Ghana and their emphasis on helping to institutionalize democracy and its related free market system. Next, they provide a description of some of the major challenges to institutionalizing democracy that scholars writing about ICT in Ghana have identified. In addition, the authors discuss several opportunities for enhancing democracy that scholars writing about ICT in Ghana have highlighted. Finally, they make a few general recommendations for mitigating the potential problems and enhancing the opportunities of the ICT revolution for Ghana as well as the entire African continent.

Chapter IX

Chin-fu Hung, National Cheng Kung University, Taiwan

China has vigorously implemented ICTs to foster ongoing informatization accompanying industrialization as a crucial pillar to drive its future economic development. The institutional and legal reforms involved were initiated and put into practice in order to meet the increasing demand for technological convergence and the negotiations for the expected entry into the World Trade Organization (WTO). The Chinese government has nevertheless long been torn by the ambivalence brought about by the Internet. It regards the Internet as an engine to drive economic growth on the one hand, and as a subversive challenge to undermine the ruling Communist Party on the other hand. As soon as ICTs were introduced and Web sites mushroomed, the Party was so determined to harness the new medium to assure the Internet's economic and scientific benefits. As a consequence, controls other than stifling ICTs would be critical for the CCP's agenda to achieve the century-long modernization process and in the meantime, consolidate its power.

Section III
Themes and Issues

Chapter X

Claudia Padovani, University of Padova, Italy
Elena Pavan, University of Trento, Italy

Political processes are undergoing profound changes due to the challenges imposed by globalization processes to the legitimacy of policy actors and to the effectiveness of policy-making. Building on a socio-political approach to governance and focusing on global information policies and networks, this chapter aims at developing a better understanding of the possibility of change in world politics nowadays, by critically analysing two innovative elements: the reality and relevance of "multi-stakeholder" practices and the growing role of information technologies as a complementary support to actors' relations. Looking at Internet Governance debates in recent years, the authors reconstruct networks of interaction connecting actors in the virtual space, and look at actors' communication modes. Thus they analyze the extent to which technological, as well as processual and cognitive innovation, shapes actors' orientations and the structures within which they interact in the specific context of Internet Governance.

Chapter XI

Mauro Santaniello, University of Salerno, Italy

The Internet Governance debate has, for a long time, been influenced by a well-defined characterization of information networks. The depiction of a decentralized network, governed on a consensual basis by distributed forms of authority, has therefore focused, as a consequence, little attention on the coding and configuration strategies of the network architecture that is implemented in a conflictual scenario by a set of parties whose interests are rarely explicated in the internet governance debate or in institutional plans and policies inspired by it. It follows that some important structures of network government are not publicly recognized as constitutive places where processes of economic, political and social shaping on technology application occur. On the contrary this chapter will be dedicated to the analysis on those geo-strategic issues relating to international flows of data and to remote control activities deployed by a small group of software houses and hardware manufacturers.

Starting from the assumption that any technology embeds the ideology, politics and culture of the society where it was created, this chapter reconstructs the specific historical and political link between the affirmation of neo-liberal paradigm, which has occurred since the 1970s in Western industrialized capitalist countries, and the dissemination of ICT. More in particular, it analyses the problem of measurement of ICT, emerged functionally to the need to identify new tools to legitimize the hierarchy of development, giving some countries the label of "most advanced" and the others of "developing" or "underdeveloped". Indeed, the measurement, acting as a scientific justification for the Western superiority, is a part of those structures of knowledge which constitute an essential element in the functioning and legitimacy of the political, economic and social structures of the existing world-system. This contribution reconstructs this methods of knowledge deployed first at the international level, within and through the work of those actors who have taken the leading role in defining the interpretative lines of the measurement of ICT: the OECD, ITU, the World Bank.

This chapter is dedicated to analyse the fabrication of networked socialities, that is to address the complex interweaving of technologies of information and communication and the manifold instantiations of sociality. Networked socialities are digital formations being produced out of the intertwining of social logics outside and inside digital spaces and society. Such contribution is organized as follows: first, it will present the theoretical frame necessary to grasp the fabrication of sociologies in our information age, drawing on some concepts elaborated by the social studies of science and technology, together with the studies of the global digital worlds. Then, it will highlight the analytical fruitfulness of this perspective by describing some digital formations, such as social network sites, virtual communities of practice, and electronic markets. Finally, it will discuss the effects and the implications of such fabrication as a re-configuration of social, the emerging post-social relationships as well as the increasing fragility of knowledge societies.

Virtual worlds are computer environments in which large numbers of human beings may interact, do useful work for each others, and build enduring social connections. For example, in World of Warcraft an estimated nine million subscribers form short-term action-oriented groups and long-term guilds, employing a variety of software tools to manage division of labor, spatial distributions, activity planning, individual reputations, and channels of communication, to accomplish a variety of often complex goals. A broader system of essentially permanent allegiances, comparable to current national governments and major corporations, frames the volatile forming and dissolving of small and medium-sized cooperative groups. New social technologies have a clear potential to supplement and render more flexible the existing structures of government, but they may also represent a significantly new departure in human social organization. The chapter will describe the diversity of information technology tools used to support social cooperation in virtual worlds, and then explain how they could be adapted to mediate in new ways between government and its citizens.

Preface

Although it is clear that the spread of the digital networks is increasingly important in world politics, there is little evidence of exactly what implications the Internet has had. Some researchers analyze the growing role of Internet in promoting freedom and changing social and political norms. Others emphasize the role of Internet in the context of globalization, and its destabilizing effects.

In order to overcome the main limits of the literature on new technologies—descriptivism and normativism—the volume adopts a perspective able to provide a meaningful framework for several issues related to the social, cultural, and political meanings and implications of the ICTs. The goal of this book is to provide recognition and a reinterpretation of the so-called "digital revolution" relating it to the processes of transformation of the current historical system. Such "digital revolution" is, in fact, a key aspect, perhaps the most important, of the contemporary systemic socio-political change. Why? And what does it mean? I think that a book developing that interpretative key, and articulating it on multiple levels of analysis, could allow to substantially advance this field of study from a theoretical and methodological point of view.

This book is divided in three sections and includes 15 chapters covering many of the important topics which are contributing to frame the historical reality of cybernetic networks within the wider contemporary systemic structure.

The first one reflects on the theoretical perspective, and, as exemple, upon a long term trend. There is no doubt; in fact, that the transformations generated by the ICTs repurpose, mutatis mutandis, something that already followed technological innovations in other phases in terms of: cultural representation, metaphors and symbols production; redefinition of borders between public authority and private subjects; concrete attribution of rights and power. Highlighting these aspects is intended to make clearer the overall framework of the other contributions.

The second section of the book, the most detailed one, analyzes the changes of the last two to three decades—the conjoncture—with a particular attention to the geopolitics of the technological innovations. This section highlights how technological innovation has been and is a strategy of reorganization of political-institutional systems, and how it goes along with specific forms of knowledge production and specific ideologies of social legitimization.

The third and final section examines some themes and issues following the re-organization of political-institutional systems and of socio-cultural practices via digital networks, from Internet governance debate and policies - a more equalitarian, institutional mechanism, or a new formula to hide inequalities? - to the tools of measurement and evaluation of the organizational practices; from ICTs (che) can be seen as destructive, reproductive as well as constitutive of forms of sociality, to "virtual worlds", computer-generated environments in which large numbers of human beings may interact, do useful work for each other, and build enduring social connections.

Let us consider now the structure and the contributions of this volume.

In the first section, the **_Longue durée,_** it will be collected chapters which analyze issues concerning the new technologies - and their socio-political implications in the contemporary world - from an historical perspective. This permits to capture relevant constitutive and transformative processes.

In _Electronic Constitution: A Braudelian Perspective_, Francesco Amoretti presents a model for the analysis of time and space structures of digital networks based on the braudelian triad of times: structure, conjuncture and event. Taking the incipit by a short description of this tri-partition, he proposes a historical method to frame the products of innovation processes such as the Internet in their wider socio-economical context.

Thus he argues that the World Wide Web (WWW), with its enormous quantity of easy-accessible and easy-produced information, is just the _evenementielle_ of a historical process that could represent a new conjuncture, with its own dynamics and phases based on the structure of capitalist world-system. The thesis at the base of this work is that Internet, rather than being a revolutionary technology that will subvert the current organization of social and economic production is a technological instrument that gives to institutions and organizations a way to re-organize their assets and processes in order to start a new conjuncture of capitalistic structures.

Most of the authors and scholars debating the transformative power of the Internet have up to day focused their attention on the WWW as the locus of a democratising and participative movement that take the technology in service of civil society. With this chapter, Amoretti shed light on the character of continuity that the Internet has regarding such traditional categories of political economy as private property, hierarchy and institutional enforcement.

Such approach conducts us to the second chapter, dedicated to one of the most important themes of the debate on digital networks: their democratic and egalitarian potentialities.

In _Old and New Rights: E-Citizenship in Historical Perspective_, Mauro Di Meglio and Enrico Gargiulo aim to offer a view on the issue of citizenship, and e-citizenship in particular, adopting a long-term perspective and questioning the basic assumptions of most of the literature on this topic, that is, its unit of analysis; in fact, the misunderstanding of the role played by citizenship derives from the attitude to analyse it mainly, if not exclusively, from a nation-state perspective. On the contrary, their analytical premise is that a different, and more satisfactory, understanding of the historical vicissitudes of citizenship requires the adoption of a world-systemic perspective, able to take into account the array of economic, political, and social long-term and large scale processes which, since at least the XVIth century, have shaped them.

From this perspective, two basic aspects of the history of citizenship clearly emerge. First, the fundamentally exclusive nature of this category, as it has emerged and developed in the history of the modern world-system. And, second, the fact that, well before the so-called "information revolution" supposedly generated by the introduction of the information and communication technologies (ICT's), technology – broadly defined as the application of "advanced" scientific knowledge to practical purposes in a particular field, and given a specific level of economic and socio-cultural development – has shaped the Western social imagination, acting, in historically different and changing forms, as an effective instrument of control and supremacy; producing asymmetric and inegalitarian effects; and providing a yardstick of the different "level of development" of the European and not-European peoples.

Taking for granted this basic asymmetry in the mastery and exercise of knowledge, the image of the European citizen has been constantly modeled on the basis of what has been considered his key features, and strengthened in his certainties of supremacy – both technological and material – over the non-citizen by the evidence of the latter's inferiority and "underdevelopment", simultaneously generating the inclusion of some and the exclusion of others. In this view, the most recent phase of the history of citizenship, his _e-_ form, seems to replicate, in new ways, both the explanations of the gap existing both between and within countries—now conceptualized as "digital divide"—and the illusory universalistic promise of an expansion of the citizenship and the rights associated to it.

In the second section, *The Conjuncture: The Geopolitics of Technological Innovations*, it is offered a group of contributions regarding policies and implementation strategies of e-government and e-democracy worldwide. The time here is that of *The Conjuncture*, the time of capitalistic world-system's reorganization. The spreading of digital technologies is one of the most relevant pillar, or better the most relevant, of the processes of world geo-political relationships. This consideration appears clear in the analysis of the role of International Organizations. The contribution of Oreste Ventrone, *International Organizations, E-Government and Development*, refers to such aspect.

Following the diffusion of e-government in the technologically advanced high income countries, international organizations, notably UN, OECD, World Bank, have promoted the implementation of e-government practices in developing countries as priority means to further good governance, democracy and development. ICTs can, indeed, be very useful in order to make public administration more transparent, accountable and participated or in connecting and networking faraway places, especially where transports and physical infrastructures are lacking or insufficient.

However, a decade later, the few researches conducted in the field show that the overwhelming majority of e-government projects in developing countries end up in total or partial failure. The best practices identified in high income countries prove difficult to be reproduced in settings that are very different in terms of organizational traditions and level of development. Despite the recognition of the need to take into account local specificities and to get the locals involved in the process, e-government in developing countries still appears essentially as a mere transfer operated by donor countries' firms with Western technologies. Moreover, as these technologies are mostly proprietary, they prevent institutions and users from developing countries to modify and adapt the tools to their particular needs and lock them in a position of permanent technological dependency

For what concerns the cost efficiency of e-government projects, it must be said that the advantages observed in developed countries, where the cost of hardware and software is more than compensated for by the savings in terms of costly human labor, are less likely to be obtained in developing countries, where the cost of labor is a small fraction of that in developing countries and the cost of ICTs is proportionally much higher.

The claim that e-government diffusion in developing countries can improve the prospects for democracy is based on a normative assumption supported by little evidence. The causality chain between e-government, good governance, and democracy (with the frequent addition of economic growth as a further stage), if at all plausible, looking at history should be probably read the other way around. In fact, some scholars consider the contribution of e-government to overall development irrelevant, if not negative, in the measure in which it diverts funds from higher priorities.

The relevance of International Organizations in defining digital policies, Action Plans, and the ideological paradigm of digital technologies revolution, certainly refers to underdeveloped countries. More complex and debated is the relation between such institutions and the more developed areas. Yet the diffusion of the digital networks is becoming a flywheel for the world-system transformation. To the United States, whose experience represents an emblematic case, it is dedicated the chapter authored by Fortunato Musella, *American Electronic Constitution: Reinventing Government and Neo-Liberal Corporatism*.

This contribution has been aimed at analyzing the strategic use of new technologies and its representations by an important world power. An evident synergy has been noted between the digital policy projects and the neo-liberal ideology wave that has traced origin in the fiscal crisis of the State in the seventies.

Thus, often presented as an occasion for reinventing national government, ICTs can be interpreted as the latest chapter of a longer-term process of reform. About four decades have transformed some political directions in true imperatives: public sector downsizing, cost-cutting in public agencies, decision-making privatization, and the principle of efficiency as a measure of collective action. If new public management has been imposed as a dominant paradigm for administrative restructuring, ICTs programs sustain reform objectives by putting emphasis on the sure advantages of technological applications.

In this country, Npm trends seem to be in continuity with some American historical tradition, in reasserting a central role of private actor in public activities and denying the dualism state-economic enterprises. Following a rich research tradition developed in the 1960s and 1970s (Kolko, 1963; Miller, 1976), Musella underlines that the political actions of federal government have been essential to the operation of the American business system, since the beginning of last century. According to the author, In the United States it has been realized a silent constitution realizing with the significant "fusion of political and economic power" so that corporations and other large scale organizations have become far more important components than the State. Digital era seems to have added a new chapter to the American corporate liberalism history, with the difference—and the aggravating circumstance—that private organizations have now more powerful instruments to control and regulate society. As remembered, besides obtaining a new role in policy making, private organization are able to intervene on the complex architecture defining the Internet rules as a sort of private law – a scenario that poses, again, the question of the limits between private interest and public functions.

Viewing to the international scene, the structure of global ICTs regime assures a quasi-monopolistic position to US private firms, while less rich states seem dependent to the power of software-hardware providers. Referring to the words of the Human Development Report, according to which «the Internet was created in the United States, but its cost slashing consequences for information and communication enhance people's opportunities everywhere» (United Nations, 2001: 95), Musella argues that the digital imperatives are still far from hiding perils of quasi-monopolistic hegemony, even if the most recent developments do not exclude that future trends could leave more space for other nations such as Europe or China.

In *The European Administrative Space and E-Government Policies: Between Integration and Competition* Francesco Amoretti and Fortunato Musella focus attention on the meaning of e-government policies in the European context.

E-government development policies represent one of the most important stages for the europeanization of national public administrations and for the creation of a "European administrative space". By providing standardization, ICTs turned out to be a crucial lever toward a greater integration within the European administrative structures and the computer-based network became a mirror—and a promise—for a new administrative set-up. In this way technology seems to constitute an essential element for the construction of the European entity, offering a premise for «cooperation mechanisms between Member States administrations, relevant national and European Union initiatives, standardisation and market initiatives, as well as research activities» (European Commission 2003a: p. 14).

A common element of the "European e-Government platform" is the attention to the development of an administrative framework favorable to business, especially through the reduction of the administrative costs, that is, the costs that the corporate sector must make in order to comply with the information obligations resulting from Government-imposed legislation and regulations. For this reason administrative reforms are included as a key element of Europe's competitiveness agenda, as they may provide user-centred services and cutting red tape (i.e. unnecessary administrative burdens), requiring that information is shared across departments and different level of government. Although the correlation between digitization of public services and a more competitive economy remains complex and elusive, wider benefits have been recognized in the introduction of new technologies.

Another important element of the European strategy indicates the interconnection between e-government initiatives and the social dimension of development. On the base of official reports, the authors underline that such result is fulfilled only through a policy convergence and a willingness to adapt regulatory frameworks in order to facilitate the mobility of citizens and businesses. By looking at some initiatives aiming at fostering such strategic objectives, it is shown the scope for the creation of Web portals designed as a single entry point for businesses, which enabled the interaction between financial actors and institutions regardless their position at local, national or national level. Although the failure of the treaty approval, it is not difficult to perceive that e-government represent a pillar of the European economic and administrative constitution, due to its contribution to the policies for efficiency as well and for social cohesion.

The chapter by Clementina Casula, *The EU and the Information Society: from E-Knowledge to E-Inclusion, in Search of Global Leadership*, considers such point, by considering the rhetoric used worldwide by policymakers in promoting the uptake of Information and Communication Technologies (ICT) to emphasize the advantages deriving for all citizens from the advent of the Information Society (IS). Access and, increasingly, uses ICT to become acknowledged as part of citizenship rights to be granted in the new society, which is said to offer unique opportunities for democratic regeneration, besides increasing competition and economic growth. Among the democratic features of the IS particularly praised are despatialisation processes, leading to a sort of 'death of distance' mainly benefitting the inhabitants of territories traditionally located in peripheral and backward areas, as well as the enlarged global market. However, research shows that the uptake of ICT varies territorially, mainly following wealth distribution, among other variables. This consideration would corroborate the view of those reading the rhetoric over IS as a facade covering the restructuring of capitalist economy at the global level and arguing that the uptake of ICT, based on an unequal model of development, further strengthens rather than reduces the territorial and socio-economic divides between centres and peripheries. The chapter confronts those two readings of the main rationale behind policymaking for the development of an IS by looking at the case of the European Union (EU). The argument is that, although global economic competition in the ICT sector seems to be the mainspring that led the EU to promote policies for the IS, social concerns are emerging as the flagship of the policy, increasingly tuned with other policies within a wider European developmental strategy, which may start up a new field on which to compete for global leadership.

As the European experience and the American one demonstrate, ICTs policies and Information-Knowledge Society initiatives have to be included in the dynamics of capitalistic world-system transformation. Yet Europe differs from the USA, as it seems to fulfill objectives such as social cohesion policies and initiatives for strengthening democracy, in order to reduce territorial and socio-economic cleavages and the note deficit of political legitimacy.

If the final results are still to be defined, the EU "vocation" to leadership is quite week. The American role of global player is only contented by China. The contribution by Chin-fu Hung, *The Politics of the Governing the Information and Communications Technologies in East Asian Authoritarian States: Case Study of China*, takes together economic and political aspects.

To date, Internet access has been expanding rapidly and extensively chiefly due to direct support and promotion by the government. China has vigorously implemented ICTs to foster ongoing informatization accompanying industrialization as a crucial pillar to drive its future economic development.

The institutional and legal reforms involved were initiated and put into practice in order to meet the increasing demand for technological convergence and the negotiations for the expected entry into the World Trade Organization (WTO). Above all, it implies that the authorities in Beijing intended to restore administrative control over the telecommunications sector from previous stages of devolution, which had resulted in fragmented governance and intensified pluralization in terms of efficient flow of information among several telecommunications service providers.

The Chinese government has nevertheless long been torn by the ambivalence brought about by the Internet. It regards the Internet as an engine to drive economic growth on the one hand, and as a subversive challenge to undermine the ruling Communist Party on the other hand. As soon as ICTs were introduced and Web sites mushroomed, the Party was so determined to harness the new medium to assure the Internet's economic and scientific benefits. As a consequence, controls other than stifling ICTs would be critical for the CCP's agenda to achieve the century-long modernization process and in the meantime, consolidate its power.

Such experiences refer to the contraposition between different uses and interpretations of technological innovation on the global scale. Participation *versus* control, democracy *versus* authoritarianism, represents different paths of the current historical conjucture. In the following cases, both regarding the African Continent, such ambivalence are even stronger. Whereas e-government and e-democracy in Western nations is a tool for resolving the perceived crisis of liberal democracies, in the developing countries they are a tool to *build* democracy, and administrative state.

The contribution by Joseph Ofori-Dankwa and Connie Ofori-Dankwa, *ICT Challenges and Opportunities for Institutionalizing Democracy in Ghana: An Integrative Review of the Literature*, is a rich description of theoretical and empirical studies regarding the potential challenges and opportunities associated with implementing ICT initiatives in developing economies.

Starting from the consideration that only recently research has focused on the ongoing ICT revolution and its potential to stimulate the institutionalization of democracy in Africa, the chapter focuses more specifically on Ghana—one of the first countries in Africa to begin to develop and implement a broad national ICT strategy. The authors incorporate several key points in their discussion. First, they provide a brief description of the global ICT revolution and its potential implications for enhancing the democratization process in countries, followed by a summary of ICT trends and policies in Ghana and their emphasis on helping to institutionalize democracy and its related free market system, and a description of some of the major challenges to institutionalizing democracy that scholars writing about ICT in Ghana have identified. In addition, they discuss several opportunities for enhancing democracy that scholars have pinpointed. Finally, they make several general recommendations for mitigating potential problems that may arise, and enhancing the opportunities of the ICT revolution for Ghana, as well as the entire African continent.

Such perils and opportunities are strongly tied to the processes of democratic institutionalization via ICTs in developing countries. However, such processes are also connected with the creation of an administrative state. Nicolas Pejout, *World Wide Weber: Formalise, Normalise, Rationalise: E-Government for Welfare State – Perspectives from South Africa*, offers an interesting interpretation of these dynamics, focusing attention on the relationships between formal and informal aspects of political action. The author argues that numerous governments, particularly those of developing countries, have to deal with challenging economic, socio-economic and political *realities*. More challenging is to deal with *unrealities*, i.e. realities that do exist but that governments can't manage because they don't know about these. These realities are real but informal: the typical example is moonlight work. They all do exist but have no official, formal, legal-administrative and statistical existence. They are "parallel" to the official-formal world of public action and stay "underground", in the shadow of public policies.

This problem of "informality" is particularly encountered by governments in developing countries. They face tremendous problems in terms of public action upon realities that they don't know of, that they can't know of, due to a lack of measuring resources and public management capacities. Various examples are: the absence of a satisfactory statistical machinery, the ineffectiveness of a formal civil status (for instance, the registry of birth), the inefficiency of tax rolls…

For governments to act upon realities, they need to know them and therefore to reveal and measure them. In other words, they need to formalise them so as to be able to control them. Governments have to normalise human activities, that is to put them into norms, into measurable and controllable frameworks. This explains, for instance, the importance of statistical machineries into the construction of nation-states.

Nowadays, governments can use an extremely powerful set of tools to formalise and normalise realities, in order to rationalise their knowledge and therefore their action upon these: information and communication technologies (ICTs). The deployment of electronic government can support a strategy of formalisation and of normalisation which aims at making a society (groups and individuals) highly visible – some might say transparent – to the power in place.

By producing formatted knowledge for the State, this ICT-based formalisation is supporting a move towards genuine rationalisation: technologies enable an extreme degree of accurateness and sophistication (data mining) so that everything and everyone can be labeled, measured, compartmentalised.

Following Foucault's analysis (1997), Pejout argues that such power of knowledge, based on the knowledge of power, can threaten democracy: full transparency of individuals to the State is impossible, due to the absolute necessity of protecting the private sphere. However, the development of the welfare State requires the administration to know most of personal data, so as to provide relevant services, for instance well-measured pensions or health care. This is all the more true when the welfare State is getting ICT-intensive, making the most of e-government to provide e-services. For such provision with efficiency and cost-recovery, the State needs to be scientific, somehow omniscient. That is why transparency of the society to the State is necessary, but to a certain extent beyond which democracy is at risk.

Most of governments in African countries are confronted with informal realities, particularly in hard socio-economic contexts. They don't have enough resources – financial, human, … – to know of realities that they nevertheless need to tackle with. That is why many of African States are focusing on ICTs and developing e-government infrastructures in order to fasten and improve their "formalisation strategy": by getting to know their society better, they can act upon it better.

According to Pejout, this philosophy drives the South African State in its impressive efforts to deploy efficient and pervasive e-government architecture, for its citizens to enjoy accurate public services and for this young democracy to be "useful" to them. By focusing on the South African case, the author underlines that the role of ICTs as tools to register, formalise and normalise, supporting the final objective of Weberian rationalisation. By considering the historical process of this strategy, across different political regimes (from Apartheid to democracy), he analyses how it is deployed within a young democracy, aiming at producing a balance between two poles: a formal existence of citizens for them to enjoy a "delivery democracy" in which they are to be transparent; an informal existence of citizens for them to live freely in their private and intimate sphere. In this tension, South Africa, given its history, is paradigmatic and can shed light on many other countries, beyond Africa. Such reflections on the African Continent, certainly the more penalized in the power recourses distribution, close the second part of the volume, dedicated to geo-political strategies and dynamics on the global scale (in the current world-system). The central idea of such section is that the ICTs have to be analysed in the reconfiguration of the world-system. This process, encouraged by the International Organizations action, id based in a neoliberal ideology (or paradigm), which brings together all Western experiences (EU, USA), and those countries that depend from Western countries (Africa). Yet Chinese case shows the limits of a neoliberal paradigm that have led e-government and e-democracy policies. Indeed the growth of Chinese power the authoritarian feature of Chinese regime weakens the equation on which neoliberalism is based: more economic growth-more democratic development.

It exists also a further level of analysis, besides the *Longue durée* and the *Conjuncture*. As demonstrates the increasing amount of research conducted on the political, cultural, and social implications of the ongoing digital revolution (transformation), such level collects relevant themes and issues featured by a more contracted "time". The contributions presented in the Third Part are aimed at focusing on some themes not sufficiently discussed in the literature on digital technologies and cyberspace.

The first two chapters of this section concern the theme the *Internet Governance*, an object of political and academic confrontation from more than ten years.

Building on a socio-political approach to governance and focusing on global information policies and networks, the chapter by Claudia Padovani and Elena Pavan, *Information Networks, Internet Governance and Innovation in World Politics*, aims at developing a understanding of the possibility of change in world politics nowadays, by critically analysing two innovative elements: the reality and relevance of "multi-stakeholder" practices and the growing role of information technologies as a complementary support to actors' relations. Looking at Internet Governance debates, they reconstruct networks of interaction connecting actors in the virtual space, and they look at actors' communication modes. Thus they analyze the extent to which technological, as well as processual and cognitive innovation, shapes actors' orientations and the structures within which they interact in the specific context of Internet Governance.

Considering the rich literature on the concept of "multi-stakeholderism", they emphasize how such term has almost become a *passé-partout*, widely adopted in political discourses, often with the implicit assumption that a consensus exists on how participatory political processes should be organized and managed. According to the authors, it is growingly evident that stakeholders' participation risks becoming a rhetoric exercise aimed at neutralising criticism through the adoption of an unproblematic consensual understanding of political life. Moreover it is crucial to take into consideration the objective constraints and necessary preconditions to full and effective participation, such as financial and knowledge resources, or the available power base on which actors define their positions in governance processes. To better articulate the multi-stakeholder notion, they suggest relating multi-stakeholderism to the very concept of diversity, to be conceived as a matter of actors involved, issues addressed, knowledge produced and, in the end, power relations.

On the base of this conceptual reformulation, they look at how organizational actors involved in Internet Governance (IG) debates at the World Summit on the Information Society (WSIS) translate their awareness of the dynamic potential offered by ICTs into an intentional strengthening of networking relations aimed at fostering new configurations of power. Thanks to the use of a research software which analyses the Web-sphere, this contribution underlines the correlation between three areas - what kind of actors are involved in the web-based "conversation" about IG, what are the prevailing issues in the IG debate, and, finally, what is the actors' capacity to (re)present and express their differences in the debate, from a geographical, linguistic and cultural point of view. If the interplay between information technology and the conduct of world affairs offers the possibility of innovation in world politics, the case of contemporary Internet Governance produces dubious results. More particularly, the most controversial aspect remains one of inclusion and exclusion. The Global South, and in particular its localities, with their languages and cultural ways of expressing different concerns and needs, have not yet found adequate space in the on-going conversations in the Web-sphere.

In spite of the expectation that actors engaged in Internet Governance, the analysis of how actors involved in Internet Governance conceive and make use of technologies does not allow a very optimistic conclusion in terms of world politics innovation through communication, at least not for the time being.

In *Who Governs Cyberspace? Internet Governance and Power Structures in Digital Networks*, Mauro Santaniello proposes a critical approach to Internet Governance. Refusing the common image of a politically neutral network administered through technical consensus and shared responsibilities, he focuses analysis on those geo-strategic issues relating to international flows of data and to remote control activities deployed by a small group of software houses and hardware manufacturers.

The role of regulatory algorithms in controlling information circulating in cyberspace is thus observed and explained according to two main dimensions: the elaboration levels upon which algorithms work, and the functional areas of cyberspace where machines are established. By following this interpretative grid, it is presented an analysis of the main control centers operating nowadays in such networks as Internet.

Particularly, power centralization trends operating on personal computers and devices alike, as well as on the Internet infrastructure and on the so-called hosting servers, are described in their historical deployment, shedding light on the political consequences of some important processes currently re-engineering digital networks' architectures. This kind analysis provide an interpretative framework to keep together some of the most controversial issues of digitalization, such as the "appliancization" of terminals, the decline of network neutrality, and the information accumulation at computing centers whose resources in terms of processing power, bandwidth and storage capacity are pushing for a monopolistic situation.

This chapter also provides an insight of relationships between information code producers and the legal code produced by territorial authorities. As the most of coding authorities are U.S. companies, in fact, the geopolitical location of a government and its international relations can lead each country to adopt a different set of cybernetic strategies: from the articulated and complex ones that are followed by China, India and other "emerging countries", to the limited and simplistic one shown for example by EU countries.

Presenting cyberspace as a conflictual scenario where companies and governments compete in order to gain control upon a wider and wider part of networks, this chapter re-contextualize the so-called digital revolution in the historical processes of capitalist world-system re-organization.

In the contribution by Diego Giannone, *Measuring ICT: Political and Methodological Aspects,* the attention is focused on the production of knowledge as expression of capitalistic system geo-culture.

Starting from the assumption that any technology embeds the ideology, politics and culture of the society where it was created and that any technical fix to its measurement represents a "political" solution, behind which they operate the more general mechanisms of reproduction of existing hegemonic powers, the author reconstructs the specific historical and political link between the affirmation of neoliberal paradigm, which has occurred since the 1970s in Western industrialized capitalist countries, and the dissemination of ICT. Neoliberalism has played a decisive role not only for the rapid dissemination of ICT, but also for their legitimacy as a criterion for measuring the progress of society. Indeed, the trajectories of neoliberalism have intersected perfectly with the incentive to produce new technological infrastructure (software and hardware), since the latter were: a) a new area of prolific development of the capitalist economy; b) an effective solution for the decrease in production costs and the acceleration both of economic transactions and of financialisation of the economy; c) an appropriate solution to the imperative of statehood more streamlined and less expensive; d) an ideological tool to reaffirm on a global scale the superiority of some countries than others. Within this process, the problem of measurement of ICT has emerged functionally to the need to identify new tools to legitimize the hierarchy of development, giving some countries the label of "most advanced" and the others of "developing" or "underdeveloped". The need to obtain data, information, and sound knowledge on the state of ICT was therefore certainly a strong motivation for the development of methods for measurement, but it is clear that the framework within which it was included transformed it into a primarily political problem and project. Taking the lesson of Wallerstein, and in the wake of the considerations already made by Gramsci, the author argues that the measurement, acting as a scientific justification for the Western superiority, is a part of those structures of knowledge which constitute an essential element in the functioning and legitimacy of the political, economic and social structures of the existing world-system. The measurement is a method of knowledge whose reform is indispensable in the battle for the realization of a hegemonic apparatus. Giannone reconstructs this process of reform of the methods of knowledge deployed first at the international level, within and through the work of those actors who have taken the leading role in defining the interpretative lines of the measurement of ICT: the OECD, ITU, the World Bank. These institutions, mostly controlled by Western countries, have worked out guidelines, selected indicators, built models of measurement that only apparently respond to the logic of a "universal universalism", instead they configure as a specific expression of a "Western universalism" and hegemony.

In *The Fabrication of Networked Socialities,* Paolo Ladri, following the approach presented by Latham and Sassen (1995a), criticizes the analyses of digital worlds dominated by a focus on technical properties, and on technology and society as if they were two separate worlds.

In order to offer a more encompassing view of the electronic constitution of society, the chapter indeed adopts a perspective which looks at the mutual *constitution of technology and society*, arguing for the appropriateness of the analytical categories of the social studies of science and technologies (in particular, from the 'actor-network' theory) in addressing the imbrications of technology and society. He argues that technologies can be seen as destructive, reproductive as well as constitutive of forms of sociality, not relying on the essence, substance, or intrinsic logic of technology but on the situated fabrication of technology *and* society. In this sense, Landri tries to expand the analysis of the forms of sociality given by Latham and Sassen, by encompassing the dystopian effects of technologies, such as the destruction of sociality inherited by the sociology of industrial society, or the post-sociality forms of post-modern reflections on the re-shaping of knowledge societies.

After highlighting the analytical fruitfulness of this perspective by describing some digital formations, such as social network sites, virtual communities of practice, and electronic markets, he discusses the effects and the implications of such networked socialities, looking at three issues.

Firstly, the fabrication of networked socialities represents an experiment in the *reconfiguration of the social.* It is a sort of laboratory for the making and the remaking of the social through digitations. Here, the new technology of information and communication does not simply reflect upon, but tries to constitute and partly stabilize forms of sociality derivative or transformative of the society. This reconfiguration is not virtual, in the sense of being potential; it has its specific materialization, its electronic space and the respective socio-technical infrastructures.

Secondly, the concept of post-social relationship seems to be able to grasp some characteristics of this emergent sociality. It implies an engagement bringing the object-centred social relationship to the forefront. Most critical theories focus on the negative, and dystopian, effects of the recent transformations. Yet, it fails to recognize the transformative and stabilizing effects of these changes. In order to address this aspect, it should probably refine the way of conceiving sociality, usually understood in reference to humans with human relationships, by taking into account the relevance of the non-human side (objects, artefacts, tools, technologies) in the social fabric (Latour, 2005). This post-social perspective helps to visualize how the modern emancipation of selves from previous social belongings (communities, social classes) has been accompanied by an increasing *objectualization of social life.*

Finally, the analyses of these forms, and the reflection on the objectualization of sociality, introduce the theme of the risks resulting from the electronic constitution of society. This issue can be addressed from different angles: in a sense, the traceability of the sociality can multiply the possibility of growing *surveillance and control* (Lyon, 2001); on the other hand, this objectualization could also reveal the *fragility of modern societies* (Stehr, 2001). The uneven and often contradictory character of digital technologies, their variability as shaped by diverse operational logics of social and cultural forms, is also the analytical focus of the final chapter.

In *Virtual Nations* William Sims Bainbridge analyses "virtual worlds", computer-generated environments in which large numbers of human beings may interact, do useful work for each other, and build enduring social connections. By giving attention to some examples of contemporary virtual worlds such as *Second Life* (SL) and *World of Warcraft* (WoW), he considers potentialities of virtual worlds in offering models of future computer-organized virtual groups, and in enhancing government operations and popular involvement in public decision-making. For example, in World of Warcraft an estimated nine million subscribers form short-term action-oriented groups and long-term guilds, employing a variety of software tools to manage division of labor, spatial distributions, activity planning, individual reputations, and channels of communication, to accomplish a variety of often complex goals. Developed for online virtual worlds, these social technologies have a clear potential to supplement and render more flexible the existing structures of government, and they may represent a significantly new departure in human social organization. They could be also adapted to mediate in new ways between government and its citizens.

Yet this may lead to the dark side of virtual world. Digital instruments could also be used in order to give governments' greater control over their citizens. Bainbridge reminds the presidential candidate Ron Paul's words, interviewed on the influential television program, *Meet the Press*, late in 2007, when he expressed concerns felt by many Americans that their nation was decaying into some form of imperialism or fascism: "We're not moving toward Hitler-type fascism, but we're moving toward a softer fascism: Loss of civil liberties, corporations running the show, big government in bed with big business". As users become more accustomed to the technical and social characteristics of virtual worlds, the worlds themselves will evolve still further, posing new challenges and opportunities for users. Yet the use of such new technologies by traditional governments represents an important element to be investigated in future research.

Acknowledgment

The idea behind this book goes back a long way. Emerging during the course of my contribution to the *Encyclopedia of Digital Government*, "Digital International Governance", it started to take shape during several workshops on electronic democracy and e-government in national and international scientific Congresses in particular, the ECPR Joint Session in Pisa, the SISP in Cagliari and Catania, Italy and the IPSA in Fukuoka, Japan. I wish to thank all the participants who with their observations, helped me to focus more clearly on the analytic perspective.

Above all, I wish to thank the individual authors of this book for their contribution and encouragement during the process. I also wish to express my gratitude to Julia Mosemann, Christine Bufton, Carole Coulson, and Jamie Snavely for the technical assistance in the realization of the project as well as IGI Global for the opportunity given me. Finally, I would like to thank Fortunato Musella and his editorial assistence whose support throughout the preparation of the book has been particularly helpful.

Francesco Amoretti
University of Salerno, Italy

«Everywhere Kings, the State, hierarchies demand obedience. [...] When Jean-Paul Sartre (in April 1974) writes that the hierarchy should be crushed, forbidding that one man be the servant of another, he states in essence, I believe, the basic truth. But is it really possible? Uttering the word Society is just like saying hierarchy. All the distinctions that have not been invented by Marx, slavery, servitude, the working class, constantly evoke chains. That they are not always the same chains does not change events to any great extent. One form of slavery ends and another is waiting to replace it. The colonies of the past have been given their freedom: all discourse affirms that this is so. Yet the chains of the Third World Countries make a deafening rattle. The people, well-fed and sheltered, adapt themselves cheerfully or in any event, have no qualms about resigning themselves to it all without a struggle».

«Partout le roi, l'État, la société hiérarchisée exigent l'obéissance. [...] Quand Jean-Paul Sartre (avril 1974) écrit qu'il faut rompre la hiérarchie, interdire qu'un homme dépende d'un autre homme – il dit à mon avis l'essentiel. Mais est-ce possible? Il semble que dire société, ce soit toujours dire hiérarchie. Toutes les distinctions que Marx n'a pas inventées, l'esclavage, le servage, la condition ouvrière, évoquent sans fin des chaînes. Que ce ne soient pas les mêmes chaînes ne change pas toujours grand-chose à l'affaire. Supprime-t-on un esclavage, un autre surgit. Les colonies d'hier, les voilà libres. Tous les discours le disent, mais les chaînes du Tiers Monde font un bruit d'enfer. De tout cela, les nantis, les gens à l'abri s'accommodent allégrement, en tout cas ils s'y résignent facilemen».

F. Braudel (1979). Civilitation matérielle, économie et capitalisme, XV- XVIII siècle – Vol. 2. Les jeux de l'échange, Paris, Armand Colin, p. 617

Section I
The Longue durée

Chapter I
Electronic Constitution:
A Braudelian Perspective

Francesco Amoretti
University of Salerno, Italy

ABSTRACT

In "The Electronic Constitution: A Braudelian Perspective", Francesco Amoretti devises a model for analysing the time and space structures of digital networks based on the Braudelian triad of times: structure, conjuncture and event. By a short description of this tri-partition, he proposes a historical method to frame the products of innovation processes such as the Internet, in their wider socio-economical context. He thus argues that the world wide Web (WWW) with its enormous quantity of easy-access and easy-made information, is just the evenementielle of a historical process that could represent a new conjuncture, with its own dynamics and phases, based on the structure of the capitalist world system. The theory underpinning this work/study is that Internet, as opposed to being a revolutionary technology that will subvert the current organization of social and economic production, is a technological tool giving institutions and organizations the means to re-organize their assets and processes so as to start a new conjuncture of capitalistic structures. Most of the authors and scholars debating the transformational power of the Internet, have to date, focused their attention on the WWW as the locus of a democratising and participative movement that puts technology at the service of the citizens/civil society. With this chapter Amoretti sheds light on the nature of continuity characterizing the Internet as regards traditional categories of political economy such as hierarchy and institutional enforcement.

INTRODUCTION

Reputable scholars believe we are looking at experiences that have never existed before. And at the origin of this new world is the cyberspace evolution. Above all in its initial phase—the 80s and the early 90s—the idea that cyberspace was infinite, borderless—and timeless—was common and shared by diverse cultural and institutional bodies, not only by the prophets of the Information Revolution[1]. Extraordinary power was attributed to this "borderless and timeless world" it could blank out history and set it in motion again based on radically new foundations (Mosco, 2004)[2].

Others believe we are looking at transformations—cultural, social, and political—that do not subvert the present. Indeed, technological innovations not only re-propose conflicts and existing gaps, they also highlight the most worrying and threatening features. Revolution or Counter-revolution? When searching for "'theoretical' literature on information technology one is apt to find empirically-disconnected speculation infused with utopian optimism or distonian cynicism" (Garson, 2003). The interpretations of cyberspace continue to fluctuate between these two extremes. Neither account is adequate. If the extensive mapping by Martin Dodge and Rob Kitchin in *Mapping Cyberspace* (2001a) and *Atlas of Cyberspace* (2001b) has demonstrated that a "borderless" and "timeless" world does not exist, and that cyberspace is characterized by specific coordinates of time and space; however it is hard to deny that nothing much has changed or is about to change: in politics as in economics and in the social practices of production, dissemination and consumption of knowledge and information. In other words: the Internet is neither a Revolution nor a Counter-revolution. To understand this model of historical analysis is needed, a model that captures and explains the different and at times (apparently) contradictory features of technological innovations.

It was exactly fifty years ago that in the Number 4 issue of the "*Annales*", the French historian Fernand Braudel published his essay destined to become a classic: Histoire et science sociales. La longue durèe. Why should that event be relevant half a century later in an article on cyberspace?[3] The reference to Braudel is explicatory, it clarifies that thanks to the Braudelian concept of time—or better—of past times, is it possible to overcome both alternatives. In terms of cultural representations, of politico-institutional changes and social practices, technological innovations follow so to speak rhythms that are temporally different, marking the life of a unique historical system, in terms of unit of analysis: the capitalist system—and the inter-state system that enabled its working and reproduction—in its dynamics of expansion. Such processes often re-propose from a new perspective, more so than people are prepared to admit, the tensions that have accompanied the history of such a system. Awareness makes for understanding, or rather, the actual terms of the issue instead of being blinded by the ideology of "newness", that if not historically contextualized, is a vague category. Therefore we shall be able to recognize new inventions—if and when they appear—only if we know how at the same time, to re-conduct them and to recompose them within the framework of the dynamic growth of this historical system. More precisely, Braudel's tripartition of historical time—structure, conjuncture, and event—is a useful interpretative tool—and an ordinate principle—for the analysis of phenomena and processes (or at least some of the most significant) that characterize cyberspace. A useful example for clarifying this concept is the fact that to date, it is well-known that most of the authors and scholars debating the transformative power of the Internet have focused their attention on the WWW—with its vast quantity of accessible and easy-to-produce information—as the locus of a democratising and participative movement that brings technology to the service of civil society. It is hard to deny the worldwide innovating lever.

But this is merely the *evenementielle* of a historical process that could represent a new conjuncture and its dynamics, of the capitalistic system, characterized by specific structures—material and ideals—of functioning.

The perspective of analysis gives an account of the state of the art of research—but also of the policies and practice—on cyberspace identifying in the diverse temporal dimensions the key elements for keeping together issues that are very distant. What is thread that links for example, the analysis of the social representations of technological innovation—in terms of metaphor, founding myths and ideologies—with the policies for technological innovation? Or, the strategies of control of cyberspace and Web politics?

In an interesting passage, anticipating by years, reflection on the embedded character of technology[4], Braudel states: *"everything is technical: violent, but also the patient and monotonous effort of men on the world outside; the rapid transformations that we usually call, a bit impulsively, revolutions, but also the slow perfecting of procedures and tools, ... the last resort, the same breadth as history and necessarily its slowness and its ambiguousness ..."* (Braudel, 1979, capp. V-VI).

Technical knowhow as a constitutive aspect of history: the meaning attributed in this work to *"Electronic Constitution"*, consequently, going beyond the juridical and normative significance. It recalls the intrinsic semantic ambiguousness of the term "constitution": not a juridical text as Lessig coherently underlines (1999) but which is neither merely the architecture of cyberspace, its Code, conceived in terms of its regulating functions and individual behavior. Electronic Constitution in the sense of a mix of processes—material and ideal—that digital technologies put in play: policies, ideologies, economic interests and individual and social practice.

The analysis of the three temporal dimensions – structure, conjuncture and event – of cyberspace has an exemplifying character. It will be necessary to verify the relevance of the perspective suggested. To better highlight, hopefully, its utility, an overview of current debate on the network and its (foreseeable) developments.

CYBERSPACE IS DEAD, LONG LIFE TO CYBERSPACE

In the space of a few years there has been a realignment—ideological and theoretical—with respect to the nature of cyberspace and its implications for society and contemporary politics. Initial doubts have been replaced by renewed thinking, self-criticism, and new certainties. With the same rapidity with which enthusiasm was spread for the virtues of new technologies, disillusionment has set in. Rather than a more equal and freer world, cyberspace seems to reproduce all the ills of the world that we know. Indeed in certain respects, it seems to contribute to making them even more acute. The fact that it was Bill Gates who heralded, with the advent and expansion of cyberspace, the realization of Adam Smith's perfect market, a "friction-free" meta-national marketplaces, where everything will be done differently, and in full freedom, should not surprise us. The community of cyber-optimists did not include only media moguls and operators of the IT sector for which reason the marriage between information technology and capitalism became, so to speak, a vital issue. Besides these intellectuals, professionals, politicians and citizens look at cyberspace as though it were a new frontier: along the "Information Superhighway" the old myths of American culture are being revived.

Amongst the most relevant from a symbolic perspective, two initiatives come into being between 1994 and 1996, becoming a constant reference for political and cultural debate for many years and not only in the Unites States.

The first is the rise of the Progress and Freedom Foundation (PFF), a think tank and lobbying organization to which we owe the document *"Cyber-*

space and the American Dream: A Magna Carta for the Knowledge Age"[5]. This manifesto takes a very laudatory view of the digital revolution. With the help of ICTs, it is possible, for the first time in the history of capitalism, to end all injustice and create a world where all are equally free to pursue their aspirations. On a political plane, the PFF's Magna Carta envisages the end of traditional politics and the birth of new. The aim is pursued re-dimensioning—annihilating—government functions and structures all to the advantage, on the one hand, of small local communities—where individuals are imagined to have more opportunities to communicate their views and to have greater access to those representing them in the political system—and, on the other, of business interests and private enterprises. Released from the control—and from the powers—of the political authority, the network space envisioned by the PFF is transparent and ubiquitous, connecting each user to every other.

This concept re-emerges in the other initiative, launched in 1996, by John Perry Barlow from Switzerland, a *"A Declaration of the Independence of Cyberspace"*[6]. The most strong statements—politically and culturally—of the Declaration are those that shape a world—cyberspace—upon which governments cannot exert any sovereignty. A new world where unlimited access is guaranteed to everyone and where everyone can express themselves freely. The First Amendment, the effective architrave of the American constitutional and juridical culture, should have found in the expansion of the Network an immense territory to safeguard, but also an extraordinary opportunity to regenerate itself. Betrayed in the real world, the dream of the Founding Fathers can thus be revived in the virtual world that is emerging. It is a new beginning, and each of us can freely choose to take part. But history cannot be erased, and, in the full flow of euphoria, is already at work to claim its revenge. This is highlighted by David Resnick (1998) in *Politics on the Internet: The Normalization of Cyberspace.*

The issue of normalization, in its crudeness, has highlighted how the diffusion of ICTs, and of the WWW in particular, led to the overcoming of the first phase characterized by egalitarian networking relations, with newsgroups, e-mail etc..., transforming "cyberspace into a mass medium which became attractive to the economic, social, and political forces that had previously ignored it" (Resnick, 1998, p. 48). Taking up again the metaphor of Locke's concept of nature where, every individual is free and equal, the American scholar maintains that "cyberspace has lost its political innocence". What happened? The transformation "from the original natural state" can be traced to the invention of currency and to the necessity for laws and policies to regulate relations and exchanges between individuals, "the Internet has its own economy...; has developed a complicated division of labor with its attendant inequalities...; and it has heard the call for laws and regulation and the protection of private property. "As virtual reality comes to mirror the real world, Cyberspace simply becomes another arena for the ongoing struggle for wealth, power and political influence" (Resnick, 1998, pp. 51-54).

The libertarian and egalitarian ethos, circulating on the Net and stoking the passions of net-citizens, belongs to the past. Rather than revolutionaire the political processes favoring forms and practices of direct democracy, the Internet is consequently, destined to be dominated by the same actors that dominate the political scene and the American society. The fortune of the digital network in the world 'off line' is such that the governments are claiming jurisdiction[7]. Paying more attention to highlighting the government's desire to regulate the Network, Resnick does not attribute the same importance to the role of the ICT corporations, seen in their capacity to contend or to share with them, such powers. Yet in commercializing processes regarding the Internet and in the (commodification) of information, this is a decisive aspect. Rather than to the critics of digital capital (Sassen, 2000) it will be an exponent of the liberal juridical culture to denounce what was happening and what risks there would be for democratic societies. Lawrence Lessig has declared, uncompromisingly, the "death of Cyberspace". Victim of its own success, the Internet is shaped more and more like a closed world and hostile towards innovation: "The return of Jefferson's ghost will end" (2000, p. 347). Transformed in a space perfectly regulated – and can be regulated – the Internet has a future made up of controls and technological devices—codes—that have force of law (Lessig, 1999). And Twin Towers were still towering over New York.

While in the United States enthusiasm was waning, in Europe Utopia was finding new sap. In the same years in which Lessig was announcing the end of cyberspace, and before the terrible events of 11 September 2001, Pierre Levy was delineating the profile of a long-term cyberdemocracy. According to the French philosopher the invention of cyberspace is part of a process of "human emancipation", and accelerates it. With the advent of a "new public space", the conditions of government are drastically redefined and new possibilities opened up – in particular on the ground of individual and collective freedom–unimaginable. The march towards a State which is transparent—made possible by the growing diffusion of digital networks—should be encouraged and sustained, not feared. Certainly, "the existence cannot be denied of a global and liberal 'empire', but one wonders whether all empires are the same" (p. 121). Even if real democracies—liberal—are not Eden on earth, they are however, the outcome of a process of social selection which is full development of human capacity. Cyberspace, from this point of view, helps the historical process to free itself from dictatorships: "dictators will capitulate to the rhythm of the expansion of cyberculture" (p. 61).

Pierre Levy's Utopia reserves a place of significance to the force of the market and to globalized capitalism. In open polemics with the no-global movements and antisystemic forces, the French philosopher argues that in the creation of the virtual agorà—a new historical object—the contribution of private firms is not only a guarantee of success, given the considerable financial resources and the professionalism involved, but also of the stability. A virtual "public" agorà, decided and sustained in other words, by governments, would perpetuate—for lack of competition—and even more important, it would be exposed to the risk of degeneration (pp. 106-107). This acknowledgement founded on the conviction that between capitalism and democracy there is no opposition: neither of principle nor of fact. Under the sign of collective intelligence, finally the "Marxist Utopia and the liberal utopia will be joined here, in this point of escape that is the conversation in the virtual communities that manage the capitalism of information" (p. 134).

This concept can result more than utopic, visionary. Above all, after 11 September, the crisis of the net-economy and the consolidating of the great info-economic monopolies, the direction the development of ICTs has taken of cyberspace seems to confirm the preoccupations of many both on the liberal side (Lessig, 2001) and on the neomarxist and the political economy of the media (Schiller, 2007) denounce an impoverishment of the democratic fabric and an accentuation of the fractures and social disparities. For both, the alliance between the States and the corporations of hardware and software has hit cyberspace hard, bending it to logic—and to interests—that suffocate the original spirit. States and corporations: in other words the institutions in opposition and/or rejected by the palingenetical myths and ideological, libertarian and egalitarian elaborations of cyberculture. This, in this battle, draws from—and is inspired by–deep currents of American tradition and culture.

MYTHOLOGY AND IDEOLOGIES OF CYBERSPACE: THE *longue durée*

"Au delà des cycles et intercycles, il y a ce... la tendance séculaire...réalité que le temps use mal et vèhicule trés longuement...: elles encombrent l'histoire, en genent, donc en commandent l'écoulement. D'autres sont plus promptes à s'effriter. Mais toutes sont à la fois soutiens et obstacles. Obstacles, elles se marquent comme des limites (au sens mathématique) dont l'homme et ses expériences ne peuvent guére s'affranchir. Songez à la difficulté de briser certains cadrei gèographiques...voire telles ou telles contraintes spirituelles: le cadrei mentaux, aussi, sont prisons de longue durée" (p. 731).[8]

Various myths, philosophies and cultural models inhabit the context of the longue durée; ideas permeating and orienting individual and collective action. Constants that demarcate the field of action and the choices of social actors, establishing the perimeter of sense. The transformations generated by ICTs re-propose, *mutatis mutandis*, what has already followed technological innovations in other phases in terms of cultural representation, the metaphors and symbols produced (Mosco, 2004). Rather than a sign of weakness, the fact that these cultural artefacts return in cycles, in concomitance with technological innovations, is a sign of their embeddedness in the societies and cultures that express them. Roots strengthen new beginnings. This is especially true in the United States, where rather than suddenly arriving bang in the middle of the American way of life, the Information Age has been underway for more than 300 years. In the interesting work *A Nation Transformed by Information. How Information Has Shaped the United States from Colonial Times to the Present* (2000) we find: "North Americans embraced information as a critical building block of their social, economic, and political world, and

invested in the development and massive deployment of the infrastructure and technologies that made all the 'hype' about the Information Age that we read about today possible" (Chandler and Contrada, from Preface)[9].

In the previous pages we have already come across some of these myths and models: Founding Fathers, the Magna Carta and the Declaration of Independence. Myths and models that return in fundamental periods in order to regenerate the Community. A glance at the history of the USA presents a very different picture of the "Information Revolution".

There is no doubt that American political culture is not homogenous, and its various trends are making their own sense of the technology. Representations and visions of virtual reality—in their political and socio-cultural implications—have borrowed from the repertory of ideas and values that have been decanted over time, giving life to the secular trends of the American nation. In this respect can the concept of cyberspace reflecting the original features of American culture and her social system, be confirmed, in other words, the country that has produced this technological innovation. However, among the diverse trends, those that can be traced (back) to "historical commitments"[10] seem the most entrenched. Of the many "historical commitments" that still dominate how Americans consider themselves and their political future, two in the context of this article are particularly important. And both were consolidated between the period from 1776 to 1836.

The first commitment made was when the U.S. issued a liberal political and constitutional order giving primacy to the protection of certain political and civil rights among its citizens. "So profound is its acceptance… that the great constitutional quarrels to follow were not so much over the validity of the Constitution as over its meaning, assuming its unquestioned validity". The second historical commitment was the belief that the only proper constitutional and political system for Americans is democracy, and that a high degree of political, social, and economic equality is necessary to guarantee its survival (Dahl, 1982, p. 235). These ideas are embedded in American culture an elaborate system of stereotypes, conceptual schema, mentality that has proven capable of coming to the surface full-blown when conditions are right.

The historical commitments obviously derive their origins from centuries old religious myths, their gnostic influence extends decisively, contaminating even the domain of political and economic action, showing—besides how useless—how improper the attempts to keep them distant and distinct are[11]. The general idea is that information technology brings about a total, revolutionary change in the social world. The rhetoric of discontinuity is not particularly sophisticated nor politically simple to classify. In its simplicity however, it justifies efficaciously, the adoption of new technology, and, conversely, stigmatizes the opponents of any technology-driven vision of institutional reform.

Above all if we consider the specific movements and initiatives characterizing the original phase—1994-1998—of public debate on cyberspace (the journal *Wired*, for example) this is certainly a prevalent aspect. It reflects the idea that "the American self is a gnostic self, its firm conviction being that authenticity derives from independence; independence that is at one and the same time, natural, unique and sovereign" (Davis, 1998). When Thomas Jefferson wrote that he had "sworn on the altar of Almighty God eternal hostility towards any kind of tyranny exerted over the minds of men", he was articulating the framework of feeling and believing that informs the American self. This framework is reversed on—and shapes—the most important secular and political documents of American history, such as the Declaration of Independence – the Charter of Civil Rights, the rhetoric of which, does not only derive from Illuminist notions of the inalienable rights of women and men. This

element of American religion contains primordial individualistic sparks, and, as far as the political structures of the United States are concerned, at times, they do not satisfy the desire for freedom intrinsic to American individuals, but on the contrary, deny that "the cornerstone stands on this gnostic ground".

The depth of this belief, together with the other belief emphasizing the sovereignty with which an individual with his needs is naturally endowed, helps to explain the strength of frontier rhetoric in American culture: a mix of freedom, self-autosufficiency and wide open spaces. A real obsession that accompanies and self-nourishes all the entire original, founding phase of cyberspace, the new "digital frontier". Consequently, it is not so curious that cyberspace has so often been understood as a world unto itself, a parallel realm. During that particular period, the contrast between cyberspace and the corporeal world, certainly in the United States but also in much of the rest of the world, is the precondition for the technological sublime: Internet levels hierarchies, decentralizes society, creates an idealized neoclassical market, and eliminates the role of intermediary institutions. As Carter (1998, p. 193) claims, "the rise of cyberspace is the apotheosis of the ideal (if it is an ideal) of individualized experience ... the appeal of the cyberspace is to autonomy: we can choose our own experiences." The e-publican rhetoric of everyday life online promise netizens a modernized mode of living beyond ordinary politics (Carter cit. in Luke, 2000). Informatic networks are creating systems of communication, exchange, and production in virtual domains of interaction beyond the constraints of material embodiment in specific territorial locations on solar time (Luke, 1998). Such visions of a more authentic autonomy on the Net are precisely what most e-publican advocates celebrate: citizens of the world, not single nations; everyone is a compatriot, nobody is a foe; physical residents of one place, virtual fraternity in all places; not cultural paralysis, social revitalization everywhere; more democracy becomes possible, new tyranny is unlikely (Luke, 2000).

This centuries old religious matrix of the ideology of cyberspace, as a recurring feature of social movements for computerization, is undoubtedly one of the strongest cultural roots. A matrix, that nourishes and is entwined with other ideas and cultural orientations. In effect, if in the context of American political culture, the concept of cyberspace reactivates the colonists' religious sense of founding a utopian 'city on a hill', this utopian project confirming at the same time, the corroding spirit towards existing structures and powers. The Magna Carta and the Declaration of Independence conceive the cyberspace as a new world, but also as an instrument for the destruction of old institutions, particularly those of government. (Agre, 2002). This suspicion of institutions is a constant of American culture, of its imaginary, and finds it roots in the criticism of the monarchy principle – in that they were centralizing authorities - during the sixteenth and seventeenth centuries. The original elitist nature of the culture of the Revolutionary leaders was weakened by the diffusion of beliefs—especially of protestant ethics—that led to a progressive democratizing of political life and American culture. In this respect, it has been maintained that "the single potentially revolutionary element of the colonial information infrastructure was the widespread, institutionalized Protestant belief in the importance of literacy for individual piety, conversion, and salvation" (Brown, 2000, p. 41) that led to an exceptionally literate public. This belief, consequently was at the origin of the democratization of information, and of the creation of an informed citizenry in all walks of life and to engage them in public affairs. If up until 1760 it normally referred only to a confined number of elite of English citizens and from the colonies, and reinforced the status quo, when the Revolutionary era began, the belief ideologically and institutionally enabled the foundations to be laid for the Information Age. After the American

Revolution the commitment to the concept of an informed citizenry became incorporated in the articles of the new state constitutions, as well as the federal Constitution. "With this objective of an informed citizenry in view, the new state constitutions encouraged information development by their declarations in favour of free speech and the press and by making provisions for schools and universities" (p. 47).

Centrally forged within political conflict, it is the commitment to legitimise the role of the new states and the Republic in creating a national marketplace for information, and the institutions to sustain it, even when the geopolitical and institutional character of the United States moved beyond that Revolutionary age. The persistence of the connection between liberty and the free flow of information, and the broad diffusion among the American people and in American public life, explains why the "commitment to democracy is included the culture of informed citizens" (p. 49).

Madison makes clear that unhindered public discussion of policy and policymakers is the very essence of republican government, and that the general and central object of the American political experiment was the transfer of absolute power from the government to the people. Free speech is a crucial means to the end of popular sovereignty (McIntosh and Cates, 1998, pp. 94-95), and to the new egalitarian society, where every man's opinion seemed as good as another's. These beliefs and ideological leanings had further implications for American political culture and for the very idea of democratic government: "In the 1790s both the Federalists and their opponents recognized the changing role popular media of communication was beginning to play in American public life… In most public writing there was a noticeable simplification and vulgarization…, (and emerges) a conception of public opinion (that) became American's nineteenth-century popular substitute for the elitist intellectual leadership of the Revolution generation" (Wood, 1982, pp. 120-125)

I don't know whether Davis' suggestive image, the myth of cyberspace is true also because its very disappearance is an intrinsic factor, "its twilight decline". If this were the case, nothing would alter the facts that history, this history, conveys to us. Despite the relative decline of mythologies, the ideologies of cyberspace will not disappear. Freedom, autonomy, democracy, markets: a cluster of ideas, values, and hopes that, precisely because they are constitutive of the western culture, continue to characterize discussions—and the political strategies—on cyberspace[12]. This continual tension between technological innovation and *longue durée*—from a perspective of ideas and cultures—contributes to throwing light on how in its expansion this innovation was accepted, justified and sustained by the institutions, the political and social forces. On this ground we can find propulsive ideals, material interests and the thirst for power on the part of States and large corporations. Thus, we are crossing the threshold of the era of the conjuncture.

THE GEOPOLITICS OF THE TECHNOLOGICAL INNOVATIONS: THE CONJUNCTURE

Un mode nouveau de rècit historique apparait,
disons le 'récitatif' de la conjoncture,
du cycle, voire de l'intercycle, qui propose à
notre choix une dizaine d'années, un quart
de siècle et, à l'extreme limite, le demisiécle...
Un temps nouveau, élevé à la hauteur
d'une explication où l'histoire peut tenter de
s'inscrire, se découpant suivant des repéres
inédit... (p. 730).[13].

Ever since the end of the 70s, and during the following two decades with even more determination, the politics and the policies for the technological innovations have become the

priority on the agenda of western governments, and worldwide (Kamark, 2004). The development of digital networks has become the main arena for the ongoing struggle for wealth, power and political influence within nations and, above all, between nations; the technological innovation for the reorganization of the states and for the redefinition of their political influence and power in the international relation system. In these geo-political processes, the role of the States, and of the inter-state system, is not declining in any way. Indeed, particularly, in relation to IT, the return of the State assumes significance in dual value terms; each connected to specific strategies of action and precise paradigms of analysis.

In the first place, the State is subject-object of IT development policies. The reform movement of state-run public institutions is executed in the use of new technologies and, in particular, of the Internet and the WWW, an extraordinary driver for growth and organizational-institutional stability. The attention addressed to this dimension explains the fortune of the neo-institutional approach to the network. *Building the Virtual State. Information Technology and Institutional Change* (Fountain 2001) highlights the greater theoretical coherence, developing and motivating the perspective: "as a revolutionary technology, the Internet—by which I mean the Internet and a host of related information technologies—provides the technological potential to influence the structure of the state as well as the relationship between state and citizen... During the 1990s alone, process redesign efforts and innovation provided evidence that IT in conjunction with government reform efforts is likely to result, over the long run, in substantial modifications of the form and capacity of the administrative state" (p. 22).

The slow and problematic reorganization of the state—via e-government policies and initiatives—is nonetheless, only one aspect of the scenario. The other concerns the role of the state in the governance of cyberspace and in the definition of its most fundamental architecture.

Processes are consequently, ambivalent: while digital networks are held an opportunity for institutional and organizational change in state administration, the latter claims its place in the contest for determining the nature of cyberspace and its Code. Both processes are collocated, finding their historico-political reason for being, in the change of direction of international political economy and world politics during the mid the 70s onwards and the following decade. Considering in detail the most significant passages of these transformations. Up to 1980, development, which had been defined as nationally-managed economic growth, was redefined as "successful participation in the world market" (World Bank, 1980, cit. in McMichael, 2004, p. 116). On an economic scale, specialization in the world economy, as opposed to replication of economic activities within a national framework, emerged as a criterion of "development". On a political level, redesigning the State on competence and quality of performance in the discharge of functions was upheld, while on an ideological plane, these were the years sanctioning the 'naturalization' of neo-liberalism, and its establishment as the hegemonic paradigm (Harvey, 2006)[14]. While during the 1980s, government reform had concentrated on deregulation, in the 1990s it focused more on the reform of core state functions and the building of state capacity. The route to development was seen to be a route of liberalization and unfiltered integration in the world economy, supplemented by domestic institutional reforms to render effective integration viable. In the 1990s innovation through IT—social and economic advancement ever more bound to technology creation, dissemination, and utilization—was at the core of the renewed focus on the role of the State and state institutions in the process. Different cultural and politico-institutional traditions contribute to explaining the features and outcome of the introduction of the digital networks in the different Countries (Rose, 2005) but are-defining the State—functions, responsibility, powers—as regards world market

priorities and logics has become strategic in the international arena, and IT a specific tool to achieve these goals. From the developed countries, the new myth is being recreated worldwide (AA.VV., 2003): in terms of organisation, in (re)directing the political agenda and public debate, i.e. in a world economy, if government—democratic and not (yet) democratic—is to have a chance, and a future, then digital networks have to be looked to and invested in (Amoretti, 2007)[15].

An ongoing parallel process with the definition of development plans and strategic policies for the development of digital networks worldwide, was the finding of new ground for political and ideological conflict for the control of cyberspace, as witnessed in recent years, by issues relative to governance structures for information and communication, and in particular Internet Governance. Criticism of non-democratic governance structures for information and communication e.g. ICANN, "became very US oriented"[16], encouraging the reinforcement of organizations such as ITU, in which the developing countries are represented (Kleinwachter, 2004). If the prospects of an "open and inclusive" approach to Internet Governance is marking time, the reason is that cyberspace has become the arena for geopolitical strategies in the capitalistic world-system. The conflict becomes more pronounced the more distance is created from the phase in which it could be affirmed with Tim Luke, that "in many respects, the Net is made in America, for America, and by America" (2000, p. 9). Over a period of a few years the scenario has changed radically. China[17], a significant example, in terms of Internet users, has surpassed the United States: circa 260 millions of users in June 2008. From the 80s, the Asian giant has devised a strategic policy for technological innovation and for the governance of cyberspace challenging powerful America on its home ground (Tsui, 2005). As part of the "Four Modernizations", besides the reform in progress, the Chinese government "has identified in the growth of the information technology sector critical economic and strategic importance, and thus, has devoted tremendous resources to its development" (Kulver, 2005, p. 199). The policies which act on the nature of cyberspace have assumed an institutional onus; for the policies that concern fundamental values in societies the destiny of the Network becomes a crucial factor (Lessig, 2001, p. 17). In other words there is in act a battle of ideas—and political strategies—of epochal importance from which outcome depends the future equilibrium of the world: the destiny of cyberspace is to be our destiny. For this reason, the contest between nations and their respective ideologies for the Network is already and even more so in the future, will be a decisive aspect of the system of international relations.

If according to some authors, the alternative is between unhindered cyberspace and controlled cyberspace (Lessig, 2004), for others the transformations of the Network are a fact: the alternative is no longer between freedom and control, but between more or less control; not least, between what models of control. If what is at stake is the future of the United States in a (cyber)space characterized by centrifugal drives and conflictual dynamics, then needs must, they will have to defend themselves from the offensive converging from the other powers, China in particular. The American decline can be contrasted with a different vision of the world, an ideology of the Network and the implementing of consequent policies. Opening up the network is in effect, "contingent, and one of the most important things is that the contingency is proportional to governance constraints that demand a specific architecture". Naturally, here it is not simply in discussion the capacity on the part of the nations to give specific shape to the architecture of the Internet in their different ways: "The point is that the United States, China and Europe are using their respective coercing powers to implement different visions of what the Internet could become… The outcome is the beginning of a technological version of the Cold War, where each faction el-

bows their way to reach their own vision of the Internet of the future" (Goldsmith and Wu, 2006, p. 184). This contribution, not sustainable from a political point of view, has without doubt the merit of having laid the cards on the table. We are far from having given life to a parallel virtual world, characterized by the existence of an open network. At the same time, we are not in the presence of a unique mechanism of centralized control through which channel all the information passes. What is happening under the surface of the Network is the development of a reality fragmented and segmented by filtering and control procedures relative to the information flows. Rather than define the Internet as "the networks of networks, it is perhaps more accurate to define it as a network of filters and chokepoints" (Deibert, 2007, p. 324). The idea has definitely declined of the immunity of the Internet from controls, the commitment of the different States and of the corporations to expand and render them more technologically sophisticated, is constantly increasing. From the mapping of these procedures, the most relevant emerging data concern their growing diffusion. If at the beginning of the New Millennium, few Countries were involved, in 2006 the interested areas were far more numerous with differences that are narrowing. In other words, not only authoritarian regimes such as China, Saudi Arabia and Iran, are up front with their capacity for filtering and control, but also many democratic Countries (Deibert et. al., 2008).

For some of them, these policies—which are policies for national security—are leading to a militarization of cyberspace. For others, national security is merely a pretext for putting in place surveillance systems that affect all the aspects of our existence (Lyon, 2007). However different their point of view, they all agree on one essential point: that such policies are the expression of an inter and intra-state contest for the determination and control of the architecture of cyberspace.

The idea that there is just one way ahead for political development—the neo-liberal way—has undergone, with the onset of the world economic crisis of recent years, a hard attack. Nonetheless, the latest developments do not suggest a radical shift: IT has the power to advance human development, and at the same time, human potential can be realized through IT and access to knowledge. The benefits of the 'Information Age' are held to be axiomatically true worldwide. This axiom however, arrives with ever greater determination "interpreted" autonomously by the players. In the new global (dis)order—that is however structurally unequal—some States (China, Russia?) and/or geopolitical regions (Ibero-American, and Europe?) intend to be protagonists. They are counting mainly on the development of digital networks as a factor of institutional change, and on the autonomous control of cyberspace.

Whether this wager will be won—and at what cost and with what implications—it is too early to say. While behind the scenes, out of sight of the citizens, the struggle goes on, and more and more technological opportunities are delivered: our message is: welcome to the marvellous world of cyberspace!

SURFACE THE WEB:
THE *événementielle*

L'événement est explosif, 'nouvelle sonnante', comme l'on disait au XVI° siècle. De sa fumée abusive, il emplit lacosciente des contemporains, mais il ne dure guère, à peine voit-on sa flemme...Le temps par excellence du chroniqueur, du journaliste...La science sociale a presque horreur de l'événement. Non sans raison: le temps court est la plus capricieuse, la plus trompeuse des durèes (p. 728).[18].

Has anything been more "explosive" over the last two decades than widespread digital networks? And has anything intrigued the press more, ever on the lookout as they are for news,

better still if it promises to be "revolutionary". During the last decade, according to data from the Website http://www.internetworldstats.com, in 2000 there were about 360 million Internet users in the world. Most of them living in developed countries. Figures in June 2008, showed that there are nearly 1.5 billion Internet users worldwide – an increase of over 300 percent in less than eight years. Much of this growth has been in the developing world. Impressive figures and percentages, which if considered together with the incredible number of technological devices, appliances and software platforms, testify an environment undergoing continual change. While markets—such as the American and Japanese—are by now in a saturation phase, technological innovation opens new horizons: mobile phones, iphones etc., having unpredictable consequences: on life-styles and consumerism, on social relations, institutions and political processes.

Some commentators speculate that on-line "content" has been taken over completely by large corporations, and/or by political power, and that "this counter-revolution would push mainstream users away from generative[19] Internet that fosters innovation and disruption, to an appliancizing network incorporating some of the most powerful features of today's Internet while greatly limiting capacity – and for better or worse, heightening its regulability" (Zittrain, 2008).

From a consumer's point of view, i.e. those who use the digital networks, it's hard to deny that surfing the Web is thriving as never before. And it is equally true, that most internet users want networks and digital devices that work (Thierer, 2008). Although increased mechanisms of control of cyberspace architecture on the part of governments and corporations are effectively in force, the empirically relevant fact remains that the number of interactive Internet sites and services is growing exponentially: blogs, social networking, shopping and consumer advice, politics, and so on. Certainly, the nature of political and party systems, political culture itself, poses a limit to the potential of digital environments. However, their use is widespread and diversified. This is evident in the first place among politicians, who consider the Internet a fundamental opportunity for their strategies of political mobilization, for acquiring consensus in the polls and for fund-raising. The analysis of political communication and interactive mass communication on the Net, records a clear rate of exponential growth emerging over the last few years. The actors in the politico-representative circuit are "scrutinized" so to speak at first hand, in their capacity to use digital media. And every polling meeting has been considered ever since the early 90s, as a new way for initiating and consolidating cyberpolitics.

It all began with the American Presidential campaign in 1992, when the future President of the United States, Bill Clinton, together with Al Gore, was the first to utilize the Internet in such a way to promote electronic democratization, "distributing press releases and general interest information over the Internet on January 20, 1993" (Whillock, 1997, cit. in Tedesco, 2004). From that moment on, there has been a race to jump on the bandwagon, with the constant fear of being left behind. In every political election cycle—both national and local—"all serious presidential candidates, the majority of U.S. congressional candidates…, and countless interest groups had established an on-line presence" (Tedesco, 2004, p. 513). For many years, these experiences often appeared rudimental "although candidates recognized the need to establish a Web presence, assessment of candidate communication via the Web indicated that candidates were not using Web tools to make candidate-public relationships stronger" (p. 514).

Started by Howard Dean in the American Presidential campaign in 2004, it appeared that a new mode of electronic strategy had been affirmed; characterized by a mix of techniques and communicational platforms that rendered the candidate-public relationship stronger and more responsible. This modality was re-launched in the

Primary Polls enabling Barack Obama to prevail over Hillary Clinton. We are thus fully immersed in the era of the Digital Campaign, based on "online venues loosely meshed together through automated linking technologies, particularly blogs and social networking applications".

Furthermore, the Internet has enabled a kind of "collective intelligence in online campaigning to emerge: a distributed networks of creators and contributors, the majority of them amateurs, can, using simple online tools, produce information goods..", particularly online video thanks to YouTube (Chadwick, 2009). Differently to the quite recent past, when the most popular Internet activities: email and instant messaging, general Web surfing or browsing, reading news, and travel information searches, election campaigns in the United States are now characterized by obsessive and continuous recalibration in response to instant online polls, fundraising driver, comments lists on YouTube video pages, and blog posts. Data are everything, and most of these data have been created by the labour of volunteers.

Government institutions have also made recourse to digital media, with initiatives aimed at the creation of a new, more participative public sphere. The spreading of e-democracy experiences—of a national nature, above all on a local scale—has resulted in a diversified map, often characterized by marked problematic issues, which makes measuring outcomes difficult (Chadwick, 2007). However, despite the variety of patterns, there is a common thread, i.e. the attention to a "Web presence" dimension: a new (virtual) space wherein to experiment with innovative skills on institution and democratic practices'. According to this perspective, the level of institutional "openness" intended as an acceptable degree of transparency and interactivity, represents the measuring unit of a government's real adherence to its basic social mission. Assuming that Website openness could be an outlet to gain entry to the institutions' working logic, the transparency and interactivity of institutional Web-sites have to be considered when analyzing different variables included in the so-called Web presence dimension. The problem is to choose the best techniques to promote or strengthen those aspects of democracy that we want to push forward.

The fundamental issue remains the formulation of the WWW, or in other words, the new opportunities for communication provided by innovative technologies. These new opportunities "are defined by the interactive capabilities of the Internet". Whether they concern the constitutional dimension (macro) the institutional (meso) or individual (micro), the perimeter—technological and procedural—within which and thanks to which it is believed possible that processes and democratic institutions can be invigorated, is that determined by the Web. The range of these opportunities is extremely wide. Personal Web sites provide the most basic interactive applications such as e-mail or Web mail. The Internet also "provides opportunities to poll constituents on particular policy issues in a timely and cost-effective manner. On-line surveys allow citizens to influence parliamentary agenda. More sophisticated mechanisms are discussion forums or public guestbooks. The political relevance of personal Web sites is also dependent upon their textual content, and upon the quality of hypertext links. Electronic voting and electronic referenda, at the constitutional level; virtual party conventions, citizen consultations on the Internet, and ways to use Internet to organize debates within—and between—political organizations at the institutional level; use of the Internet for the purpose of political communication and political participation at the individual level (Zittel, 2004a, pp. 239-240).

Indeed, the citizens: as mere individuals or organized in groups and movements. Of course, there are limitations to the potential of the Internet to contribute to a more informed society, and to a more participative politics. And the more serious attempts at analysing political participation via digital networks have take into consideration that the Internet is a multifaceted phenomenon (Polat,

2005). The plethora of online political groups and activism—also at individual level—in any event signals the potential of the Internet for political engagement and mobilization.

According to some authors, the expansion of cyberspace seems to favour the emergence of a fairly broad-based civil society in electronic space, particularly in the Net, which often assumes the nature of a movement of "resistance against overarching powers of the economy and of hierarchical power" (Sassen, 2000, p. 200). For other authors, the diffusion of digital networks also contribute to affirming national political identity in the form of " virtual community" characterized by the sharing—online—of sentiments, hopes and memories (Eriksen, 2006).

While technological innovations open new frontiers of research, at the same time, the approach remains essentially unchanged. The attention addressed to this level of analysis—and experience on-line—reflects on the methodology used. The comprehension of the événementielle does not require the recourse to a historical and systematic approach but rather, to empirical studies, that are often impressionistic and not really accurate. Analyses on Web content, surveys and opinion polls carried out on quite limited samples or samples that are not exactly representative are not the exception but the rule in an area of research which, erupting within the short time span of a few years, follows technological innovation wherever it establishes itself. The temporal horizon linked to the issue under investigation is a horizon which is contracted and decontextualized. As has been written, "the explorative nature of most of these case studies and the lack of a common relevant theoretical focus does not allow for cumulative knowledge and for a general conclusion regarding the impact of new digital media on democracy" (Zittel, 2004, pp. 71-72).

Even if there is greater awareness that the challenge is as much one of institutional design as it is about the adoption of the latest technology, the new generation technological artefacts are still considered in the same way as in the past, from the same point of view and asking the same questions that have always been asked, discovering their threatening side and their extraordinary virtues. YouTube, MySpace, SecondLife, FaceBook, and so on. With many of the leaders starting their own YouTube channels; or which, like Avatar, project themselves into a virtual world which everyone often at a price – can own. That this (virtual) reality can satisfy the appetites of the big software corporations is not important, if the institutions, policies and citizens can explore new highways? (Winograd & Hais, 2008). We are merely observing another of the many ambiguous scenarios history is steeped in.

CONCLUSION

Compared to the early years, the founding years of the Network, the cultural climate has profoundly changed, and there is clear awareness in its materialness the virtual world expresses the tensions – and the aspirations – of the world off line. Indeed it is a constituent part. The Utopian vision of a worldwide agorà which would revitalize democratic processes has now faded and, this awareness has been transformed into accusations denouncing the promises made and not kept by (cyber)democracy. Responsible for the deception, are the market and political forces, accomplices, the masses of bored and indifferent citizens. There is no doubt that the cyberspace has undergone a marked transformation. However, like every great transformation, it projects bright spots and shadows. We would need to understand its origins, sense and direction. Many eminent scholars have dedicated themselves to this issue producing broad range analysis, both theoretical and empirical and descriptive, such as Manuel Castells (2000) and Yokai Benkler (2006). More interested in macro social dynamics—the reproduction based on new foundations, of social disparities—the former; more inclined to stake, starting from what the

Internet is today, on the liberal policies promised, the latter, the two researchers share a basic assumption: that the diffusion of digital networks produce an effective socio-cultural and economic change. And the emerging model—"informationalism" or "networked information economy"—leads to the decline of centralized and hierarchic organizations to the advantage of organizations in (network) which are decentralized and horizontal: in the economic field and in the political field (network state). But just how new is it; how new is the "networked information economy"? A legitimate question, from a Braudelian point of view. Away with digital networks, the capitalist system has been re-organized, not turned upside down: we are in a phase of conjuncture. All the more reason for highlighting the fact if the emerging forms are the result or are placed in relation to, globalization, which is by now in crisis (McMichael, 2008).

And the destiny of States? Projected towards the XXI century, the emerging models are likely to place the state in a position whereby a radical transformation is imperative, a new form: a (network state) or in institutional systems that are congruent with the logics of the "networked information economy". Even though deprived of their palingenetic myths, the Information Revolution is confirmed by these analyses, to be a real revolution. Utopia is not relevant here. Rather, it is a political prediction on the basis of how the Network really works, and not as we imagined it ten years ago. In cyberspace it seems rather as though the mechanisms and principles that the two authors believe are on "the way to extinction": hierarchical control and the centralization of mechanisms of command, have been reinforced. With the complicity of the "invisible but heavy hand", of the hardware and software corporations, which, worldwide, condition the destinies of entire countries, rendering "the political development on a national scale for most States…impossible to achieve whatever method is used, and in those few cases where it is possible, the benefits will necessarily be obtained at the expense of some other area (Wade, 2003).

We have not crossed the threshold of the XXI century on padded feet. But we should take care not to be deaf to the clinking of chains which, somewhere in the world, are making a "deafening rattle". And from idleness, fear, or merely because we are overcome by technological wonders, we are not willing to (ac)knowledge them.

REFERENCES

AA.VV. (2003). *The global course of the information revolution: Recurring themes and regional variations.* Rand Publications, www.rand.org

Agre, Ph., E. (2002), Cyberspace as American culture. *Science as Culture, 11*(2), 171-189.

Amoretti, F. (2007). International organizations ICTs policies: E-democracy and e-government for political development. *Review of Policy Research, 24,* 331-344.

Benkler, Y. (2006). *The wealth of networks: How social production transforms markets and freedom.* CT: Yale University Press.

Braudel, F. (1958). Histoire et sciences sociales. La lungue durée. *Annales E.S.C., 4,* 725-753.

Braudel, F. (1979). Civilisation matérielle, économie et capitalisme (XV-XVIII siècle). *Les Structures du quotidien: Le possible et l'impossible.* Paris, Librairie Armand Colin.

Brown, R. D. (2000). Early American origins of the information age. In Alfred D. Chandler & James W.Contrada (Eds.), *A nation transformed by information: How information has shaped the United States from colonial time to the present.* UK: Oxford University Press.

Chandler, A. D. & Contrada, J. W., (Eds.), (2000). *A nation transformed by information: How information has shaped the United States from colonial time to the present.* UK: Oxford University Press.

Chadwick, A., (2007). *Internet politics: States, citizens, and new communication.* UK:Oxford University Press.

Chadwick, A. (2009). The 2008 Digital Campaign in the United States: The Real Lesson for British Parties, http://www.andrewchadwick.com. Accessed online on the 12 January 2009.

Dahl, R. A. (1982). On removing certain impediments to democracy in the United States. In Robert H. Horwitz (Ed.), *The moral foundations of the American republic,* Second Edition. Charlottesville: University Press of Virginia, .

Davis, Erik (1998).*Techgnosis: Myth, magic & mysticism in the age of information.* New York: Crown Publishers, Inc..

Davis, J. R. and Post, D. G., (1996). Law and borders: The rise of law in cyberspace. *Stanford Law Review, 48*

Davis, R. (1999). *The Web of politics: The Internet's impact on the American political system.* New York/UK: Oxford University Press.

Deibert, R. J. (2007). The geopolitics of Internet control: Censorship, sovereignty, and cyberspace. In A. Chadwick & Ph.N.Howard (Eds.), pp.323-336, *Routledge Handbook of Internet politics.* London: Routledge.

Deibert, R., Palfrey, J., Rohozinski, R. & Zittrain, J. (Eds.) (2008). *Access denied: The practice and policy of global Internet filtering.* Cambridge, MA: MIT Press.

Drezner, D. W. (2004). The global governance of the Internet: Bringing the state back in. *Political Science Quarterly. 119*(3), 477-498.

Formenti, C. (2008). *Cybersoviet.* Milan, Italy: Raffaello Cortina Editore.

Eriksen, Th. H. (2006). *Nations in cyberspace.*

Ernest Gellner lecture, London School of Economics, 27 March. Accessed online on the 12 of Octrober 2008 at http://www.philbu.net/media-anthropology/eriksen_eseminar.pdf

Garson, David G. (2003). Toward an information technology research agenda for public administration. In David G. Garson (Ed.), *Public information technology: Policy and management issues.* Idea Group.

Goldsmith, J. & Wu, T. (2006). *Who controls the Internet?* UK: Oxford University Press,Inc.

Kluver, R. (2005). US and Chinese policy expectations of the Internet. *China Information, 19*(2), 229-324, Special Issue.

Harvey, D. (2006). Neo-liberalism as creative destruction. *Geogr.Ann., 88* (2), 145-158.

Kamark, E. (2004). *Government innovation around the world.* Research Working Papers Series. John F. Kennedy School of Government, February.

Kleinwachter, W. (2004). Beyond ICANN vs ITU? How WSIS tries to enter the new territory of Internet governance. *Gazette, 66*(3-4), 233-251.

Lessig, L. (1999). *Code.* New York: Basic Book.

Lessig, L. (2000). The death of cyberspace. *Washington and Lee Law Review, 57*(2).

Lessig, L. (2001). *The future of ideas.* New York: Random House, Inc.

Lessig, L. (2004). *Free culture.* Available for free under a Creative Commons licence, at http://www.free-culture.cc/freecontent

Lévy, P. (2002). *Cyberdémocratie.* Paris:Editions Odile Jacob.

Lévy, P. (1994). *L'intelligence collective. Pour une anthropologie du cyberspace.* Paris: Editions La Découverte.

Luke, T. W. (1998). The politics of digital inequality: Access, capability and distribution in cyberspace. In Ch. Toulose & Timothy W. Luke (Eds.), *The politics of cyberspace.* New York/London: Routledge.

McIntosh, W. V. & Cates, C. L.(1998). Hard travelin': Free speech in the age of the information super highway. In Ch. Toulose & Timothy W. Luke, (Eds.), *The politics of cyberspace.* New York/London: Routledge.

McMichael, Ph. (2004). *Development and social change: A global perspective.* London: Sage.

Mosco, V. (2004). The digital sublime: Myth, power, and cyberspace. Cambridge, MA/London: The MIT Press.

Polat, R. K. (2005). The Internet and political participation: Exploring the explanatory links. *European Journal of Communication, 20*(4), 435-459.

Resnick, D. (1998). Politics on the Internet: The normalization of cyberspace. In Ch. Toulose & Timothy W. Luke, (Eds.), *The politics of cyberspace.* New York/London: Routledge.

Rose, R. (2005). A global diffusion model of e-governance. *Journal of Public Policy, 25*(1), 5-27.

Sassen, S. (2000). The impact of the Internet on sovereignty: Unfounded and real worries. Accessed online on the 20 of Octrober 2008 at http://www.coll.mpg.de/pdf_dat/sassen.pdf

Schiller, D. (2007). How to think about information. Urbana and Chicago: University of Illinois Press.

Tedesco, J. (2004). Changing the channel: Use of the Internet for communicating about politics. In Lynda Lee Kaid (Ed.), *Handbook of political communication research.* Mahwah, NJ: Lawrence Erlbaum Associates.

Thierer, A. (2008). *The Internet isn't dying.* Accessed online on the 20 of September 2008 at http://www.pff.org

Tsui, L. (2005). Introduction. Sociopolitical Internet in China. *China Information, 19(*2), 181-188, Special Issue.

Wade, R. H. (2002). Bridging the digital divide: New route to development or new form of dependency? *Global Governance. 8,* 443-466.

Whittaker, J. (2004). The cyberspace handbook. London/New York: Routledge.

Winograd, M. & Hais, M. D.(2008). *Millennial makeover: MySpace, YouTube, and the future of American politics.* NJ: Rutgers University Press.

Wood, G. S. (1982). The democratization of mind in the American revolution. In Robert H. Horwitz (Ed.), *The moral foundations of the American republic,* second edition. Charlottesville: University Press of Virginia.

Zhao, Y. (2004). Between a world summit and a Chinese movie: Visions of the 'Information Society'. *Gazette.* 66(3-4), 275-280.

Zittel, Th. (2004). Digital parliaments and electronic democracy. Iin Rachel K. Gibson, Andrea Rommele & Stephen J. Ward (Eds.), *Electronic democracy. Mobilisation, organisation and participation via new ICTs.* London/New York: Routledge.

Zittel, Th. (2004a). Political communication and electronic democracy: American exceptionalism or global trend? In Frank Esser & Barbara Pfetsch (Eds.), Comparing political communication: Theories, cases, and challenges. UK: Cambridge University Press.

Zittrain, J.(2008). *The future of the Internet and how to stop it.* CT: Yale University press.

ENDNOTES

[1] Cfr. Davis and Post (1996).

[2] Cfr. special issue of The Information Society, 18, 2002.

[3] The Internet is the most visible of the new technologies, and it is used interchangeably with the World Wide Web, the Net, cyberspace, and the information superhighway, among others.

[4] Cfr. Paolo Ladri's contribution in this volume.

[5] Cfr. Vincent Mosco (2004, pp. 105-115). The Magna Carta at: http://www.pff.org/issuespubs/futureinsights/fi1.2magnacarta.html

[6] Barlow's Declaration at: http://homes.eff.org/~barlow/Declaration-Final.html

[7] "The regulation of Cyberspace is part of the process of normalization – of transforming a marginal frontier into a populous settled territory of advanced industrial society" (pp. 57-60).

[8] Way beyond historical cycles and mid-cycles … an age-long tendency emerges… different realities that time abhors and drags along for far too long… : realities that encumber, confuse and cause history to stumble, at the same time, determining its path. Other realities crumble more easily: but paradoxically, they are both history's support its obstacles. As the latter, their nature is marked in terms of limits (in a mathematical sense) from which mankind and human experience can in no way escape. Imagine for instance, the difficulty of breaking up or dividing particular geographical confines ....or indeed, this or that spiritual restriction: even mentalities or outlooks can be long and enduring prisons (p. 731).

[9] "Well, American history is shaped by information. But such resources, from the origins to the twentieth century, vary substantially. The creation and evolution of the national

postal system, the electric telegraph, the telephone, the electronics, and so on, always are technological innovations, new power sources, and a device of social and cultural development".

10 The expression "historical commitment" "pertains to periods in our history in which some alternative possibilities seemed open to the principal historical actors, who, however, were in conflict over the relative desirability of the alternative they perceived" (p. 234).

11 Cfr. Davis, Erik (1998). Techgnosis. Mith, Magic & Mysticism in the Age of Information. Crown Publishers. In particular, Chap. IV, Techgnosis, American-Style.

12 According to Are "… does it suffice to ask whose interests the cyberspace ideology serves; although it has certainly been a conscious and successful component of an industry lobbying strategy, the idea of cyberspace has deeper roots and comes from more directions than the average campaign of corporate public relations".

13 A new type of historical narrative appeared, what we could call the "narrative" of the conjuncture, of the cycle, or of the mid-cycle if you like. We are faced with the choice of a decade, a quarter of a century or at the other extreme, half a century…A new kind of time, raised to the dignity of explicatory criteria within which history attempts to find its place, dividing itself up temporally, on the basis of new points of reference… (p. 730)

14 "Neoliberalism is in the first instance a theory of political economic practices which proposes that human well-being can best advanced by the maximization of entrepreneurial freedom within an institutional framework characterized by private property rights, individual liberty, free markets and free trade. The role of the state is to create and preserve an institutional framework ap-

propriate to such practices" (Harvey, 2006, p. 1).

15 Debate is ongoing in terms of the boundaries and the nature of the policy community helping to consolidate the myth, conferring on it an extraordinary force of diffusion and cultural embedding: an intricate network of actors and interests – professionals, corporation managers, intellectuals, Nation-States, local elites – governed by the International Organisations, the core protagonists of this community of mythmakers (Mosco, 2004).

16 "After the burst of the .com bubble and the terrorist attack of 11 September 2001, the broader political and economic environment for internet governance changed dramatically…ICANN turned from a project on 'cyberdemocracy" into a instrument for 'cybersecurity'. …The original principles remained the same, but ICANN 2.0 became a little bit less a self-regulatory body and a little bit more a 'public-private partnership organization" (p. 240).

17 Kluver writes that "The Chinese emphasis on "informatization" particularly an emphasis on telecommunications, began in earnest in the early 1980s. During this period of economic reform, IT became targeted as a "key strategic industry" (p. 301).

18 An event is explosive, "resoundingly new" as was said of the Sixteenth century; its deceiving smoke enfolds contemporary consciences but it is destined not to last, its flame is seen with difficulty……Time par excellence, narrated by reporters or journalists… Social science shies away almost with horror when events occur and not without reason: brief time is the most capricious, the most deceiving of them all (p. 728)

19 By "generative" Zittrain means technologies or networks that invite or allow tinkering and creative experimentation.

Chapter II
Old and New Rights:
E–Citizenship in Historical Perspective

Mauro Di Meglio
University of Naples "l'Orientale", Italy

Enrico Gargiulo
University of Naples, Italy

ABSTRACT

This chapter offers a long-term perspective on citizenship, questioning one of the basic assumptions of most of the literature on this topic, that is, the nation-state as unit of analysis. Through the adoption of a world-systemic perspective, two basic aspects of the history of citizenship stand out. Firstly, the fundamentally exclusive nature of this category, as it emerged and developed over the history of the modern world-system, since at least the "long 16th Century". And, secondly, that well before the so-called "information revolution" of the last decades, "technology" has shaped the Western social imagination, acting, in various and changing historical forms, as an effective instrument of control and supremacy, producing asymmetric and inegalitarian effects, and providing a yardstick of the different "levels of development" of Western and non-Western peoples. In this view, the most recent phase of the history of citizenship, his e-form, seems to replicate, in new ways, the explanations of the gap existing both between and within countrie—now conceptualized as "digital divide"—and, at the same time, the illusory universalistic promise of an expansion of the citizenship and the rights associated to it.

During their travels, each time the Spanish encountered a native individual or group they read to the Indians a statement informing them of the truth of Christianity and the necessity to swear immediate allegiance to the Pope and to the Spanish crown. After this, if the Indians refused or even delayed in their acceptance—or, more simply, their understanding—of the *requerimiento*, the statement continued:

I certify you that, with the help of God, we shall powerfully enter into your country and shall make war against you in all ways and manners that we can, and shall subject you to the yoke and obedience of the Church and Their Highness. We shall take you and your wives and your children, and shall make slaves of them, and as such shall sell and dispose of them as Their Highness may command. And we shall take your goods, and shall do you all the mischiefs and damage that we can, as to vassals who do not obey and refuse to receive their lord and resist and contradict him (quoted from Helps, 1900, pp. 264-267).

Usually, the Spanish did not wait for the Indians to reply to their demands: «*After they had been put in chains, someone read the **Requerimiento** without knowing their language and without any interpreters, and without either the reader or the Indians understanding the language they had no opportunity to reply, being immediately carried away prisoners, the Spanish not failing to use the stick on those who did not go fast enough*» (Todorov, 1984, p. 148).

The reading of the *Requerimiento* wasn't necessary to the Spanish in order to perpetrate inhuman violence against the native peoples they confronted. The proclamation, rather, «*was merely a legalistic rationale for a fanatically religious and fanatically juridical and fanatically brutal people to justify a holocaust*» (Stannard, 1992, p. 66).

The practice of the *requerimiento* reveals, in a clear and dramatic way, a crucial historical reality: well before the information and electronic "revolution", technology—that is, the employment of "advanced" scientific knowledges in order to achieve specific ends, given a certain level of economic and socio-cultural development—shaped in an ambiguous and dangerously contradictory way the European imagination. More specifically, within this imagination, two particular aspects of technology, the language and the law, marked the boundaries of the Western man, the citizen

of the Old World. The first aspect, the language, represented in fact a barrier to any process of communication. As we have seen, the Requerimiento was not usually translated. On the other side, the second aspect, the law, represented a seemingly non-violent instrument of supremacy, which was nonetheless able to produce effects of tremendous violence: its formally symmetric and egalitarian aspect hardly concealed its despotic, asymmetrical and anti-egalitarian substance.

The use of law mediated by language, therefore, constituted a proper technology. It contributed, scientifically and methodologically, to the achievement of specific ends, and, at the same time, it offered a yardstick of the "level of development" obtained by the European peoples. Given this asymmetry in the control of knowledge and in its use, the image of the European citizen was modeled in its main traits and reinforced in its certainties by its supremacy—technological even before than material—over the non-citizen, and by the demonstration of the inferiority—the underdevelopment—of this non-citizen. The trajectories of the inclusion of the former and of the exclusion of the latter were *simultaneously* formulated in terms of Western literacy and legal technology. Europeans regarded *their's own* language and legal system as the only existing ones and, theoretically, as the only possible ones.

The list of rights of the European citizen progressively expanded over the centuries, enriching and fortifying his image. In its current phase, the trajectory of citizenship, heavily characterized by informatization, has produced, according to the view of many scholars and experts, a new character: the e-citizen, whose emancipatory potential has been often magnified. It has been argued that, thanks to the information and electronic resources, the participation to the collective choices, from being a nostalgic utopia of the ancient Greek polis, can now become a concrete reality even within modern states.

At the same time, the trajectories of exclusion implied by the e-citizenship, and which can be summed up in the so-called "digital divide", are considered, once again, as a consequence of a cultural and material backwardness, as the result of the—at least temporary—failure in adapting to the Western standards. If this gap should be closed, "development" would mechanically be exported and shared on a global level. Just like 500 years ago, the West is today offering, through its technology, its own image. "Informatization for all!": this is the path to equality and well-being.

From this point of view, the structure of the European supremacy has maintained some constant features over time. In these terms, the digital divide is not something new, which emerged in the last decades, but the result of long-term historical processes. Through a rather complex process, the juridical equality as a veil able to conceal the substantial inequalities has been joined by the rhetoric of informational equality. In this chapter, we will examine the most relevant steps of this process, and will try to point out, through the vicissitudes of citizenship, the thread which links the origins of the modern world-economy to the present of the historical capitalism.

1. A METHODOLOGICAL PREMISE: CITIZENSHIP AND THE UNIT OF ANALYSIS

In step with the optimistic and now classical view presented by Marshall (1950), most of the analyses on the history of citizenship tended, and are still inclined, to emphasize its inclusivity, at the expense of its exclusive dimension. This has been possible also thanks to the specific perspective from which this institution has been studied, one which assumes that the nation-state is the proper unit of analysis. State-centred and atomistic perspectives are quite common in the social sciences, whose epistemic structures and programs have been closely attached to, and shaped by, the experience of modern-state formation. As a consequence of this methodological

nationalism, most analyses share the assumption that «*the nation/state/society is the natural social and political form of the modern world*» (Wimmer & Glick Schiller, 2002, p. 303).

In fact, if considered from a state-centric perspective, citizenship appears as the natural link between a nation-state and its members: each state has its own citizens, and this is the natural state of things. And conventional political thought has treated the nation-state as the naturable and inevitable site of citizenship. As Linda Bosniak has pointed out, «*citizenship has been conventionally assumed to be a national enterprise; it has been assumed to be an institution or a set of social practices situated squarely and necessarily within the political community of the nation-state. Given this assumption, there has not seemed to be much of anything to talk about*» (Bosniak, 2001, p. 237).

Given there premises, most scholars focused on how this link is shaped. As a consequence, the debates around citizenship deal with its features, the rights that it grants and the ways through which these rights may be claimed; while the questions most frequently asked about citizenship concern its effectivity in contemporary societies, or, in other words, the—active or passive—nature of the citizen. From this standpoint, the route of citizenship appears like a glorious path of progressive inclusion, which, by means of an increasingly generous system of rights, leads to equality among the citizens.

But if, in place of a state-centric perspective, we adopt a world-systemic one, the picture appears quite different. From this angle of vision, citizenship looks like a somehow arbitrary link between a state and some of the people who live within its boundaries. This link, now, appears no longer as something natural, but as an artificial fact, while the questions concerning how to concretely defend citizens' rights are replaced by questions concerning how to have access to these rights, who holds them, and which are the reasons why these people are considered as citizens.

At the same time, in order to understand the concrete political dynamics of historical capitalism it's necessary to go beyond a perspective that assigns to the state(s) and the interstate system a political role which is independent from the economic sphere and that, as a consequence, conceives the "state" and the "market" as dichotomous terms. Nevertheless, we don't mean to deny the importance of the first term within the logic of capitalism. Indeed, as Braudel has pointed out, capitalism «only triumphs when it becomes identified with the state, when it is the state» (Braudel, 1981, p. 64).

Rather, it's important to achieve a better understanding of the role of the state(s) within the capitalist world-system. In fact, the close relationship between political power and the market makes it necessary to link the study of the states with that of the context in which the interstate system has risen and has evolved. This context can be characterized as a world-system, that is «a social system ... that has boundaries, structures, member groups, rules of legitimation, and coherence. Its life is made up of the conflicting forces which hold it together by tension, and tear it apart as each group seeks eternally to remold it to its advantage» (Wallerstein, 1974, p. 347).

According to Wallerstein, up till today only two kinds of world-systems have existed: "world-empires" and "world-economies". The concrete boundaries of both are defined by a single division of labor. However, while world-empires are characterized by the existence of a single political system over most of its area, world-economies are characterized by the existence of a multiplicity of political entities (Ibidem, p. 348).

Capitalism is thus constituted by a world-economy which has survived for 500 years without transforming itself into a world-empire: it «*has been able to flourish precisely because the world-economy has had within its bounds not one but a multiplicity of political systems*» (Idem). Considered as an *historical system,*

capitalism can be identified with «*that concrete, time-bounded, space-bounded integrated locus of productive activities within which the endless accumulation of capital has been the economic objective or 'law' that has governed or prevailed in fundamental economic activity*» (Wallerstein, 1983, p. 18).

As a result, the state cannot assumed as the proper unit of analysis in the study of historical social change, since «*capitalism as an economic mode is based on the fact that the economic factors operate within an arena larger than that which any political entity can totally control. This gives capitalists a freedom of maneuver that is structurally based. It has made possible the constant economic expansion of the world-system, albeit a very skewed distribution of its rewards*» (Wallerstein, 1974, p. 348).

Once we drop a state-centric perspective and adopt instead a world-systemic one, the path of citizenship – now conceived as a world-wide phenomenon – appears under a different light: within the capitalist world-economy, the lengthening of the catalogue of the rights of citizenship, as well as the extension of the citizen's set, that may be noticed "inside" the economically and politically "core countries" appear closely linked with the worsening of the life conditions—and with the constriction of citizenship—that can be observed "outside" them. In this sense, the transformation of the "subject" into the "citizen" which has concerned some areas of the capitalist world-economy went along a parallel process whose outcome has consisted in the establishment of new subjects in different areas of the *same* world-system.

The function of citizenship, therefore, cannot be adequately understood if considered exclusively as a chapter of each single state's domestic policy. It must be rather reinterpreted in the light of the role performed by this institution within the world-economy: that of a legal instrument—better if equipped with socio-cultural rhetorical statements—used as an instrument of control within political and economic processes that generate at the same time inclusive and, above all, exclusive effects[1].

In the light of these suggestions, exclusion appears as the central element in the history of citizenship. The path of this institution, in fact, especially along the centuries following the French revolution, has been characterized by an uninterrupted conflict among different groups, some of them aiming at narrowing the number of citizens, and others, instead, aiming at widening it (Wallerstein, 2003, p. 651).

Citizenship has therefore made legal a worldwide system of privileges. Throughout the history of the modern world-system, the assignment of certain rights to some groups of people has gone hand in hand with an increase of the groups which have been excluded: «*The halfway house of citizenship—the inclusion of some and the exclusion of others—served precisely to appease the most dangerous strata of the countries of the core zones, the working classes, while still excluding from the division of the surplus value and political decision making the vast majority of the world's populations*» (Wallerstein, 1998, p. 21).

Therefore, the entire trajectory of citizenship—and not only its latest development—can be adequately understood only if considered within a world-systemic perspective. Worldwide economic, political, and social processes have shaped such a path from the beginning. In this sense, the role of citizenship has always been supra-national, even before the European citizenship was born.

We don't mean to deny the inclusive dimension of citizenship, nor we mean to consider it like a sheer "ideology". What we intend to emphasize is that, in the history of citizenship, emancipatory projects, although effective and significant, have always been accompanied by opposite projects. In other terms, the progressive and inclusive dimension of citizenship has never been detached itself from its mirror dimension, which, on the contrary,

is regressive and exclusive. The inseparability of these two dimensions is understandable only if it is traced back to the role they have performed within the capitalist system: they are both necessary to its functioning.

2. THE EUROPEAN EXPANSION AND THE ORIGINS OF CITIZENSHIP

As the foregoing theoretical and methodological remarks have pointed out, a long term and large scale perspective is needed in order to reconstruct the processes of simultaneous inclusion and exclusion which have given shape and substance to the history of citizenship. As we have already mentioned, the importance of the European expansion for the entire path of citizenship already become manifest through the use of law as a technological device able to include and, at the same time, to exclude. It is now time to turn our attention in more detail to the historical trajectory of citizenship.

The early stage of the modern era, the "long sixteenth century", provided the backdrop for the emergence of a new protagonist—the individual-citizen—and of the institutional framework within which he took his first steps – the state. If compared to the pre-modern citizen, the individual-citizen is characterized by the particular link that ties him with the authority: «the citizen is the subject which obeys the sovereign and obtains in return security from the internal and external enemy» (Costa, 2005, p. 24). A "corporatist" metaphor of citizenship is replaced by an individualistic one. By virtue of the latter, citizenship is no longer defined by an organicistic membership to a community, but by the relationship of subordination towards the sovereign. However, the form of citizenship enclosed within this metaphor only guarantees a partial emancipation, since the subject continues to live within the body of the citizen (Mezzadra, 2004, p. 3). The submission to a specific authority, in other words, is the price that the individual has to pay in order to free himself by the oxbow of the community.

If within the borders of the state the individual-citizen obtains only a partial autonomy, outside of these borders his condition can however radically change. More specifically, he gains his autonomy also through the comparison with a different kind of individual, to which the status of citizen is denied. This kind of individual, the non-citizen, is a subject of a particular kind: although he formally doesn't belong to the state, he is however submitted to the control of the latter. The control, in this case, doesn't take shape as an "internal" political-legal link between a sovereign and his citizens, but as a relationship grounded on the power and the conquest "from the outside".

In the process of citizenship's building, therefore, the external expansion of the states has been no less important than their domestic policy. By means of the latter, in fact, it has been possible to establish a lay relationship between the citizens and the political order. However, this relationship has kept the citizen in a partially subordinate position. Through the external expansion of the states, then, it has been possible to build, on the difference with the non citizen, a citizen's image within which the subordination made partially room to autonomy. In this sense, we can assert that the representation of the citizen as an autonomous individual, in order to be strengthened, has needed a projection of every state towards other states and other populations external to it.

In the course of the "long Sixteenth century", such a projection has been guaranteed by the European expansion, which has provided the citizen with a one-time chance to represent himself as such and not as a subject. This chance, more specifically, has been provided by the encounter with before unknown populations. After this encounter, the European individual has started to give himself a new physiognomy exactly, once he set himself against his contrary, the "savage".

But the discovery of the savage has raised a completely new problem for the citizen of the Old World: a clear and unambiguous definition of this character had to be offered. In the Sixteenth-century, Europe «*had very little knowledge and still less understanding of the people beyond its borders, [and] there were very few terms with which to classify men. [...] In European eyes most non-Europeans, and nearly all non-Christians, including such 'advanced' peoples as the Turks, were classified as 'barbarians'*» *(Pagden, 1982, p. 13-14). According to many interpretations, therefore, the peoples of the New World would have busted «within the European's space of experience with all the radicality of the absolute other», a radicality able to «turn pale the image of all the traditional **others**»* with which the Europeans had come in contact during the previous centuries (Scuccimarra, 2006, p. 167). The differences between the European and the savage were so striking as to call into question the fact that the second were members of the same species (Todorov, 1984, p. 5 and Lévi Strauss, 1997, p. 73-74).

But these interpretations may be considered not completely exact: «*No Spanish man of the early Sixteenth century would have questioned the belonging of the Americans to the mankind – namely to the progeny of Adamo. It's true that someone spoke about the bestiality of Indians, but not to affirm that they belong to a different zoological species*» (Gliozzi, 1976, p. 287). Rather, the bestiality of Indians was brought back to their *intellectual* and *moral* incapability to receive the evangelical preaching: not a natural and primal data, but a fruit of the sin. Moreover, the bestiality of the Indians consisted in disobeying the *natural laws*; and this infringement made them deserving a punishment (Ibidem, p. 293).

Historically, however, the question has been quickly solved: with the Papal bull *Sublimis Deus* of 1537, Paolo III expressly excluded that the inhabitants of the Indies had to be considered as animals, and invited Christendom to try to redeem them. The acknowledgment of the humanity of the Indians is not at all surprising if we consider what's was at stake: evangelization was the main formal justification of the conquest, consecrated by Pope Alessandro VI in *Inter Caetera*; if it would have failed, also the control on the New World claimed by the Castilian Crown would have broken down. Any evangelization of subhuman subjects, in fact, would have been impossible (Cassi, 2007, p.106). The inhumanity of the Indians,[2] therefore, promoted the private interests of the *conquistadores* but damaged the material interests of the Church and the Spanish Crown, oriented towards an exploitation of the workforce more than towards a destroyer robbery of the local population (Gliozzi, 1971, p. 4).

However, even once accepted the human nature of the savage, another problem arose: the definition of the social and legal status of such a subject. The attempts to obtain an acceptable definition reached their apex with the contention between Bartolomé de Las Casas and Juan Ginés de Sepúlveda[3]. The thesis of the latter, grounded on a radical interpretation of Aristotle's political theories, maintained the natural inferiority of the Indians and their being outside of the reason. As a consequence of this inferiority, the inhabitants of the New World were represented as subjects intrinsically extraneous to the political and civil live – namely as "natural slaves". Las Casas, instead, claimed that men, as sons of God, could be holders of *dominium* independently from grace and faith. As a consequence, their slavery wasn't natural but rather accidental.

The thesis of Las Casas had been anticipated by another theologian and jurist of the early Sixteenth century, Francisco de Vitoria. Through the strengthened categories of the Aristotelian thought, he brought back into unity the Old and the New World on the base of mankind's common rational character and common moral root. Thus, the Spanish theologian accorded to the Indians

the same rights of the Europeans (Pagden, 1982, p. 143).

But after having denied the anthropological diversity and encompassed the otherness of the savage, de Vitoria turned the difference into subordination, putting the new subject into a pre-existent hierarchy of status and powers (Costa, 1999, p. 124). According to the Spanish theologian, therefore, the savages were rational but, however, they remained in a position of inferiority if compared to the Europeans. As a consequence, the latters had to direct them like parents do with their children. De Vitoria, in fact, pursued the vindication of the Spanish conquest, and he did so by means of arguments grounded on the freedom of missions for Christians[4]:

The norms conceived by de Vitoria, as well as the more general strategy intended to include the Indians within the Spanish empire's legal system, turned these people into docile and obedient subjects (Merker, 2006, p. 32). The relation between *conquistadores* and Indians, at least formally, from external became internal: compulsion was replaced by a political link, shaped by a strongly paternalistic attitude.

The European legal technology is now completely deployed: by means of this formal inclusion, a common constituent—a sort of *general equivalent* through which compare European citizens and Indians subjects, without, however, putting them *de facto* on the same plane, but, on the contrary, crystallizing their differences—was founded. In other words, the formal acknowledgment made still more solid the substantial inferiority of the Indians when compared with the citizens of the New World. From this point of view, the spuriously liberal, and only in the abstract, universalistic nature of the rights theorized by the Europeans becomes evident: «*as a matter of fact, only the Spanish can exert these rights—moving away, occupying, dictating the laws of the unequal trade—while the Indians are uniquely passive parts and victims*» (Ferrajoli, 1997, p. 16).

3. BIRTH OF THE OWNER-CITIZEN

As we have already suggested, the guarantist nature of some of the legal measures introduced by the Spanish soon revealed itself only seemingly so: the guarantee instrument had become a vehicle of injustice. With the inclusion of the America's inhabitants within their own system of rights, Europeans did nothing else but to reassert their *property* on these subjects, now considered as mere labour-force.

Property, therefore, became a crucial element of citizenship. Notwithstanding, another organizational passage in the process of colonization and, more in general, in the dynamics of the capitalist system was necessary so that property could become central in the definition of the citizen. This organizational passage overlapped with the temporal passage from the Sixteenth to the Seventeenth century and with the theoretical passage from the problems of the state—of the "reasons of state"—to the attention towards the "peoples", the productive forces and the worldwide market, that is, to the growing perception of the existence of a world-economy (Lentini, 2003, p. 67).

A geographical postponement of the discourse on citizenship occurred with the passage from the state to the productive forces: England replaced Spain in the process of citizen's building. The reasons of such a postponement may be better understood if we compare the Spanish and the British systems of rule.

In Spain, the external projection of the state showed a strong weave between religious and strictly economic motivations. This weave was already evident during the process of Christianization of the internal unfaithful persons, which represented a factor of national cohesion for the Spanish and the Portuguese citizens and a good training for colonialism (Merker, 2006, p. 20). More in general, Spain exerted its sovereignty in an authoritarian way, showing a clear will of realizing a hierarchical and patriarchal government (Bailyn e Wood, 1987, p. 22), a sort of «bureaucratic empire controlled by lawmen» (Reinhard, 2002, p. 61).

Since the beginning of the conquest of America, Spanish monarchs attempted in all ways to obstruct the feudal tendencies that previously had threatened their power within their own territory (Elliott, 1982, p. 80). Fernando and Isabella, in fact, conceived America as a kingdom which was inextricably tied with the Crown of Castile. Consequently, the governmental jurisdictions within American territory were subordinated to the Crown itself; moreover, the settlers enjoyed a scarce autonomy, while the American natives, as we have already seen, were turned into direct subjects of Spanish monarchy (Bailyn e Wood, 1985, p. 26). As subjects, the natives, for order of the Queen, couldn't turned into slaves[5].

The conquerors and the settlers, therefore, considered the Crown and its defending attitude towards the Indians an obstacle to their activities and to the defence of their interests. Thus, a continued tension between the interest of the Crown—not uniquely dictated by humanitarian and Christian-inspired reasons, but also by political-economical ones—to take care of the well-being of the Indians subjects and the double necessity, on one hand, of satisfying the conquerors and the settlers offering them just the labour force of the Indians, and on the other hand of producing resources for the insatiable fiscal machine (Reinhard, 2002, p. 71).

Spanish imperial experience, as we have seen, has strongly emphasized the role of the state. English empire, instead, has never had an effective structure of government: the authority of the Crown and of the British Parliament was fragmented in half a dozen of not governmental organisms which were not coordinated among them; moreover, this authority was much superficial, rarely proceeding beyond naval stores and dock customs (Bailyn e Wood, 1985, p. 25). Also the English penetration within the western hemisphere presented characteristics which were contrary to the Spanish one: starting from the discovery of America and through the first fifty years of the Sixteenth century, while Spain conquered and exploited wide areas of central and southern America, England didn't anything in order to claim its pretensions on northern America. During the kingdom of Henry VIII, both the Crown and the private citizens showed a scarce interest for the lands of America.

However, during the kingdom of Elisabeth I, new economic needs forced towards the search for new ways out for the capitals. These ways out were located in the overseas lands and in the development of new markets (see Hill, 1977, in particular Chapter IV), whose conquest was entrusted, in good part, to private companies holder of special grants by the Crown (Arrighi, 1994).

If the primacy of Spain was attributable to religious motivations, the primacy of England was attributable to their technical and scientific skills and to the "lay" relations that such skills were able to promote (cf. Merker, 2006, p. 34). The myth of a "commercial civilization" went hand in hand with the myth of an industrious and technically superior Britain[6]: a society which was centred on the *diversity* of goods and men was regarded as superior to another society which, instead, was centred on the political and religious uniformity. With the myth of commercial civilization, the ideology of technological superiority, although differently dressed, has appeared again: the rules of trade[7], strictly centred on property, were a new and more evolved form of knowledge adapt to be exported outside of Europe in order to civilize the savages. The relationship between the latters and the English, as a consequence, was characterized by a strong pragmatism: the conquerors adjusted the image of the Indians on the base of their own purposes: «*A basic rule was that any given Englishman at any given time formed his views [of the savages of America] in accordance with his purposes. [...] In short, like the most modern architects, the Englishmen devised the savage's form to fit his function*» (Jennings, 1975, p. 65).

The difference between the Spanish and the English systems of rule can be observed also with regard to slavery. About the Spanish empire, the relationship between slave and citizen was mediated by religious and political components. As a consequence, the choice—however only partially respected—for the prohibition of the slavery as regards the Natives (with the exception of the ones which refused Christian religion) gave rise to a multilevel system of exclusion: black men has occupied the most external level; Indians the intermediate level; while the population of Spanish origin the innermost one. Anyhow, the ones which occupied the intermediate level, as subjects of the Crown, has benefited, at least formally, from a guaranteed by law protection. With regard to English empire, instead, things has gone differently: «*From the standpoint of the British government Indian slavery, unlike later Negro slavery which involved vital imperial interests, was a purely colonial matter*» (Williams, 1944, p. 8). For such a reason, this kind of slavery was never declared illegal, neither it was that extensive in the British dominions (Ibidem, p. 9).

As it has become evident till now, the differences between the Spanish and the British systems of rule may in part explain the shaping of different features of citizenship. The first system, limiting the autonomy of the settlers and turning the American natives into formal subjects, has kept citizenship within the narrow horizon of the state. Absolute sovereignty, despite the local decentralization of power that has partially enforced by the Crown, has been the predominant characteristic of the citizenship's model expressed within the Spanish system: a communitarian and organicistic kind of membership as regards the settlers-citizens and a paternalistic recognition of their belonging to the state-structure as regards the natives-subjects. In such a context, the image of the citizen has been hardly separated from that of the subject, inasmuch as the scarce autonomy of the first one has made the difference with the second one more static than dynamic, grounded more on the status than on the role. The relation between the individual and the political order, in other words, has been shaped in its vertical dimensions, but not in its functional characteristics; while the relation between the "out-and-out" citizen and the native has always been *mediated* by the state, being lacking a link between these two different kinds of individuals which was *independent* from the Crown.

Spanish imperial system, therefore, with its emphasis on the role of the state, has contributed to fix the relation between the political order and the citizen/subject without, however, fixing the contents of this relation. British system, instead, with its growing attention towards the economic role of the productive forces, has brought with itself a significant novelty for citizenship: the entrance of property in its semantic field. Through property, the citizen has more and more parted from the subject.

The owner-citizen is an individual who is able to provide for himself, who is independent and autonomous; and this autonomy, apart from being the essence of civil citizenship, constitutes the premise of the political one. The link between property and political autonomy, more in detail, is one of the most interesting and illuminating threads in the history of citizenship. Through this link, the process of shaping of the owner-citizen clearly shows its symbiosis with the consolidation of the capitalist economic-system, since citizenship has inherited the possessive and unequal nature of this system. The owner-citizen, in fact, free from economic concerns, is the only one who may actively participate to the communitarian life. Participation, consequently, is more and more depending on economic capacities: a privilege for few more than a benefit for all. The enlargement in the catalogue of citizen's rights, thus, is reserved to a minority. More citizenship rights, in short, but only for few citizens.

The case of French revolution, about, is exemplary[8]. Here, two categories of citizens have been immediately created: passive and active. Siéyès said that «*natural and civil rights are rights "for whose maintenance and development society is formed". These are passive rights. There also exist political rights, "those by which society is formed". These are active rights*» (Wallerstein, 2003, p. 651).

4. THE RISE AND DEMISE OF SOCIAL CITIZENSHIP

After the French revolution property gained a great importance in the definition of the citizen. Its exclusionary nature, nevertheless, was the main limit of this category: property made citizenship a privilege reserved to an excessively restricted number of persons. The growth of social conflict and of the claims carried on by the most disadvantaged classes has pushed then the dominant classes to make citizenship more inclusive, at least for some categories of individuals.

The process of citizenship's enlargement, then, has shown itself, in a particularly significant way, in the end of Nineteenth century, with the birth of the welfare state[9]. By means of social security, the states have put in practice national hegemonic projects that aroused an inter-class loyalty into the people. Proletarian internationalism, as a consequence, went in a crisis when the masses became able to claim social protection from their states.

About this question, Edward Carr spoke of "socialization of the nation", making reference with this phrase to the centrality of the economic requests expressed in those days by the masses. As a consequence of the socialization of the nation, the task of domestic policy ceased to be only the safeguard of order and became also the wellbeing of nation's members. To defend salaries and jobs became a priority of domestic policy, even when this priority conflicted with other nation's interests. The socialization of the nation, thereby, had in the nationalization of socialism its corollary (Carr, 1945).

At the end of XIX Century, therefore, a vicious circle between domestic conflict and international conflict became manifest: internal and external, social and national protectionism moved in the same direction[10] (cf. Arendt, 1958; Hobsbawm, 1987; Polanyi, 1944; and Silver and Slater, 1999). With the birth of the welfare state the growth of the *internal side* of citizenship—namely the extension of the rights reserved to citizens— had a counterpart in the worsening of the forms of *external* exclusion[11]. The language of citizenship has been tightly entwined with the language of nationalism, insomuch as to give rise to the actual confusion between the two categories (Gallissot, 2001, p. 269-278). From that moment, the separation between insiders and outsiders started to be grounded no longer exclusively on sex, property and race, but predominantly on national membership (Balibar e Wallerstein, 1988): in the name of nation, some persons have been considered different from others and, as such, "deserving" different rights.

This new centrality of national citizenship—centrality that has to do with the gradual spatial enlargement of citizenship and with the process of capitalism's expansion—caused remarkable changes in the criterion of discrimination between citizens and not citizens. This criterion became independent from the *residence* in a given national territory. As a consequence, the concept of "foreigner" assumed a different meaning: the link between national sovereignty and nationalism has transformed the stranger into an *outsider*. Also previously foreigners were considered as such, but in the beginning of the Twentieth Century they started to be identified as a group "a part" that the state, on the base of its institutional legitimation, had the power of exclude from civil society (Sassen, 1996).

The strong relation between domestic and international politics is also proved by the social policies enforced during the First World War: the states committed to the wellbeing of its members just because they represented a precious resource

in order to obtain the final victory. But the reverse side of this increased protection has been the exclusion of the non-citizens from its enjoyment. According to Carr, the closing of the frontiers during the post-war years made even more clear the link between the economical nationalism and the socialization of the nation (Carr, 1945). Moreover, the costs of the extension of citizenship have also been transferred, by means of explicitly imperialistic economical and military policies, toward the peripheral areas of the world—namely toward the colonies—creating by this way new excluded persons. National politics, therefore, has merged with social policies more strongly than ever, giving rise to a "double-faced Giano", socialist and nationalist at the same time.

Later on, nevertheless, claims for inclusion have started to come also from the peripheral areas of the world. There, the complaints of the marginalized groups have become stronger; while the risk of a world revolution has become every day more concrete. If during the XIXth Century had been possible simply to transfer outwards the costs of the internal conflict, in the course of XXth Century the question has become more complicated.

At the beginning of this century, after the first World War, United States have attempted to respond to the claims for inclusion through a reorganization of the international order. The project of president Woodrow Wilson was the creation of the Society of Nations, an international institution able to level the differences and keep a lasting peace among the states (Ventrone, 2004, p. 12). A decisive role in the formulation of Wilson's plan has been played by Lenin and its incitement to a worldwide revolution. To this incitement in favour of solidarity within the international proletariat and against imperialism, Wilson responded in fact with a call in turn internationalist, but reformist and not revolutionary (Barraclough, 1967), putting the self-determination of peoples and an universalist vision of freedom and democracy in the middle of his discourse. The typical characteristics of the owner-citizens/individual member of a democratic system were extended, with his call, to the international system; while the then rising welfare state (practically beginning in the United States) has started to configure itself as an instrument of development for the "backward" countries (Wallerstein, 1995, p. 107).

The "global" project of a political integration foresaw, in embryo, a *worldwide* social policy. At the base of this policy there was the experience of the *New Deal*[12], and more specifically its organizational aspects. The New Deal, in fact, has been innovative on a institutional plan more than a economical-political and socio-political one. For the United States, the institutional engineering has been the strong point between the two world wars, which enabled their hegemonic transition. The power of American capitalism was based on new organizational forms, which were on a global scale and, above all, qualitatively different compared to other forms of "organized capitalism". These organizational forms, having reference to the *corporate liberalism*[13], were «a compromise between a certain way to mean liberal principles in a entrepreneurial age and the new ideological need of the giant corporation (Lentini, 2003, p. 290).

The Society of Nations, nevertheless, hadn't succeeded to go out from a eurocentric vision. United Nations, on the contrary, had explicitly choosen to include non european peoples within them (Claude, 1956, p. 87). Only with World War two, the birth of the United Nations and the process of decolonization, therefore, the programme of a global welfare has reached maturity. The strategy applied to realize such a program has been a worldwide extension of the *idea* of social citizenship by means of a real extension of the—now decolonized—international system.

Modernization theory has been the rhetoric through which this double extension has been possible. Such a theory, which has covered a key role within the American strategy for a worldwide hegemony, appears as a real ideology (cf. Latham,

2000; McMichael, 2000; Ventrone, 2004 and Di Meglio, 2008): it gained importance in the context of the Cold War and of the struggle among the superpowers for the dominance on the Third World; its role has consisted in eliminating the new chances for the Soviet and Chinese expansion that the collapse of colonialism was making possible (Latham, 2000, p. 2-3). The social function of modernization was «*to give a pattern of gradual and not-revolutionary development, grounded on the trust on progress, on rationality and on industrialization, on the base of an euro-centred equation that assimilated modernization and westernization*» (Di Meglio, 1997, p. 17).

The contents of modernization theory were based on the recognition of the claims for an autonomous development expressed by the single states. Within the American plans, moreover, the path to development, although hauled by the states, would request international aid. Thus, the worldwide extension of the welfare state—or, better said, the "project" for its extension—performed a fundamental role in the modernization theory. The "welfare project", in fact, shared the substantive characteristics of this theory: a linear idea of the progress, the faith in development as economical development and the centrality of the *state*, assumed both as the mode of organizing political power and as the privileged—if not exclusive—unit of analysis in the study of processes of social change.

The progressive development of the international system has rendered citizenship an increasingly important institution: every person has been imperatively collocated within this system, becoming in this way a citizen. The building of a world of states, in other words, has had as a result the building of a world of citizens. The relationship between the citizen and the political authority, from then on, has become—except some categories of individuals[14]—exclusive: being citizens means being citizens of *one* state.

The development and welfare state policies, therefore, assume a different meaning if seen with the eyes of the centre or of the periphery of the world. In the centre, they have meant a partial inclusion of the low classes into the political life and the productive system; at the periphery, they have meant only the illusion of a comparable process of inclusion. The extension of citizenship, in other words, has been formal but not substantial, since the new *de jure* citizens have realized to be, *de facto*, citizens which are provided with not guaranteed rights. If globally viewed, these policies have clearly delineated, above the distinction between citizens and not citizens which is internal at every state, a systemic distinction between "big leagues" and "bush leagues" citizens. Hence, by means of social citizenship it has been possible to reach a compromise – the inclusion of a few and the exclusion of the others – that has permitted, within the core and richer countries of the world-system, to pacify the most dangerous classes through the medium of the welfare state and the political rights, while the remaining part of the world population, despite—or maybe it would be better to say "through"—the ideology of a worldwide welfare, was, by-fact, excluded by the sharing of the wealth produced.

The inclusivity of citizenship, therefore, has been in many cases more apparent than real. The universalism of social rights has proved to be an absolutely "particularistic" universalism: only few categories of persons have benefited from a really rich and inclusionary citizenship.

5. FROM MATERIAL TO VIRTUAL MEMBERSHIP: THE CRISIS OF THE WELFARE STATE AND THE RISE OF THE E-CITIZENSHIP

During the 1970s, the scene of the welfare state changed. The so-called "golden age of capitalism" started to show some cracks, revealing by this way the signs of an hegemonic crisis. In the 1980s, these signs became even more evident. In the beginning of this decade, in fact, the shift of

capital from production and commerce to finance accelerated, giving life to a new financial expansion and, as a consequence, to the polarization of wealth within the world-system (Silver and Slater, 1999, p. 207).

The outcome of these socio-economical changes has been the following: in the entire world, the competition for mobile capital has become increasingly intense, inducing the dismantling of legal and economical measures established in the past in order to contrast unemployment; similarly, the projects oriented to development have been dropped in favour of IMF-imposed structural adjustment and austerity programs aimed at making Third World countries solvent in the world financial markets (McMichael, 1996). In other words, what become manifest was that, despite the expansion of the welfare state in western countries, the world-system wasn't structured in such a way as to offer a level of welfare able to satisfy the requirements asserted by the Universal Declaration of Human Rights of the UN.

Since the 1970s, therefore, the welfare state started its decline and lost its international legitimacy as the main systemic-wide instrument of control of justice and security. If, during the apogee of the welfare state, modernization theory represented the dominant rhetoric, with the crisis of the welfare state the "globalization project" replaced the "development project" (Ibidem). Within the idea of globalization, the development is still exclusively intended as economic growth. But, contrary to modernization theory, that offered national solutions to national problems, the idea of globalization, although continuing to identify the impediments to development with factors internal to the single states, imposes solutions based on the external orientation of the national economies and exalts the benefits, for poor countries, of interdependence under the discipline of the market (Ventrone, 2004, p. 122).

The measures of structural adjustment have represented a strong attack against social citizenship, in particular against the social rights of the poorer classes. For the persons belonging to these classes, the just started path of social citizenship has been all of a sudden interrupted. In the peripheral areas of the world-system, as a consequence, it has meant a tacit confession of the impossibility of a real development (Silver and Slater, 1999, p. 211).

In the "globalization project", as a corollary of the structural adjustment, the control of the mobility of persons has played a fundamental role. Hence, the history of citizenship is also the history of its "contrary", namely the history of the condition of the non-citizen, of the foreigner, and of the ways in which this condition has been imposed to some people, denying them the possibility to have access to citizenship. The history of citizenship, in other words, is the history of the strategies through which the mobility of people, and in particular of the workers, has been limited.

However, at the end of the XX century, along with the crisis of the welfare state, another crucial event has occurred: the birth and consolidation of a new type of citizen, the "electronic" citizen. This new form of membership, more than being a new phase in the process of expansion of rights, presents itself as an alternative way to get access to the same rights. The electronic citizen, in other words, represents a new opportunity to exercise some rights—civil, political and social—which are already part of the equipment of every citizen.[15] In this sense, this new kind of membership could become a more direct form of managing the resources assigned to the citizens.

The main instrument for the building of the e-citizenship is represented by the Internet. Its mass diffusion across the population has led many to speculate about the its potential effects on society at large. Enthusiast have heralded the potential benefits of this technology suggesting that it will reduce inequality by lowering the barriers to information allowing people of all backgrounds to improve their human capital, expand their social networks, search for and find

jobs, and otherwise improve their opportunities and enhance their life chances. In contrast, others caution that the differential spread of the Internet across the population will lead to increasing inequalities, improving the prospects of those who are already in privileged positions while denying opportunities for advancement to the underprivileged (Hargittai, 2003).

More specifically, within the scholars there is considerable disagreement—especially in the United States—about whether inequalities in access and use are increasing or decreasing across different demographic categories. Some argue that with time the majority of the population will be online and no policy intervention is necessary to achieve equal distribution of the medium across the population (Compaine, 2001). Others emphasize the increasing differences among various segments of the population at large (Dickard, 2002).

The issue of the digital divide—that is, the issue concerning the inequalities in the access to the new communication technologies, and in particular to the Internet—is relevant in order to understand the dynamics of e-citizenship. Usually, the phenomenon of the digital divide is considered as a dichotomic one. Some scholars, nevertheless, offer a more complex view of the nature of the digital divide. Mark Warschauer, for example, questions the fact that this phenomenon would have a bipolar nature, suggesting, instead, that it would show a complex gradation with varying degrees of marginalization. Moreover, he underlines how this concept tends to focus attention on causality running from lack of access to diminished opportunities and distracts from causal complexity, which at a minimum would seem to require notions of cumulative and circular causation. In conclusion, Warschauer argues that the concept of digital divide, as it is commonly meant in the social sciences, tends to imply digital solutions through provision of hardware and software and overlooks the critical and complex set of linguistic, educational, and institutional resources that are vital in providing meaningful access to ICT (Warschauer, 2003).

Besides the nature of the digital divide, equally important is the issue of power and of its management within a scenery strongly characterized by the new technologies. The technological developments driving the current changes, in fact, can only be understood when placed within the political context of an unequal and changing pattern of power relations.

More in general, about the real nature of the e-citizenship scholars are split: although many optimists see the widespread access to computer-linked networks as a way of spreading information and knowledge to many more citizens, and thus sharing political and economic influence even more widely, others believe that access to those networks will simply be laid over the same old patterns of geographic and economic inequality. Some scholars, therefore, stressing the emancipatory possibilities offered by the electronic citizen, see in its virtual nature a strong element of change, arguing that cyberspace offers the chance to "ordinary" people to construct new identities which can free them from the imposed classifications of class, race, gender or disability associated with material space and place (Haraway, 1991; Barlow, 1996). Other scholars, instead, stress how the emancipatory possibilities of the e-citizenship clash with the material conditions within which this institution is shaping itself: «*The apolitical and frequently deterministic accounts of the information society ideologues, by envisioning the future through their virtually constructed realities, may well be guilty of overlooking the material impoverishment of large numbers of the world's population by those both better equipped to take advantage of ICTs and also use it for the protection of their privileged position; a social and economic process which has much in continuity with previous epochs*» (Loader, 1998, p. 8).

The systematic exclusion of some groups of people, framed in a *longue durée* perspective, shows therefore the gap between the "virtual" and the "material" dimension of the electronic citizenship. The material exclusion from the produced wealth and from the benefits connected to this wealth tends to deny the hypotheses which are intended to show how the new technologies, spontaneously and automatically, would be able to level off the inequalities: their stratification, instead, seems more likely than their normalization (Sartori, 2006, p. 141).

In this sense, also the ideology of globalization associated with the use of the Internet[16] clashes against the reality represented by the inequalities. The Internet, in fact, is a global phenomenon, but its development, from an economic and social point of view, is tightly tied to the conditions of every single country (Kogut, 2003). The new technologies, thus, more than helping less developed countries to achieve the degree of development of the western ones, may reinforce their dependency on the West (Wade, 2002). Social divisions and distinctions, in fact, have historically remained largely untouched by the massification of a whole range of computer-based technologies, and the Internet could be no different. Low-income groups, therefore, «*become the victims of a powerful triumvirate: a developing cocktail of selfish technology for the economically stable, a propensity for the middle ground to adopt a more reactionary political stance with regard to public investment in information infrastructure, and the continuing invisibility of the social and economic value of public information sources in the information-knowledge chain*» (Haywood, 1998, p. 22-23).

CONCLUSIVE REMARKS

The electronic dimension of citizenship, therefore, reveals a strong continuity with the previous steps in the trajectory of this institution. In fact, it shows a tight interweaving of elements and regressive elements. Since the beginning of the citizen's history, technology has played a role of absolute prominence, but not in the direction of equality: «*Over two hundred years on from the world's first iron bridge, technology has failed to deliver the levels of release from poverty and drudgery that its proponents anticipated. [...] Only the most dement of optimists can expect internetworking to challenge the basis of this kind of society. The differentials run deeper than the deepest cable and they stand every chance of being made worse by the tendency for the "haves" to observe the world from insulated, screen-based cells rather in among the community outside their front door. The opportunities that easy access to information can bring have never been distributed evenly among the members of any community, rich or poor, large or small. What is probably more important is that despite the convergence of whole range of new technologies, easy access to the information that can really empower and liberate people still looks likely to be the preserve of an affluent minority*» (Ibidem, 26).

From this point of view, Internet as an "eraser of difference" is only a myth (Holderness, 1998, p. 36). The virtual nature of e-citizenship, in fact, clashes against the materiality of the conditions in which it has been established. Thus, to be effective for all citizens, including those left outside of the electronic club, this kind of citizenship should be followed up by action in the real space which they inhabit. But this may be difficult: «*Tucked away in virtual worlds, we will be less organised, less streetwise and thus less effective in the real world because we will have forgotten how it works*» (Haywood, 1998, p. 29-30). As a consequence, the utopian dream of a universalistic citizenship, despite its recent electronic form, is still far from becoming true.

REFERENCES

Alber, J. (1982). *Vom Armenhaus zum Wohlfahrtsstaat. Analysen zur Entwicklung der So-zialversicherung in Westeuropa*. Frankfurt/Main: Campus Verlag.

Andrews, G. (Ed.). (1991). *Citizenship*. London: Lawrence and Wishart.

Arendt, H. (1958). *The origins of totalitarianism*. New York: Meridian books.

Arrighi, G. (1994). *The long twentieth century: Money. power and the origins of our time*. London: Verso.

Balibar, E., & Wallerstein, I. (1988). *Race nation classe. Les identités ambigües*. Paris: Éditions la Découverte.

Barlow, J. P. (1996). Declaration of the independence of cyberspace. *Cyber-Electronic List*. 8 February.

Barraclough, G. (1967). *An introduction to contemporary history*. London: C.A. Watts & Co.

Bailyn, B., & Wood, G. S. (1985). *The great republic: A history of the American people*. Lexington. D.C: Heath and Company.

Beiner, R. (Ed.). (1995). *Theorizing citizenship*. Albany: State University of New York Press.

Bosniak, L. (2001). Denationalizing Citizenship. In T. A. Aleinikoff & D. Klusmeyer, (Eds.), *Citizenship today: Global perspectives and practices*, (pp. 237-252). Washington, DC: Carnegie Endowment for International Peace.

Braudel, F. (1981). *Afterthougths on material civilization and capitalism*. Baltimore: Johns Hopkins University Press.

Burley, A. M. (1993). Regulating the world: Multilateralism. international law and the projection of the new deal regulatory state. In J. G. Ruggie (Ed.), *Multilateralism matters: The theory and praxis of an institutional form*. New York: Columbia University Press.

Carr, E. (1945). *Nationalism and after.* London: Macmillan.

Cassi, A. A. (2007). *Ultramar. L'invenzione europea del Nuovo Mondo.* Roma-Bari: Laterza.

Claude, I. Jr. (1956). *Swords into plowshares: The problem and progress of international organization.* New York: Random House.

Compaine, B (Ed.). (2001). *The digital divide: Facing a crisis or creating a myth?* Cambridge, MA: MIT Press.

Costa, P. (1999). *Civitas. Storia della cittadinanza in Europa. Vol I. Dalla civiltà comunale al Settecento.* Roma-Bari: Laterza.

Costa, P. (2005). *Cittadinanza.* Roma-Bari: Laterza.

Dickard, N. (2002). *Federal Retrenchment on the Digital Divide: Potential National Impact.* Washington, DC: Benton Foundation.

Di Meglio, M. (1997). *Lo sviluppo senza fondamenti.* Trieste: Asterios.

Di Meglio, M. (2008). *La parabola dell'eurocentrismo. Grandi narrazioni e legittimazione del dominio occidentale.* Trieste: Asterios Editore.

Elliott, J. H. (1981). *Imperial Spain 1469-1716.* London: Edward Arnold Limited.

Ferrajoli, F. (1997). *La sovranità nel mondo moderno.* Roma-Bari: Laterza.

Gallissot, R. (2001). Nazionalità. In R. Gallissot A. Kilani, & A. Rivera (Eds.), *L'imbroglio etnico in quattordici parole chiave* (pp. 269-278). Bari: Dedalo.

Gargiulo, E. (2008). *L'inclusione esclusiva. Sociologia della cittadinanza sociale.* Milano: Franco Angeli.

Girotti, F. (2002). *Welfare state. Storia modelli e critica.* Roma: Carocci.

Gliozzi, G. (1971). *La scoperta dei selvaggi. Antropologia e colonialismo da Colombo a Diderot.* Milano: Principato.

Gliozzi, G. (1976). *Adamo e il nuovo mondo. La nascita dell'antropologia come ideologia coloniale: dalle genealogie bibliche alle teorie razziali.* Firenze: La Nuova Italia.

Haraway, D. (1991). A cyborg manifesto: Science. technology and socialist-feminism in the late Twentieth century. In D. Haraway (Ed.), *Simians. cyborgs and women: The reinvention of nature.* London: Free Association Books.

Hargittai, E. (2003). The digital divide and what to do about it. In D. C. Jones (Ed.), *New economy handbook.* San Diego, CA: Academic Press.

Haywood, T. (1998). In B. D. Loader (Ed.), *Global networks and the myth of equality: Trickle down or trickle away?.* (pp. 19-34).

Helps A. (1900). *The Spanish conquest in America.* London: John Lame, vol. 2.

Hermes, J. (2006). Citizenship in the age of the Internet. *European Journal of Communication, 21*(3), 295-309.

Hill, C. (1967). *Reformation to Industrial Revolution.* Middlesex: Penguin Books.

Hobsbawm, E. (1975). *The age of capital. 1848-1875.* London: Weidenfeld and Nicolson.

Holderness, M. (1998). In B. D. Loader, *Who are the world's information-poor?* (pp. 35-56).

Hopkins, T. K. (1982). The study of the capitalist world-economy: Some introductory considerations. In T. K. Hopkins & I. Wallerstein *World-systems analysis theory and methodology* (pp. 9-38). Beverly Hills, CA: Sage.

Isin E. F., & Turner B. S. (Eds.). (2002). *The handbook of citizenship studies.* London: Sage.

Jennings, F. (1975). *The invasion of America. Indi-*

ans. colonialism. and the cant of conquest. Chapel Hill: University of North Carolina Press.

Kogut, B. (Ed.). (2003). *The global Internet Economy.* Cambridge, MA: The Mit Press.

Lentini O. (2003). *Saperi sociali. ricerca sociale 1500-2000.* Milano. FrancoAngeli.

Lévi-Strauss, C. (1997). *Tristes tropiques.* New York: Random House.

Loader, B. D. (1998). *Cyberspace divide: Equality, agency and policy in the information society.* London/New York: Routledge.

McMichael, P. (1996). *Development and social change. A global perspective.* Thousand Oaks, CA: Pine Forge Press.

Marshall, T. H. (1950). *Citizenship and social class.* Cambridge: CUP.

McAlister, L. N. (1985). *Spain and Portugal in the new world. 1492-1700.* Minneapolis: University of Minnesota Press.

Merker, N. (2006). *Europa oltre i mari. Il mito della missione di civiltà.* Roma: Editori Riuniti.

Mezzadra, S. (2004). *Cittadinanza : soggetti. ordine. diritto.* Bologna: CLUEB.

Ntia (National Telecommunications and Information Administration). (1999). *Falling through the Net: Defining the digital divide.* July. www.ntia.doc.gov.

Pagden, A. (1982). *The fall of natural man: The American Indian and the origins of comparative ethnology.* UK: Cambridge University Press.

Pagden, A. (1995). *Lords of all the world. Ideologies of empire in Spain. Britain and France c. 1500-1800.* New Haven-London: Yale University Press.

Pagden, A. (2003). Human rights. Natural rights. and Europe's imperial legacy. In *Political Theory, 31*(2), 171-199.

Polanyi, K. (1944). *The Great tansformation.* New York. Holt: Rinehart & Winston Inc.

Reinhard, W. (1996). *Kleine Geschichte des Kolonialismus.* Stuttgart: Kröner Verlag.

Ritter, G. A. (1991). *Der Sozialstaat. Entsehung und Entwicklung im internationalen Vergleich.* München: R. Oldenbourg Verlag GmbH.

Rosanvallon, P. (1992). *Le sacre du citoyen: histoire du suffrage universel en France.* Paris: Gallimard.

Sartori, L. (2006). *Il divario digitale. Internet e le nuove disuguaglianze sociali.* Bologna: il Mulino.

Schurmann, F. (1974). *The logic of world power: An inquiry into the origins, currents. and contradictions of world politics.* New York: Pantheon.

Schmitt, C. (2003). *The* Nomos *of the earth in the international law of the* Jus Publicum Europaeum. New York: Telos Press Publishing.

Scuccimarra, L. (2006). *I confini del mondo. Storia del cosmopolitismo dall'Antichità al Settecento.* Bologna: il Mulino.

Silver, B. J., & Slater, E. (1999). The social origins of world hegemonies. In G. Arrighi & B. J. Silver (Eds.), *Chaos and covernance in the modern world system* (pp. 151-216). Minneapolis: University of Minnesota Press.

Smelser, N. J., & Baltes, P. B. (Eds.). (2001). *International encyclopedia of the social and behavioral sciences.* Oxford: Elsevier.

Stannard, D. E. (1992). *American holocaust.* New York: Oxford University Press.

Todorov, T. (1984). *The conquest of America. The question of the other.* New York: Harper & Row.

Ventrone, O. (2004). *Globalizzazione. Breve storia di un'ideologia.* Milano: FrancoAngeli.

Wade, R. H. (2002). Bridging the digital divide: New route to development or new form of dependency?. In *Global governance, 8*, 443-466.

Wade, R. H. (2005). Failing states and cumulative causation in the world system. In *International Political Science Review, 26*(1), 17-36.

Wallerstein, I. (1974). *The modern world-system. Vol. 1. capitalist agriculture and the origins of the European world-economy in the sixteenth century.* New York: Academic Press.

Wallerstein, I. (1983). *Historical capitalism.* London: Verso.

Wallerstein, I. (1995a). *After liberalism.* New York: The New Press.

Wallerstein, I. (1998). *Utopistics: Or, historical choises of the twenty-first century.* New York: The New Press.

Wallerstein, I. (2003). Citizens all? Citizens some! The making of citizen. *Comparative Study in Society and History, 45*, 650-679.

Warschauer, M. (2003). *Technology and social inclusion: Rethinking the digital divide.* Cambridge, MA: MIT Press.

Williams, E. (1944). *Capitalism and slavery.* London: Deutsch.

Wimmer, A., & Glick Schiller, N. (2002). Methodological nationalism and beyond: Nation-state building. Migration and the social sciences. *Global Networks, 2*(4), 301-334.

Zolo, D. (Ed.). (1994). *La cittadinanza. Appartenenza. identità. diritti.* Roma-Bari: Laterza.

ENDNOTES

[1] For a broader and more detailed analysis of the worldwide forms of exclusion connected to citizenship and of the political and economic reasons which underlie these forms see Gargiulo, 2008.

[2] The debate on the humanity of Indians is summarized in McAlister, 1986.

[3] For a close examination of this contention see Pagden, 1982, in particular chapters 5 and 6. About the contention between de Las Casas and Sepúlveda considered within the broader discourse on the human rights see Pagden, 2003.

[4] See about Schmitt, 2003: 114.

[5] On the exceptions to this principle see McAlister, 1986, p. 207.

[6] These myths have strongly influenced the representation of the British colonies as entities radically different from the French and the Spanish colonies (Pagden, 1995, 128).

[7] Also the Netherlands is a very good example of the "technological-commercial superiority" ideology. On the "paternalist multiculturalism" practised by Dutch in their colonies see Merker, 2006, particularly Chapter III.

[8] On the French revolution see Rosanvallon, 1992: 473-474.

[9] On the birth of the welfare state see, among others, Alber, 1982 and Ritter, 1991.

[10] The policies enforced by Bismarck are a good example of this dynamic. Also the social reforms enforced in Great Britain by the liberal party after 1906 are also a product of the compromise between the interests of the masses and the interests of the nations.

[11] On the history of the imperialistic and colonialist politics see Reinhard, 1996.

[12] On the significance of New Deal as a model for the American hegemonic projects after World War two see Schurmann, 1980 e Burley, 1993.

[13] For a more precise definition of corporate liberalism and on the differences between corporate liberalism and other forms of organized capitalism see Lentini, 2003: 286-292.

14 These exceptions are represented by the individuals which are holder of a double citizenship and by the stateless individuals, who lack citizenship.

15 On this aspect, see the "Information Rights", already ratified by the article n. 19 of "The Universal Declaration of Human Rights": «Everyone has the right to freedom of opinion and expression; this right includes freedom to hold opinions without interference and to seek, receive and impart information and ideas through any media and regardless of frontiers». Other rights, as those to the access and the accessibility, make explicit reference to the inclusion/exclusion mechanism which characterizes the information system and, more generally, the telecommunications system. Seemingly, these rights are inextricably tied to the new computer and communicative technologies. In this sense, they would seem to characterize themselves as rights "of new generation."

16 On the relation between globalization and the Internet see Warschauer, 2003.

Section II
The Conjuncture:
The Geopolitics of
Technological Innovations

Chapter III
International Organizations, E-Government and Development

Oreste Ventrone
University of Naples Federico II, Italy

ABSTRACT

Following the diffusion of e-government in the high income countries, international organizations, notably the UN, OECD, World Bank, have promoted the implementation of e-government practices in developing countries. However, the little researce conducted in the field show that the overwhelming majority of e-government projects end up in total or partial failure. Despite the recognition of the need to take into account local specificities and to get the locals involved in the process, e-government in developing countries still appears essentially as a mere transfer operated by donor countries' firms with Western technologies. Moreover, as these technologies are mostly proprietary, they prevent institutions and users from developing countries to modify and adapt the tools to their particular needs and lock them in a position of permanent technological dependency. The causality chain between e-government, good governance, and democracy, if at all plausible, looking at history should be probably read the other way around. In fact, some scholars consider the contribution of e-government to overall development irrelevant, if not negative, in such that it diverts funds from higher priorities.

E-government, which initially developed in the United States in the 1990s, and gradually spread to other technologically advanced countries, has rapidly become globalized thanks, primarily, to the substantial involvement of international organizations in the elaboration, dissemination and promotion of government practices based on the utilization of information and communication technologies (ICTs), notably the Internet, worldwide.

The United Nations (UN), with resolution 73/1998 of the Plenipotentiary Conference by its agency, the International Telecommunication Union (ITU), gave way to consultations amongst UN agencies which led to the resolution being discussed in ITU's Council session of 2001 and adopted by the General Assembly in its 90th plenary session and which endorsed the holding of the World Summit on the Information Society (WSIS), held in two phases: in Geneva from 10-12 december 2003 and in Tunis in 2005. This resolution (56/183) stated "the urgent need to harness the potential of knowledge and technology for promoting the goals of the United Nations Millennium Declaration", according to the conviction of

[...] the need, at the highest political level, to marshal the global consensus and commitment required to promote the urgently needed access of all countries to information, knowledge and communication technologies for development so as to reap the full benefits of the information and communication technologies revolution, and to address the whole range of relevant issues related to the information society, through the development of a common vision and understanding of the information society and the adoption of a declaration and plan of action for implementation by Governments, international institutions and all sectors of civil society. (United Nations, 2001)

Alongside the unique nature of the organizing agency (ITU) as the only UN agency based on the cooperation between governments and the private sector, the fundamental novelty of WSIS was the official participation, for the first time, of civil society[1] at a UN summit. This was indeed regarded as an 'historical event'.

The number of international organizations that have joined the debate on e-government - conceived as a crucial set of tools to foster development - includes the World Bank, International Monetary Fund (IMF), Organization for Economic Cooperation and Development (OECD), World Trade Organization (WTO), G8 and various other regional or sectoral multilateral and private institutions.

The World Bank, for instance, dedicated its 1998/99 World Development Report (WDR), *Knowledge for Development*, to the role of knowledge in the development process, deemed more important than capital itself. The WDR recognizes the difficulties involved in closing the knowledge gap between developing[2] and high-income countries as the latter constantly push the knowledge frontier outward and the gap in knowledge creation remains greater than differences in income. At the same time, it notes that

developing countries need not reinvent the wheel - or the computer, or the treatment for malaria. Rather than re-create existing knowledge, poorer countries have the option of acquiring and adapting much knowledge already available in the richer countries. With communication costs plummeting, transferring knowledge is cheaper than ever. Given these advances, the stage appears to be set for a rapid narrowing of knowledge gaps and a surge in economic growth and human well-being (World Bank, 1999, p. 2).

The Bank has since launched a new Web site and a number of initiatives and publications concerning e-government and, more generally, e-development, thus rapidly becoming one of the most influential institutions on the subject, not least because of its direct funding capability. However, ten years after these first steps, the overall balance of e-government in development is a mixed one, while technical implementation remains problematic and even its very logic still appears highly controversial.

TECHNICAL IMPLEMENTATION

E-government in high-income countries has developed gradually, through a process of progressive, step-by-step integration and networking of already established principles and practices, with a certain degree of proximity between designers and users. However, even in these countries, the process has seldom been a smooth and straightforward one. In developing countries, the technical implementation of e-government projects is likely to face even more, and different, challenges.

Technology is frequently considered as somewhat neutral, something that can be isolated and replicated without affecting other dimensions, but this is seldom the case. Technologies carry with themselves an inscribed "vision of (or prediction about) the world" (Akrich, 1992, p. 208, cit. in Heeks, 2002b). "This 'world-in-miniature' includes inscriptions of how processes will be undertaken; of the values people will have; of the structures in which they are to be placed; and so on. Technology must therefore be seen not in a reductionist manner as a separate dimension, but in a systemic manner as a group of related dimensions" (Heeks, 2002b, p. 104). Heretofore, the transfer of technologies for e-government has been based primarily on the reproduction of best practices identified in, and patterned on, experiences of high-income countries. The transposal of these practices, however, has proved to be problematic and is rarely crowned with success. After all, having been developed in different organizational contexts, stemming from specifical needs and priorities, a high rate of failure should be anticipated. Moreover, independent evaluations of the outcomes of e-government projects in the developing world, and especially analyses that take into account more than a single case, are scarce. This can be attributed, according to Heeks, to the fact that "Those who have the will to evaluate - such as academics - often lack the resources and capacity. Those who have the resources - such as the donor agencies - often lack the will to evaluate" (Heeks, 2002b, p. 102).

Just to name a few of the difficulties involved in the technical implementation of e-government, developing countries administrations are gener-ally already using some form of information technology, with some agencies more advanced than others and plus the ensuing problems of compatibility. "In sum, the IT readiness within the government administration is uneven: it is a matter of technology (old and new platforms coexisting and new ones being implemented all the time, often independently from the e-government projects); the *de facto* independence and autonomy of the Ministries; the different practices in systems implementation; sometimes inappropriate user involvment and training; the need for a deep culture change towards the new ways of working, and so on" (Ciborra, 2005, pp. 264-265)[3]. The provision of services is further conditioned by the number of dependencies involved and by the possibility (or the lack of) of unifying payments in a 'one stop shop' fashion. But what is possibly the single most important obstacle to the implementation of e-government projects is the resistence opposed by the 'angry orphans', those who are to be substituted by electronic processing or those who have to learn and adapt to the new processes.

CONTEXTUALIZING E-GOVERNMENT IN DEVELOPMENT

Bruno Lanvin's preface to *The E-government Handbook for Developing Countries* begins with a series of statements which summarize the most common positive assumptions about the relationship between e-government and development: "The process of globalization may very well entail both a reduction of income disparities among countries, and increasing income inequalities within countries. [...] On both fronts, e-government will be a powerful tool to help all types of economies (developed, developing and in transition) to bring the benefits of the emerging global information society to the largest possible part of their respective populations". E-government can directly affect the "cost effectiveness in government and public operations", determining "significant savings in areas such as public procurement, tax collection and customs operations, with better and continuous contacts with citizens, especially those living in remote or less densely populated areas". It can also have important indirect effects, such as "greater transparency and accountability in public decisions, powerful ways to fight corruption, the ability to stimulate the emergence of local e-cultures, and the strengthening of democracy" (World Bank, 2002, p. 2).

GLOBALIZATION

The Bretton Woods institutions[4] have been among the first and most active institutions in promoting globalization. Until the 1970s, under the label of modernization, mainstream development theories saw this process as the replication of the idealized model of the Western path. The main ingredients of this recipe included rapid industrialization (regularly used as a synonym of development) through import substitution, substantial control of the flow of commodities and, in particular, capital across frontiers and an interventionist role of the state in identifying and promoting top priorities.

In the World Development Report of 1980, the World Bank (1980) radically redefined the development concept—as successful integration in the world market—creating the conditions for the emergence of a new vision, known as the 'Washington consensus', which predicated export-led growth, liberalizaton of commercial and financial flows, minimal state, privatization and control of inflation as the prime instrument of sound monetary policies. Development was always seen essentially as economic growth but, compared to modernization models, which provided national solutions to national problems, globalization models, while still identifying the main impediments to development in national factors, imposed solutions based on outward orientation of national economies, emphasizing the advantages of integration and interdependence under market discipline.

This new approach, articulated and structured during the 1980s and 1990s, which even pushed itself as far as questioning the need and scientific dignity of development theory, has become the object of heavy criticism from a growing number of scholars; initially outsiders, and later also very influential ex-insiders, like Joseph Stiglitz, formerly chief economist of the World Bank itself. Even more significant was the fact that development history and theory became the ideological battleground of an unequal struggle between the Bretton Woods institutions, with their strong western (especially US) neoliberal bias, on one side, and, on the other, the United Nations Development Programme (UNDP) and United Nations Conference on Trade and Development (UNCTAD), which tended to reflect more the point of view of developing countries. On the opposite front to this new version of the neoliberal orthodoxy, UN agencies acquired a new vitality in the 1990s. During this decade, two important challenges to the Washington consensus took form. The first was the construction of a new approach for a sustainable human development, undertaken by the UNDP, which took as a starting point the United Nations Children's Fund (UNICEF) report *Adjustment with a Human Face*'s critique to the mainstream and further elaborated it in the *Human Development Reports*, starting in 1990. The second emerged from the analyses produced in the perspective of *late comers*, and can be considered the result of the convergence between the positions of United Nations Economic Commission for Latin America and the Caribbean (ECLAC) and those of the studies on the development of East Asian countries contained in the United Nations Economic and Social Commission for Asia and the Pacific (ESCAP) report *Restructuring the developing economies of Asia and the Pacific in the 1990s* and further articulated in UNCTAD's *Trade and Development Report*s (TDR). The

TDR of 1997, titled *Globalization, Distribution and Growth*, for example, told a different story about globalization and inequality.

The big story of the world economy since the early 1980s has been the unleashing of market forces. The deregulation of domestic markets and their opening up to international competition have become universal features. The 'invisible hand' now operates globally and with fewer countervailing pressures from governments than for decades. Many commentators are optimistic about the prospects for faster growth and for convergence of incomes and living standards which greater global competition should bring. However, there is also another big story. Since the early 1980s the world economy has been characterized by rising inequality and slow growth. Income gaps between North and South have continued to widen. In 1965, the average per capita income of the G7 countries was 20 times that of the world's poorest seven countries. By 1995 it was 39 times as much.

And since then, despite the spectacular rise of China and India, the situation has not changed very much, with one billion people living on less than one dollar a day.

Historically, there is no evidence of the power of free markets to promote late comers' economic growth and convergence. Invariably, the countries that have achieved high and sustained economic growth have done so through a skillful combination of tariff and non-tariff barriers together with various forms of sectoral strategic policies. The United States, for example, is currently the most vocal advocate of free markets, but in the nineteenth century, when it was the main emerging economic power on the world stage, it selectively imposed very high tariff barriers - between 25 and 60 percent depending on the sector, with an average level of 40 percent (Bairoch and Kozul Wright, 1996, pp. 19-20). It was indeed Alexander Hamilton, the first US Secretary of the Treasury, who systematically outlined for the first time the

'infant industry' argument, in his *Reports of the Secretary of the Treasury on the Subject of Manufactures* (1791). He argued that foreign competition and the 'forces of habit' could have hindered the development of internationally competitive US firms, unless their initial losses were guaranteed by government aid. This aid could take the form of import duties or, in rare cases, even the prohibition of import (Chang, 2003). Reproposed by authors such as Adam Muller in Germany, Jean-Antoine Chaptal and Charles Dupin in France the argument in favor of protectionism found its most accomplished formalization in Friedrich List's works. Exiled in the United States from 1825 to 1831, he could confront German backwardness, studied in his continental experience, and the problems of American development. During that time, more precisely in 1827, he was commissioned to write a series of papers advocating protectionism for the United States and formulated a first version of his critique of the free trade paradigm. Later collected in a volume titled *Outlines of American Political Economy*, these works earned him a reputation in America and the approval of James Madison (Lentini, 2003).

List's approach to the development of late comer countries culminated in the publication of his *magnum opus, The National System of Political Economy* (1841). Opening with an historical reconstruction of national cases, the book expounds List's conception of organizational development as the economic and institutional behavior of entrepreneurial groups united by a political bond that somehow comes ahead of and determines the degree of competitiveness. Moreover, it is strongly emphasized the strategic nature of the kind of division of labor imposed and of the underlying economic policies. This way of conceiving the world, by then already established as 'school' reached its apex with the Smithian paradigm, a paradigm which List polemically labeled as 'cosmopolitical', as it does not take into account the national peculiarities of development processes. However, List's critique of Smith, at times harsh, conceals a substantial adhesion, which, at the same time, constitutes an integration and further elaboration of Smith's thought. In his reading of the *Wealth of Nations*, List re-examines the process of paradigm construction and emphasizes how the British, in order to advance their interests in the world division of labor, imposed, through Smith, the idea of 'free trade' (cosmopolitism) while the vast majority of late comers need an intermediate phase of 'protection' to allow for the competitive development of their activities to take place.

This attitude towards development and competitive integration in the world market, examined by List as an historical constant in the development of every core country, is identified not only as the economic behavior of the French starting from Colbert, but also as a basic guideline starting from the Venetians and the Dutch ending with the British, which have made of it the inherent principle of their success. Therefore, the free-trade doctrine, displayed by Britain in the phase of its primacy on the world markets, serves essentially the interests of the core countries, while, at the same time, harms all the others (*ibidem*).

It is a very common clever device that when anyone has attained the summit of greatness, he kicks away the ladder by which he has climbed up, in order to deprive others of the means of climbing up after him. In this lies the secret of the cosmopolitical doctrine of Adam Smith, and of the cosmopolitical tendencies of his great contemporary William Pitt, and of all his successors in the British Government administrations. Any nation which by means of protective duties and restrictions on navigation has raised her manufacturing power and her navigation to such a degree of development that no other nation can sustain free competition with her, can do nothing wiser than to throw away these ladders of her greatness, to preach to other nations the benefits of free trade, and to declare in penitent tones that she has hitherto wandered in the paths of error, and has now for the first time

succeeded in discovering the truth (List, 1885, pp. 295-6).

Ironically, during World War II, when the terms of trade had reversed, after a period of generalized withdrawal from the world market the US turned the liberalist mantra against its former, most authoritative priests, the British, forcing them to dismantle their colonial empire and to open up to free trade once again.

EFFECTIVENESS AND COST EFFICIENCY

E-government practices can indeed be of great benefit in situations in which the virtualization of communication and services helps to overcome great distances, physical impediments, or where transportation is lacking, costly and/or of poor quality. In these cases, of course, especially in the logic of public service, considerations about costs give way to the essential nature of the needs to fulfill. An argument often advanced in support of e-government is the effect of disintermediation that it can produce. Allegedly, by reducing the passages and dependencies of the administrative process, e-government can make the latter more rapid and straightforward and, as long as it reduces the number of people implied in the process itself, it can also counter corruption minimizing the opportunities these people have to take advantage of their position to act as gatekeepers and arbitrarily impose bribes. At the same time, the disintermediation effect is not likely to be obtained automatically by e-government implementation. This could be possible only in the case of countries where basic services, like Internet connectivity, computer literacy and ownership are widely diffused. In many developing countries, however, services are of poor quality or lacking altogether, levels of computer literacy are very low and a personal computer can become so only for a small fraction of the population. Hence, intermediation persists, only the intermediaries change, from public officials to those who own and/or are able to operate a computer. However, as Heeks suggests, for e-government to be effective, "citizens will have to rely on reintermediation models that insert a human intermediary between the citizen and the growing digital infrastructure of e-government. Where insitutionally based, these can be thought of as 'intelligent intermediaries' that add human skills and knowledge to the presence of ICTs" (Heeks, 2002a, p. 105).

As far as cost efficiency is concerned, while it can be said that successful implementation of ICTs in government practices can significantly reduce the costs in administrative operations of high-income countries (for the very reason of the high income of these countries), in most developing countries, where the cost of labor is much lower, the cost advantages of ICTs, even in cases of successful implementation, are likely to be lower, if not reversed, compared to the use of human labor. Of course, the kind of technology used can make a huge difference. Proprietary hardware and software can be very expensive for developing countries.

This relationship is neatly demonstrated by comparing licence fees with a country's GDP per capita [...]. As is quickly apparent, in developing countries, even after software price discounts, the price tag for proprietary software is enormous in purchasing power terms. The price of a typical, basic proprietary toolset required for any ICT infrastructure, Windows XP together with Office XP, is US$560 in the U.S.. This is over 2.5 months of GDP/capita in South Africa and over 16 months of GDP/capita in Vietnam. This is the equivalent of charging a single–user licence fee in the U.S. of US$7,541 and US$48,011 respectively, which is clearly unaffordable. Moreover, no likely discount would significantly reduce this cost, and in any case the simple fact that a single vendor controls

any single proprietary software application means that there can never be a guarantee that any discount offered is intended to be sustained for the long term, rather than as a temporary measure used to tempt consumers into a lock–in situation at which point in time the discount can be reduced (Ghosh, 2003).

Before the application of the Agreement on Trade-Related Aspects of Intellectual property rights (TRIPs), the countries of sub-saharian Africa, in one of the poorest areas of the world, paid around 24 billion dollars annually (FOSSFA, 2004, p.7, cit. in May, 2006a, p.123) for the use of proprietary software. However, until recently, the cost of software was a less relevant issue in many developing countries because of generalized software piracy, which, in turn, has helped establish the monopolies of proprietary software, especially Microsoft, that runs on 90% of the world's computers.

Nowadays, following the Agreement on Trade-Related Aspects of Intellectual Property Rights, the prospects for developing countries seem to have worsened as the WTO has assumed the role of safeguard of the system of protection of intellectual property. Before signing the Agreement, many countries had little or no regulation on the subject and given the disproportionate number of patents and copyrights owned by developed countries, it is reasonable to expect an increase in the net transfer of royalties from developing to developed countries. Moreover:

Even after the Doha modifications TRIPs leaves in place a much more restrictive environment for technology transfer than the older industrialized countries enjoyed during the early stages of industrialization and the new industrialized countries of East Asia enjoyed during theirs. Recall that Japan, Taiwan, and South Korea were each known as 'the counterfeit capital' of the world in their time. And the US in the nineteenth century, then a rapidly industrializing country, was known - to

Charles Dickens, among many aggrieved foreign authors - as a bold pirate of intellectual property (Wade, 2003, p. 626).

Free and open source software (FOSS) can be an appealing alternative, especially for developing countries. The Linux operating system is free and its source code is open, which means that the software can be freely modified and adapted. Being free, Linux can cut down the cost of hardware if bundled in a new computer instead of Windows and can be customized to local needs. But current Linux releases, in spite of big improvements in the user interface, still appear less user-friendly than its proprietary competitors from Microsoft and Apple. There are still compatibility issues with some peripherals and applications and while there are very good free applications for most of the common tasks, several commercial software companies do not release Linux versions of their programs, e. g. you can download and use OpenOffice for free or buy Sun Microsystems' StarOffice but you cannot run Microsoft Office on Linux. This can be a problem in the private sector, where foreign contractors often require their local suppliers to use the same software platforms as their own. These problems of interoperability seem to trap Linux in a catch-22, because it cannot become widely used unless its applications become popular, but its applications will become popular only if Linux is widely used.

For e-government and the public sector more generally, however, FOSS seems to be a viable alternative for a number of reasons. Now that IPRs enforcement of IPRs becomes mandatory even for the least developed countries, under the pressure exerted by Western countries and firms - "Microsoft, for example, only agreed to set up a software production facility in Egypt on condition of tougher government legislation against piracy" (Heeks, 1999, p. 19) - it is likely for the public sector, and e-government along with it, to be one of the areas economically most affected by the enforcement of IPRs. In addition, the pos-

sibility to independently modify the open source software allows its adaptation to local needs, for example, local language version development, which, depending on the size of the markets involved, can be overlooked by commercial software houses (Bridges.org, 2004). According to Vinay, professor of computer science at the Indian Institute of Science:

Free software allows teachers and students to look into the software and not just treat it as a mystical black box. Children like to play with things, tear them apart and (if we are lucky!) put things back together. Free software encourages such exploration, allows interaction with other children (without inducing any guilt of being a 'pirate'!), and learn to understand large complex programs. Dexterity in creation and not in usage is crucial if a developing country like India has to create its own niche. Or else, we will merely be followers (Vinay, 1999, p. 6).

In fact, free software allows developing countries to develop their own software sector and governments can lead the way adopting FOSS in the first place. This could help further the diffusion of these technologies without resorting to direct, subsidized, industrial policies, towards which international financial institutions (IFI) have shown stubborn hostility (Wade, 1996; Woo-Cumings, 1999).

Compatibility issues associated with business partners' technologies have the least influence on governments in developing countries, because they exchange relatively little data with more established groups and systems. However, because governments are the biggest single user of hardware and software in most developing countries, any decisions they make about adopting a specific operating system can have a powerful secondary compatibility effect on the OSs subsequent spread and dissemination nationwide (Kshetri, 2004, 80).

Last but not least, governments around the world have expressed concern about the potential threats of proprietary software on national security. Because of its hidden protocols, it can be very difficult to know what the software is doing or whether sensible data is being inappropriately shared. In order to avoid these risks many countries are enacting laws and providing guidelines to promote the diffusion of Linux. "By mid-2002, Latin American countries such as Brasil, Mexico, Argentina, and Peru had proposed bills mandating the use of open source in government organizations" (Kshetri, 2004, p.80).

On the issue of free vs. proprietary software, international organizations can play a crucial role. For instance, The United Nations Educational Scientific and Cultural Organization (UNESCO) actively promotes FOSS through its Free Software Portal, which 'provides a single interactive access point to pertinent information for users who wish to acquire an understanding of the Free Software movement, to learn why it is important and to apply the concept' and, together with the New Zealand Digital Library Project[5] and the Belgian NGO Human Info[6], has launched the Greenstone software suite, a tool for building and distributing digital library collections for the free dissemination of knowledge. But multilateral aid can also have serious drawbacks, spurring a dynamic similar to drug dependence.

Microsoft and other major suppliers also have been very active in supporting their products' position within multilateral informational development programs. In January 2004, at the annual World Economic Forum, Microsoft announced a one billion dollar grant (cash and software) to fund a program with the UN Development Program (UNDP) to bring computers to local communities in developing countries. Likewise, Microsoft had also signed agreements with the New Partnership for African Development and the United Nations High Commissioner for Refugees. The involvement of Microsoft in these projects prompted a largely negative reaction from FOSSFA[7], with one discussion list correspondent suggesting that the UNDP was promoting 'technological slavery' through the use of Microsoft products rather than supporting the development of local programming skills (May, 2006b, p. 155).

DEMOCRACY FROM CYBERSPACE?

In the last three decades, starting with the Structural Adjustment Programs, international donor agencies have emphasized the role of the public sector in developing countries. Postcolonial states were considered overly bureaucratic, inefficient and corrupt. But, if in the 1980s they placed the emphasis on the size of the public sector, i. e. on the shrinking of its scope through deregulation, privatization and unhindered integration in the world economy, later on the focus of their intervention shifted to the qualitative aspects of government action (Amoretti, 2007, p. 333). Some of the main donor agencies and international financial institutions, like the IMF and World Bank, have increasingly made their aid conditional on reforms promoting 'good governance', whose conception is based on the Western neoliberal consensus around New public management (NPM) as an agenda for policy reform. The UN *World Public Sector Report* (2005) addresses this approach claiming that "it can be argued that this issue seems to fall into the trap of predicating all 'development' on the need to follow a single path or trajectory: the trail that has been blazed by 'advanced Western democracies'. Such a view is open to the charge of ethnocentricity. This transfer of Western blueprints and models as if they were relevant to all times and places has aptly been labelled 'institutional mono-cropping'. Crucially, such an approach is likely to miss the opportunities for improvement that emerge from tapping local problem-solving capacities". It would rather be advisable "to suggest a somewhat prudent scenario in which innovation has to await the slow, attentive development of support systems and institutional frameworks that Western countries have developed over centuries" (p. 61). As a matter of fact, the institutional reforms proposed by this model in order to transform a developing country into a developed one only tell us what it should look like at the end of the process, but they don't tell us anything about the process itself (Kahn, 2002).

In the vision of Bretton Woods institutions good governance coincides with a model of service delivery, a minimal state that treats the citizen as a customer. For example, a World Bank report (2005) states that "a major step towards" the creation of "the 'right' enabling environment" for the development of ICTs "is to establish clear and transparent governance structures and respect for the rule of law", and the recipe for regulatory reform includes "encouraging market-based approaches and ease of market entry; promoting business confidence and clarity; enhancing transactional enforceability; ensuring interoperability (of systems, standards, networks, etc.); and protecting intellectual property and consumer rights" (p. xv).

The idea that e-government could help achieve good governance and promote democracy and development appears very controversial. Given that it is constructed on an abstract model, which does not take history into account, this idea is based on a synoptic illusion. The countries where e-government is well established are also, generally, those which are the most developed, democratic and transparent in government action. However, the transformation of the simultaneous presence of these factors in correlation, or even in a causality chain, constitutes a naive and arbitrary, if not deliberately ideological, theoretical choice. This is comparable to the transformation of the relationship between neoliberalism and economic growth in causal nexus, with the first determining the second, which is central to the ideology

of Bretton Woods institutions. Just as it cannot be claimed that the recently developed countries have achieved growth following the *laissez-faire* recipes of neoliberalism, similarly, it cannot be argued that these countries have distinguished themselves with high levels of transparency, good governance or democracy.

Moreover, in the few countries that have experienced high levels of economic growth, the state has been far from minimal. It has invariably assumed an interventionist role, from the agrarian reforms in Taiwan (Wade, 1990), to various forms of industrial policy (Amsden, 1989; Chang, 2002) or, as in South Korea and China, heavily subsidized supercomputer projects. These policies have spurred a host of studies on 'the developmental states' (Woo-Cumings, 1999). These states have been generally undemocratic, and, as in the case of Thailand, also very corrupt, and this is not just the case for the recently developed countries. According to Chang (2002), today's leading economies, like UK, USA and Italy, in earlier times had relatively low levels of institutional development compared to the countries that are at a comparable level of development today. Even when the levels of transparency seem to be higher in less developed countries, as in some post-communist nations, a closer look can reveal a different reality.

Statistical analysis confirms our research hypotheses that both the level of democracy and cultural legacies affect openness of cabinet-level Web sites in OECD and post-communist countries. Regression analysis shows that both historical legacy and the level of democracy affect openness of electronic governments. Democracy is a major factor affecting variation of openness of cabinet-level Web sites in OECD and post-communist countries. The Polity and Freedom House indexes of democracy are positively associated with the openness of cabinet-level Web sites. The level of economic development is also positively associated with the openness index. (Katchanovski & La Porte, 2005, pp. 677-678)

Therefore, if the e-government > good governance > democracy > development causality chain exists at all, it is more likely to work the other way around.

CONCLUSION: WITHER E-GOVERNMENT?

A tremendous amount of money has been spent in developing countries on putting government services online. However, the results so far have been disappointing, as most e-government projects in developing countries have ended in failure (Heeks, 2003).

As Wade (2002) points out, the movement towards e-governance is particularly worrisome. Developed countries' governments and international donor agencies make their aid conditional on good governance, to be achieved through the creation of integrated ICT infrastructures in the public sector. Recipient governments are thus steered towards Western firms, which are the only subjects capable of providing the needed know-how and technologies. The difficulties experienced by Western countries are brushed aside and developing countries are induced to implement ICTs for e-governance simply to get more aid. "They then tie themselves to the standards of the ICT suppliers that their aiders choose and enter an open-ended dependency on these suppliers for the continued functioning of their public administration" (p. 461).

In the worst case scenario, as Ciborra notes (2005), e-government could even be turned into an instrument for the rich metropolitan states to 'govern at a distance' the potentially dangerous, weak, borderland states, through the control of sophisticated methodologies and technologies. As already shown by the Chinese government, far from being a tool that automatically promotes democracy, ICT can be a terrific instrument of oppression; from big brother-like sophisticated systems of video surveillance to Internet censorship through content filtering, which in some cases is willfully supported by Western firms eager to secure greater access to China's fast growing market.

This said, e-government should not be considered as a value in itself. The ifs and hows of its introduction in developing countries should be carefully evaluated case by case, as the impact on development can be irrelevant or, even worse, counterproductive. In countries fraught with problems of day-to-day survival, it could indeed divert badly needed funds from higher priorities.

REFERENCES

Akrich, M. (1992). The description of technical objects. In W. E. Bijker, & J.Law (Eds.), *Shaping technology/building society* (pp. 205-224). Cambridge, MA: MIT Press.

Amoretti, F. (2007). International organizations ICTs policies: E-democracy and e-government for political development. *Review of Policy Research*, *24*(4), 331-344.

Amsden, A. H. (1989). *Asia's next giant: South Korea and late industrialization*, New York: Oxford University Press.

Bairoch, P., & Kozul Wright, R. (1996). *Globalization myths: Some historical reflections on integration, industrialization and growth in the world economy* (Discussion Papers, No. 113). Geneva: UNCTAD.

Bridges.org (2004). *Straight from the source: Perspectives from the African Free and Open Source Software Movement.* Retrieved December 3, 2007, from http://www.bridges.org/files/active/1/straight_from_the_source_may04.pdf

Chang, H. (2002). *Kicking away the ladder: Development strategy in historical perspective.* London: Anthem Press.

Ciborra, C. (2005). Interpreting e-government and development: Efficiency, transparency or governance at a distance?. *Information Technology & People, 18*(3), 260-279.

ESCAP (1990). Restructuring the developing economies of Asia and the Pacific in the 1990s. New York: United Nations.

Free and Open Source Software Foundation for Africa (FOSSFA) (2004). *FOSSFA Action Plan 2004–2006.* Nairobi: FOSSFA.

Ghosh, R. A. (2003). Licence fees and GDP per capita: The case for open source in developing countries. *First Monday, 8*(12). Retrieved November 18, 2007, from http://firstmonday.org/issues/issue8_12/ghosh/index.html

Hamilton, A. (1791). *Report of the Secretary of the Treasury of the United States, on the subject of manufactures: Presented to the House of Representatives, December 5, 1791.* Philadelphia, PA: Childs and Swaine.

Heeks, R. B. (1999). Software strategies in developing countries. *Communications of the ACM, 42*(6), 15-20.

Heeks, R. B. (2002a). E-Government in Africa: Promise and practice. *Information Polity, 7*(2/3), 97-114.

Heeks, R. B. (2002b). Information systems and developing countries: Failure, success and local improvisations. *The Information Society, 18*(2), 101-112.

Heeks, R. B. (2003). *Most e-Government for development projects fail: How can risks be reduced?* (IDPM i-Government Working Paper No. 14). UK: University of Manchester, Institute for Development Policy and Management

Kahn, M. H. (2002). State failure in developing countries and strategies of institutional reform. *Proceedings of the ABCDE Conference*, Oslo, 24-26 June.

Katchanovski, I., & La Porte, T. (2005). Cyberdemocracy or Potemkin e-villages? Electronic governments in OECD and post-communist countries. *International Journal of Public Administration, 28*(7), 665-681.

Kshetri, N. (2004). Economics of Linux adoption in developing countries. *IEEE Software, 21*(1), 74-81.

Lentini, O. (2003). *Saperi sociali, ricerca sociale, 1500-2000,* Milan: Franco Angeli.

List, F. (1885 [1841]). *The national system of political economy.* London: Longmans, Green and Company.

May, C. (2006a). Escaping the TRIPs' trap: The political economy of Free and Open Software in Africa. *Political Studies, 54*(1), 123-146.

May, C. (2006b), The FLOSS alternative: TRIPs, non-proprietary software and development. *Knowledge, Technology, & Policy, 18*(4), 142-163.

Schware, R. (Ed.) (2005). *E-development: From excitement to effectiveness.* Washington, DC: The World Bank Group.

UNCTAD (1997). Trade and development report: Globalization, distribution and growth. Geneva: United Nations.

UNDP (1990). Human development report: Concept and measurement of human development. New York: Oxford University Press.

UNICEF (1987). *Adjustment with a human face.* Geneva: United Nations.

United Nations (2005). World Public Sector Report 2005: Unlocking the human potential for public sector performance. New York: United Nations.

United Nations (2001). *General Assembly Resolution 56/183 - 90th plenary meeting.* Retrieved November 9, 2007, from http://www.wsis-pct.org/resol-56-183.html

Vinay, V. (1999). What is Free Software?. *Resonance*, 4(4), 1-6.

Wade, R. (1990). *Governing the Market: Economic Theory and the Role of Government in East Asian Industrialization.* Princeton, NJ: Princeton University Press.

Wade, R. (1996). Japan, the World Bank, and the art of paradigm maintenance: The East Asian miracle in political perspective. *New Left Review*, I/217, 3-36.

Wade, R. (2002). Bridging the digital divide: New route to development or new form of dependency? *Global Governance*, 8(4), 443-466.

Wade, R. (2003). What strategies are viable for developing countries today? The World Trade Organization and the shrinking of 'development space'. *Review of International Political Economy*, 10(4), 621-644.

Woo-Cumings, M. (Ed.) (1999). *The developmental state.* Ithaca, NY: Cornell University Press.

World Bank (1980). *World Development Report, 1980.* New York: Oxford University Press.

World Bank (1999). *World Development Report: Knowledge for Development.* New York: Oxford University Press.

ENDNOTES

[1] According to WSIS' definition, civil society includes "All civic organisations, associations and networks which occupy the 'social space' between the family and the state except firms and political parties; and who come together to advance their common interests through collective action. Includes volunteer and charity groups, parents and teachers associations, senior citizens groups, sports clubs, arts and culture groups, faith-based groups, workers' clubs and trades unions, non-profit think-tanks, and 'issue-based' activist groups. By definition, all such civic groups are non-government organisations (NGOs)". See http://www.wsis-pct.org/wsis-info.html.

[2] The countries referred to here as developed (or high-income) and developing are, respectively, OECD and non-OECD countries.

[3] Here the author refers in particular to the Jordan case. His considerations, however, can be extended to most developing countries. See also, for Africa, Heeks, 2002a.

[4] World Bank and International Monetary Fund (IMF).

[5] www.nzdl.org

[6] humaninfo.org

[7] The Free and open Software Foundation for Africa(FOSSFA) was launched in february 2003 in Geneva, Switzerland, in order to promote the use of FOSS and the FOSS model in African dvelopment. See www.fossfa.net.

Chapter IV
American Electronic Constitution:
Reinventing Government and Neo-Liberal Corporatism

Fortunato Musella
University of Naples Federico II, Italy

ABSTRACT

The chapter is dedicated at analyzing the strategic use of new technologies in the United States. An evident synergy has been noted between the digital policy projects and the neo-liberal ideology wave that has traced origin in the fiscal crisis of the State in the 1970s. About four decades have transformed some political directions in true imperatives: public sector downsizing, cost-cutting in public agencies, decision-making privatization, and the principle of efficiency as a measure of collective action. If new public management has been imposed as a dominant paradigm for administrative restructuring, ICTs programs sustain reform objectives by putting emphasis on the sure advantages of technological applications. In addition to this, administrative reforms seem to be in continuity with some American historical tradition, in reasserting a central role of private actor in public activities and realizing a significant "fusion of political and economic power". Digital era seems to have added a new chapter to the American corporate liberalism history, with the difference—and the aggravating circumstance —that private organizations have now more powerful instruments to control and regulate society. New technological instruments seem to be used essentially to produce a neo-liberal interpretation of government activities.

INTRODUCTION

The more widespread interpretations of the strategic use of new technologies start usually from their democratic promises. With specific reference to the United States, it has be argued that digital policy may intervene as a remedy for the three evils bothering the society: «poor communication between general public and decision-makers in the political system; a lack of political participation, either caused by structural or functional deficits in the political system; and a negative effect of mass media both on the political system in general and on political participation in particular» (Hagen, 2003). Literature on information technologies is generally concentrated on the themes of public involvement, besides appearing empirically disconnected and infused of optimism (Garson, 2003). For instance Grossman states that interactive technologies «make it possible to revive, in a sophisticated modern form, some of the essential characteristics of the ancient world's first democratic polities» (Grossman, 1995: p. 48, in Needham, 2004, p.43)[1]. Also an entry of the *International Encyclopedia of the Social & Behavioral Sciences* has been dedicated to the concept of "e-democracy", confirming that it «refers to the use of information technology to expedite or transform the idea and practice of democracy» (Street, 2001). No room is left to other critical questions of the ICTs adoption: how new technologies contribute to change power relationships inside and outside the State? Do they impact on existing disparities among groups and geopolitical areas? What kind of political program they help to promote?

The following chapter moves from the hypothesis that new technologies have a deep impact on political structures and representations, even if it does not probably produce a return to Greek polis. The Internet presents a systemic impact on several areas between market-politics, so that it seems to justify the reference to a sort of constitutive function[2]. Indeed it seems to contribute to a strategy of reorganization of political institutional

systems at the national as well as at global level on several areas. Also starting from the statement according to which the so-called "information society" may represent an obstacle for democracy, a true "false-friend" (Agre, 1999), several authors have confirmed that new technologies encourage new social, political and economic structures: «the change brought about by the networked information environment is deep. It is structural. It goes to the very foundations of how liberal markets and liberal democracies have coevolved for almost two centuries»[3] (Benkler, 2006: 1). On other side other scholars have underlined that representations—and myths—on cyberspace, and in association with them images of information age, globalization and e-democracy, may act as a powerful instrument to justify concrete political choices and depoliticize speech (Mosco, 2004: 16). Although the Internet seems to deal with a «story about how ever smaller, faster, cheaper, and better computers and communication technologies help to realize, with little effort, those seemingly impossible dreams of democracy and community with practically no pressure on the natural environment» (Mosco, 1998: 59), it produces relevant changes in power relations even tending to conceal such implication. The use of new technologies, far to constitute a neutral and forced option, seems to be part of an ampler neo-liberal program.

More specifically, in this article it will be considered the case of the United States. It will be devoted attention on the intersection between the development of digital policies digital revolution and the diffusion of a new administrative approach: the so-called "new public management". A confluence featured by three key points: the bipartisan agreement to reinvent government also on the base of the potentialities provided by new technologies, the consequent articulation of the public-private relationships, and, last but not least, the strategic value of new technological supply on global scale. Although the rhetoric of direct involvement of citizenship has furnished a source of legitimacy for political intervention in ICTs area, numerous analyses present a different picture: new instruments seem to be used essentially to produce a neo-liberal interpretation of government activities. The expression "work better, cost less" has represented the most waving flag of the program of public administration reform and federal downsizing: in a few years the principle of efficiency has become one of the most relevant measure of political action. Consequently, technologies have been mainly used for achieving improvement in terms of speed of public services delivery and reduction in costs[4].

On the other side the citizen has been considered as a passive recipient, in the sense that service users have been given little opportunities to influence the set of choices on offer[5]. Public participation is conceived only as a general goal, with a few specific reference to how new technologies can really act to realize citizens' participation in public decision making. A question commonly note in literature as the "managerial bias" of electronic policy[6] (Chadwick, 2003a): in the United States as well as in other geo-political areas, the main objective of the introduction of new technologies remains essentially tied to the scope of embracing a more efficient administrative behavior.

If the new public management has become in the last years a dominant world-spread paradigm, the specific position of the United States in the global system let also to acknowledge an "Us specificity". From such point of view, ICTs may be interpreted as an instrument for maintaining economic hegemony. Indeed, differences in rates of technological progress determine wide divides between the most developed and underdeveloped nations. Digital technologies influence flows of investments, goods and global services in the global market, on the base of disparities in opportunity to access the Internet. They also create new forms of dependency, as the United States corporations are the principal software and hardware providers. As

it will be shown, after the announced beginning of the "information revolution", new fractures seem to be added to the old ones.

STRATEGIC USE OF NEW TECHNOLOGIES: A BIPARTISAN PROGRAM FOR REINVENTING GOVERNMENT?

The use of new technologies as a political strategy began in the United States during the Clinton administration[7] (1993-2001). From the beginning it was clear that reasons for ICTs adoption have been searched mainly in market considerations. Jane Fountain (2007) has recently reminded that the "Internet euphoria" has been inserted in a context in which Us economy indicators seemed particularly favorable: national unemployment rates and inflation were low, and the federal budget was briefly in surplus. Consequently new technologies have been mainly interpreted as an occasion of further economic growth.

First interventions in the field of digital policies have accompanied the birth of e-commerce[8], as reforms initiatives through new technologies have mainly concerned service provision for business. Producing a disintermediation effect between service providers and users, the e-commerce suggests a limited role for the State that is devoted mainly to assure a regulatory framework for free competition in marketplace. A next step has been consisted in the transfer of e-commerce experiences to the area of e-government[9], since the first one has been supposed to act as a stimulus for public administration change, transferring concepts and systems from private to the public sector (Wimmer, Traunmüller & Lenk, 2001). It has introduced a new administrative logic, an "entrepreneurial spirit" able to transform public sector towards the adoption of a more customer-oriented attitude to public services.

The use of new technologies has been considered as part of the broader program of new public management, which in a few years has modified modalities of actions of public agencies in the new as well as in the old Continent. Digitalization and back-office restructuring have represented instruments for the "reinventing government", an expression used to indicate the attempt to focus the public sector on results in terms of efficiency, effectiveness, and quality of service. The term e-government itself has been coined by the Us Program with reference to a process of government transformation – as reported in the *National Performance Review.* Briefly defined, the new approach was «a radically new way of doing business in the public sector»[10] (Osborne and Gaebler, 1992: xviii).

A policy document entitled "Reengineering Through IT" has been also published in September 1993, to make clear the "customer-orientation" of Us electronic government. Enthusiast tones have been used in order to present information and communication technology as the essential infrastructure for the government of the 21st century. In a rhetoric vein, e-government has not been considered as a modality for automating the old, worn processes of government, but as a way to rethink, in fundamental ways, how governments work growth.

After four years, such layout has been echoed by the text "*Access America*", in which the main digital policy objective has been posed in allowing any American citizen to transact business with the government by electronic means, in an integrated system of administration. Garson was right when he stated that mainly e-government represents a method for bringing business model into the public administration, in order «*to mark a new era of greater convenience in citizens access to governmental forms, data and information*» (Garson 2004, 14). Although Al Gore in his introduction to "Access America" has clarified that ICTs does not dealt only with the provision electronic services, because change may concern the whole administrative architecture, nevertheless the necessity of a cost-saving and of a better procurement system appears the central questions[11].

Under the Bush administration, interpretations of the political use of new technologies did not substantially change. A policy document published in February 2001, "*A Blueprint for New Beginning*", proposed three primary objectives for government reforms. With a very concise formula government has supposed to become citizen-centered, not bureaucracy-centered; results-oriented, not process-oriented; and market-based, actively promoting, not stifling, innovation and competition (Executive Office of the President 2001a: p. 179). They are points of government reform capable to become a successful conceptualization for political communication strategy also in other countries.

Democratic promises have not been forgotten. Yet, although it has affirmed that service provision and information dissemination represents only the first step in e-government, the road for the implementation of a more participative model was not clear. Us digital program remained consumerist in its language and managerial in its vision, driven by «an understanding of the technology as offering a means to do more of the same, in a quicker fashion, rather than on unleashing its inherently and democratic potential» (Needham, 2004: 50).

Therefore new public management program seems to constitute a bipartisan program. Nevertheless some acute observers have noted some differences between the two Administrations on the base of the changed economic scenario and political perspectives. The discriminating elements are «first, the need to reduce ICTs costs during a much more constrained budgetary environment; second, a desire to evaluate and consolidate a plethora of disconnected, grassroots reinvention efforts which had produced a fragmented e-government landscape; third, heightened awareness of security and privacy challenges, post-9/11» (Fountain, 2007: 10).

Two of such desiderata are tied to the importance of the effects of cost-cutting in digital policy evaluation[12]. Indeed, the main unit of measurement of implementation of ICTs strategy has often coincided with the productivity of investments, a point of view that has led to consider unsatisfactory the performance of new technologies initiatives during Clinton Administration. As noted in an official document «IT has contributed for 40 percent of the increase in private-sector productivity growth, but the $45 billion the U.S. government will spend on IT in 2002 has not produced measurable gains in public-sector worker productivity» (Executive Office of the President, 2001b: 22). Bush Administration has aimed at overcoming some limits of past administration by starting from economic considerations. The ICTs spending has been rethought, so that programs and systems that have not brought significant savings have been cut. However significant space has been left to projects focusing on privacy and security, and promoting e-authentication to protect online transactions.

Another core element concerns the role of the United States in the international system after the 9/11 events. Despite the rhetoric of participation have often accompanied the so-called digital revolution, after terror attacks a change of direction can be observed in terms of "culture of openness". For instance Garson underlines that many agencies have limited freedom of information:

The Nuclear Regulatory Commission pulled its entire web site. The U.S. Geological Survey removed maps of open water spaces. The Environmental Protection Agency (EPA) eliminated data on toxic waste sites needed by community groups to identify chemical hazards. The Department of Energy removed information on environmental impacts of nuclear plants and information on which communities are traversed by trucks carrying hazardous materials information previously used by public interest groups (Garson, 2005: 395).

Bush Administration has systematically sought to limit disclosure of government records while expanding its authority to operate in secret[13]. According to an official report «taken together, the Administration's actions represent an unparalleled assault on the principle of open government» (Government Reform Minority Office, 2004). However, despite of some differences concerning the complex relationship between the use of new technologies and the principle of administrative transparency, Clinton and Bush Administrations seem to show similar positions on the objectives of bureaucratic reform and of privatization of public administration.

U.S. STATE HISTORICAL ROOTS AND NEW PUBLIC-PRIVATE RELATIONS

New public management tends to see the public-private dichotomy as essentially obsolete, corresponding to a philosophy of «*generic management* [that] argues that all management has similar challenges and hence should be resolved in similar ways in public - and private- sector organizations» (Peters and Pierre, 1998: 229). A conception that fits well with the traditions of American context, usually depicted as a state-less society where private sector, extolled as the model of efficiency and good management, gains more space to define and implement policies[14] if compared with the Old Continent. As it can be easily argued, although during the past two centuries scientific political analyses have dealt primarily with the preoccupation with the state, American scholars were the first «to put the stamp of approval on the proliferation of private governments as a counterpoise of public authority» (Eells, 1962: 9). Indeed, compared with European *Rechtstaat* ideals, «American society has grown as a dispersed and diffused power structure, with many decision-making centers, both public and private»[15] (Id.: 13).

In such tradition it put down its roots the recent emphasis in Us administrative reforms on the role of public actors, that expresses an additional indication of the homage paid to private sector. As remembered, *The National Performance Review* expresses a clear opposition toward hierarchies through its formula «let the managers manage». The union between managerial paradigm and e-government has been defined as a (quasi)-perfect marriage: it has been observed that both the concepts are often used in discussions about modernizing government addressing the same problems, including lack of accountability, under-performance and diminished level of legitimacy (Homburg, 2004). As it has been partly shown, they also propose similar solutions to problems of increasing public sector efficiency, sharing the neo-liberal conception of the role of the State.

Richard Heeks (2002) has confirmed that ICTs impact on the four classical 'pillars' of the Npm agenda: efficiency, accountability, decentralization and marketization. It represents another reaction to the public crisis that has emerged since the 1970s. Due to the increasing and unsustainable public expenditure, to the excessive centralization and remotiveness of governmental activity, and to its inadequate performance, a new vision has been affirmed which in its crudest form states that "market good, government bad", putting emphasis «on the economic efficiency of markets, of the forces of competition and of individual decisions» (Id.: 10). The new managerialism has also striven to conduct to a sort of "rolling back of the state", with the replacement of public agencies with private institution, assigning to public sector the function of market support. Digital programs can be considered as the latest manifestation of a longer-term process of reform (Id.: 20).

The conceptual overlapping between Npm paradigm and new technologies programs is strictly tied to a process of convergence between public and private areas. Both of them foster a new way of thinking public sector activity on the base of the way in which private actors operate.

The consumer sovereignty, the fact that public organizations strive to produce outputs more in line with what citizens want, constitutes a central issue. Indeed, although new public management and technology-based reforms are difficult to be defined, it can be easily identified their main characteristic in the change from input to output orientation (Schedler & Proeller, 2000: 5), focusing on how governments have or are supposed to adopt new methods or IT-tools in order to provide better and faster services to citizens and businesses.

Some initiatives can be cited as a notable example of private-public integration[16]. For instance the portal "firstgov" (http://www.firstgov.gov), providing enormous amounts of information and services from the U.S. government in one place, has been acknowledged as one of the best practice in the field of digital experiences. It has been launched in 2000, under the Clinton Administration, in order to collect about 27 million federal agency web pages about specific topics or for customer groups. The portal follows an e-business model by offering an horizontal view of government – one that minimizes the "agency" aspect of services and information and capitalizes on the "content" aspect, or the subject of the information need. The important element here is that such initiative can be realized only on the base of a strong public-private partnership and has been designed to enable interactions and transactions at different levels. Indeed, according to the document *"E-government Strategy"*, produced by the Presidential Office of Management and Budget (2002). Firstgov represents the primary online delivery portal for government-to-citizen (G2C) and government-to-business (G2B) interactions.

Another point of interest in U.S, digital policy has been represented by a new valorization of private actors in policy making. If it is true that the basis model of interaction between the state and citizens that underpins the ICTs practices is managerial in its essence, also public-private relationship reflects such Npm nature. Indeed the State as a facilitator of economic activity stimulates more participation of stakeholders in public activities. This leads to consider the new category—and rhetoric—of governance, which interprets public activity as a negotiated process amongst private and public actors[17]. For instance a new administrative procedure, called electronic rule-making, signals the adaptation of the political-institutional action to such principle. Rule-making process is also known as a «notice and comment method», because it follows three steps: announcement, comment and publication (Coglianese, 2004). The agency publishes a notice containing both a proposed law and a purpose explanation of such regulatory action under consideration. The interested public is invited to send comments and proposals via e-mail during a fixed time period, and the agency will analyze and take into account such comments in its final version (Brandon and Carlitz, 2003). The citizen participation is not only a stimulus for such regulatory activity, but it also affects directly the final version of law. In any proposal introduction, people can find not only the reasoning leading to such decision, but also an explanation of changes included in the last version, as specific answers provided to issues raised from citizens. An administrative culture, bringing together a decision-making process and new computer-mediated technologies, seems therefore aimed at a specific goal: the increase of private actors involvement[18].

Such more or less consultative experiences are part of a general trend. Probably decision-making support tools for public officials constitute only the most visible initiatives in the field of digital experimentations. Yet the public-private integration is realizing on a different, and more pervasive, frontier, and concerns the private actors influence on cyberspace. Private firms determine the combination of software, hardware, and network design that substantially defines the nature of cyberspace, and on which it will be based public activities[19]. Most decisions on the Internet governance are being made by private players and are being embodied in the definition of a technical code that allows for access to the Internet and provides its myriad of applications[20]. An example may be identified in the question of sovereignty over the Internet's protocols (Transmission Control Protocol/Internet Protocol, TCP/IP), whose design has not established by traditional governments, notwithstanding they exert a profound effect on the range of conduct that can occur in cyberspace (Kalir & Maxwell, 2002: 6). In addition to this, the design of many relevant portals of public interest is often outsourced to Us corporations such as NIC, Accenture and IBM (Brewer et al., 2006: 475).

Thus, if the overlapping between private-public activities occurs in several policy areas, digital era presents a new weaving among competencies. Private actors may establish and control how public agencies operate and interact with citizens. It can also be argued that they have gained a sort of regulatory function. As Lawrence Lessig (1999: 59) states in his note contribution on the laws of cyberspace: «Architecture is a kind of law: it determines what people can and cannot do. When commercial interests determine the architecture, they create a kind of privatized law». The rise of private governance arrangements on the Net represents the most important chapter of the decision-making privatization. The Us vision on such point has been clearly expressed. Governments should regulate the Internet only when necessary: «through government played a role in financing the initial development of the Internet, its expansion has been driven primarily by the private sector. [...] The private sector must continue to lead» (White House, 1997: 2).

UNITED STATES AND THE INTERNATIONAL PERSPECTIVE

As a brief history of the digital networks shows, the Internet has been created and developed with an international projection. When it was born in 1969, under the name of Arpanet, it replies to military objectives. The primordial web has represented a decentralized structure of communication able to resist to nuclear attacks thanks to its basic structural features. Between 1969 and 1991 other private and governmental digital networks arose which presented similar principles of data transmission and protocols. However, only after 1991 the term Internet has been used to design all current networks connected each others. What was a small-scale experimental system of links among U.S. academic institutions has become a gigantic global network continuing to expand in terms of size and scope.

In the last decades the Internet has seemed lost its military purposes. Yet it preserves an important function in the system of world economy competition, as access to the world wide web seems to determine inclusion or exclusion in an integrated system of communication and commercial transactions[21] (Wade, 2002). The most economically advanced states present also the most advanced

ICTs developments, taking consequently more advantages from it. The United States represent the most evident example for such statement. For instance Netchaeva (2002) reminds that the Us are currently acknowledged as leader in most studies and surveys, in terms of economic resources dedicated to ICTs development (about $200 million a year), quality of public portals, public service provision (34 percent of all government sites have inline services).

On the other side, although the rhetoric of use of new technologies outlines their potentialities for the future of developmental countries[22], several studies have shown that not just e-government applications, but also information systems in general have often failed in less rich world. According to Heeks (2003) the majority of e-government-for-development projects fails, estimating that 35% are total failures, 50% are partial failures, and only 15% are successes. It can be concluded that the ability of developing countries to reap the potential benefits of the "digital imperative" is still limited, since they are largely hampered

Table 1. Regional e-government readiness ranking

World Regions	2008	2005	2004	2003
North America	0.8408	0.8744	0.8751	0.8670
Europe	0.6490	0.6012	0.5866	0.5580
South and Eastern Asia	0.4290	0.4922	0.4603	0.4370
South and Central America	0.4838	0.4643	0.4558	0.4420
Western Asia	0.4857	0.4384	0.4093	0.4100
Caribbean	0.4480	0.4282	0.4106	0.4010
South & Central Asia	0,3628	0.3448	0.3213	0.2920
Oceania	0.4338	0.2888	0.3006	0.3510
Africa	0.2739	0.2642	0.2528	0.2460
World Average	*0.4514*	*0.4267*	*0.4130*	*0.4020*

Source: Global E-Government Readiness Report 2005 (United Nations, 2005: 30) and Global E-Government Survey. From E-Government to Connected Governance (United Nations, 2008: 22).

by the absence of relevant political, social and economic premises[23].

Note reports produced by the United Nations (2005, 2008) show significant disparities in the use of new technologies in different geo-political areas. They place countries of North America and Europe in the leadership position in the world in terms of e-government readiness, a quantitative index that measures the capacity and willingness of countries to use e-government (Tab. 1). More particularly, in 2005 the United States reach an index of 0.9062, representing the world leader. In that year the World e-government readiness average was 0.43, while African countries present a very low value (0.27). Three years later, in 2008, Scandinavian countries surpassed the United States as the leader: a result that probably reflects the reduction in ICTs investment during Bush Administration. Nevertheless, most developing countries do not present significant improvements in e-government rankings.

Such disparities are based on relevant gaps in terms of infrastructures. The United States are at the top of the degree of penetration of new technologies in society (Table 2). As the Internet

World States show, North American population constitutes only the 5.1% of world population. Yet it represents the 18.8% of world users, with a ICTs penetration of 70.9%. Asia provides very different results. The above four billion Asian inhabitants constitute the 56.6% of world population, with a penetration of only 12.4%. Such continent has known a usage growth during the period 200-2007 (303.9%), and it presents potentialities for further developments. Yet the ranking position of this big giant is completely unsatisfactory. It can be concluded that there is an evident disparity in ICTs access and among regions and countries of the world. Despite their initial efforts, the majority of developing countries is way behind achieving any meaningful economy-wide benefits of the information society. Digital divides are so persistent that a long list of paradoxes can be presented. As Rahman points out: the total Internet bandwidth in Africa is equal to that of the Brazilian city of Sao Paulo; the total Internet bandwidth in all of Latin America is equal to that in Seoul Republic of Korea; as a proportion of monthly income, Internet access in the United States is 250 times cheaper than in Nepal and 50 times cheaper than

Table 2. World Internet usage and population index

World Regions	Population	Population % of World	Internet Usage	% Population (Penetration)	Usage % of World	Usage Growth 2000-2007
North America	334.659.631	5.1	237.168.545	70.9	18.8	119.4
Oceania/ Australia	33.568.225	0.5	19.243.921	57.3	1.5	152.6
Europe	801.821.187	12.1	343.787.434	42.9	27.2	227.1
Latin America/ Caribbean	569.133.474	8.6	122.384.914	21.5	9.7	577.3
Midle East	192.755.045	2.7	33.510.500	17.4	2.7	920.2
Asia	3.733.783.474	56.5	461.703.143	12.4	36.6	303.9
Africa	941.249.130	14.2	44.234.240	4.7	3.5	879.8
World Total	*6.606.970.166*	*100.0*	*1.262.032.697*	*19.1*	*100.0*	*249.6*

Source: Internet World Stats, http://www.internetworldstats.com/stats.htm, November 30, 2007.

in Sri Lanka; and in the United States, 54.3 per cent of citizens use the Internet, compared to a global average of 6.7 per cent. In the Indian sub-continent, the proportion is 0.4 per cent (Rahman, 2006 quoted in United States 2008: 116).

Notwithstanding e-government policy initiatives have gained international validity by the donor community as a catalyst for relevant reforms, and the global favor dedicated to new public management principles and methods, disparities among different geopolitical areas do not seem to be mitigated by the intensive use of new technologies. Indeed such trends seem to contradict Oecd assumptions, which in some official reports has asserted that ICTs are capable to contribute to development goals, by reducing transaction costs and helping to deliver information services. Indeed the concept of good governance, mainly developed in the framework of United Nations Development Program (UNDP) since the mid 1990s, has been strongly connected to the condition of liberalizing market, on the base of the statement that only a competitive and not discriminatory market order is conducive to economic growth. Thus ICTs programs typically carry a political agenda of "government by the market" (Wade, 2002: 449).

On the contrary ICTs projects seem able to create new forms of dependency[24]. The adoption of property software make developing countries more vulnerable to the power of the providers of key ICT goods and services, which are mainly concentrated in the United States and continue to dominate the global market. As Carmel notes (1997), the United States benefits from several factors that sustain its advantage in this industry. A long list permits to better understand how Us firms have gained such position: skilled labor, favorable capital conditions, sophisticated customers, close association with hardware vendors, a competitive marketplace, geographic concentrations, first-mover advantage, a strong intellectual property regime, and English as the software *lingua franca*[25]. Conditions conduct-

ing to a quasi-monopolistic regime which bring developing countries very far from the idea of a digital «*open architecture networking in which any type of network anywhere can be included*» (Oecd, 2005: 29).

CONCLUSIVE REMARKS

Numerous analyses on the political implication of the introduction of new technologies deal with simplistic—and often enthusiastic—assumptions. It may be acknowledged that contributions in such field tend to converge in a rhetorical discourse generally entailing an overarching vision of popular empowerment. Such activity of myth making seems to «build on political, structural and semantic dynamics unfolding in the cultural domain, promoted by left-wing political entrepreneurs in strange alliance with the forces of high-tech capitalism, mainly amongst academic quiescence» (Lusoli, 2006: 27). Based on deterministic conception of technological development, it tends to assume that decisions on information system, as a combination of software, hardware and network, are merely technical (Brewer et al., 2006). And to deny that, as a powerful vehicle for political change, ICTs programs exert a structural impact on market-politics arena.

This contribution has been aimed at analyzing the strategic use of new technologies and its representations by an important world power: the United States of America. An evident synergy has been noted between the digital policy projects and the neo-liberal ideology wave that has traced origin in the fiscal crisis of the State in the seventies. Thus, often presented as an occasion for reinventing national government, ICTs can be interpreted as the latest chapter of a longer-term process of reform (Heeks, 2002). About four decades have transformed some political directions in true imperatives: public sector downsizing, cost-cutting in public agencies, decision-making privatization, the principle of efficiency as a measure of collective action. If new public management has been imposed as a dominant paradigm for administrative restructuring, ICTs programs sustain reform objectives by putting emphasis on the sure advantages of technological applications.

In this country, Npm trends seem to be in continuity with some American historical tradition, in reasserting a central role of private actor in public activities and denying the dualism state-economic enterprises. Some authors have sustained that the political actions of federal government have been essential to the operation of the American business system, since the beginning of last century (Kolko, 1963). For instance Miller (1976) has described the silent constitution realizing with the significant "fusion of political and economic power" so that corporations and large scale organizations have become far more important components than the state. Digital era seems to have added a new chapter to the American corporate liberalism history, with the difference—and the aggravating circumstance—that private organizations have now more powerful instruments to control and regulate society. As remembered, besides obtaining a new role in policy making, private organization are able to intervene on the complex architecture defining the Internet rules as a sort of private law. A scenario that poses again the question of the limits between private interest and public functions.

Turning attention to the international scene, the introduction of new information technologies appears even more controversial. Notwithstanding many reports produced by international organizations present optimistic position on the ICTs opportunities for development countries, empirical studies reveal a completely different picture. The structure of global ICTs regime assures a quasi-monopolistic position to Us private firms, while less rich states seem dependent to the power of software-hardware providers. In addition to this, the definition of the Internet governance rules does not represent a participative process,

being dominated by Us private corporations. Thus new divides have been added to the old disparities among geo-political areas. Although in the words of the Human Development Report «*the Internet was created in the United States, but its cost slashing consequences for information and communication enhance people's opportunities everywhere*» (United Nations, 2001: 95), digital imperatives are still far from hiding perils of quasi-monopolistic hegemony. Yet this does not exclude that future trends could leave more space for other nations such as Europe or China «*to shape the Internet's architecture in different ways [... and use] their coercive powers to establish different versions of what the Internet might be*» (Goldsmith and Wu, 2006: 184).

REFERENCES

Agre, P. E. (1999, Octomber). *Growing a democratic culture*. Paper presented at the Media in Transition Conference. Cambridge: Massachusetts Institute of Technology.

Benkler, Y. (2006). *The wealth of the networks. How social production transfirms markets and liberty*. New Haven and London: Yale University Press.

Brandon, B., & Calitz, R. (2002). Online rule-making and other tools for strengthening our civil infrastructure. *Administrative Law Review, 54*(4), 1421-1478.

Brewer, G. A., Neubauer, B. J., & Geiselhart, K. (2006). Designing and implementing e-government systems. Critical implications for public administration and democracy. *Administration and Society, 38*(4), 472-499.

Calise, M. (2002). Corporate authority in a long-term comparative perspective. Differences in institutional change between Europe and the United States. *Rechtstheorie, Beiheft 20*, 307-324.

Carmel, E. (1997). American hegemony in packaged software trade and the culture of software. *The Information Society, 13*(1), 125-142.

Chadwick, A. (2003). Bringing e-democracy back in: Why it matters for future research on e-governance. *Social Science Computer Review, 21*(4), 443-455.

Chadwick, A., & May, C. (2003b). Interaction between States and citizens in the age of the Internet: "E-government" in the United States, Britain and the European Union. *Governance: International Journal of Policy, Administration and Institutions, 16*(2), 271-300.

Chandler, A. D., & Cortada, J. W. (Eds.). (2000). *A nation yransformed by information: How ynformation has shaped the United States from colonial times to the present.* New York: Oxford University Press.

Coglianese, C. (2004). Information technologies and regulatory policies: New directions of digital government research. *Social Science Computer Review, 22*(1), 85-91.

Cordella, A. (2007). E-government: towards the e-bureaucratic form? *Journal of Information Technology, 22*(3), 265–274. Retrieved January 15, 2008 from http://www.palgravejournals.com/jit/journal/v22/n3/full/2000105a.html#bib53

Eells, R. (1962). *The government of corporation.* New York: The Free Press of Glencoe.

Executive Office of the President (2001a). *A Blueprint for new beginnings.* Washington.

Executive Office of the President (2001b). *The President's Management Agenda.* Washington.

Executive Office of the President, (2002). *E-government strategy. Implementing the President's Management Agenda for e-government.* Washington.

Fountain, J. E. (2007). Bureaucratic reform and e-government in the United States: An institutional perspective. National Center for Digital Government Working Paper, No. 07-006, to be printed in A. Chadwick & P. N. Howard (Eds.), *The handbook of Internet politics.* New York: Routledge.

Forman, M. (2003). *Memorandum for the chief information officers. Procedures for requesting funds from the e-government fund.* Executive Office of the President, Office of Management and Budget, Washington DC.

Fukuyama, F. (2004). *State-building: Governance and world order in the 21st century.* Ithaca: Cornell University Press.

Garson, D. G. (2003). Toward an information technology research agenda for public ddministration. In D. G. Garson (Ed.), *Public information technology: Policy and management issues.* Hershey, PA: IGI Publishing.

Garson, G. D. (2005). Patriotic information systems: Evaluating Bush Administration information policy. *Social Science Computer Review, 23*(4), 395-400.

Goldsmith, J., & Wu, T. (2006). *Who Controls the Internet?: Illusions of a Borderless World.* Oxford: Oxford University Press.

Government Reform Minority Office (2004). *Secrecy in the Bush administration.* U.S. House of Representatives, Washington DC.

Kalir, E., & Maxwell, E.E. (2002). *Rethinking boundaries in cyberspace. A report of the Aspen Institute.* Washington DC.

Kolko, G. (1963). *The triumph of conservatism: A reinterpretation of American history, 1900-1916.* New York: The Free Press of Glencoe.

Habermas, J. (1986). *Between facts and norms. Contributions to a discourse theory of law and democracy.* Cambridge: MIT Press.

Hagen, M. (2003). *A typology of electronic democracy.*

Retrieved March 11, 2007 from http://www.uni-giessen.de/fb03/vinci/labore/netz/hag_en.htm

Hagen, M. (2004). Electronic government in the United States. In Eifert, M. & Püschel, J.O. (Eds), *National electronic government: Comparing governance structures in multilayer administrations* (211-240). London: Routledge.

Heeks, R. (2002). Reinventing government in the information age. In Heeks, R. (Ed.), *Reinventing government in the information age. International Practice in IT-Enabled Public Sector Reform* (9-21). London: Routledge.

Heeks, R. (2003). Most eGovernment-for-development projects fail: How can risks be reduced? *iGovernment Working Paper Series*. Paper no. 14. Retrieved March, 15, 2008, from unpan1.un.org/intradoc/groups/public/documents/NISPAcee/UNPAN015488.pdf

Homburg, V. (2004). E-government and Npm: A perfect marriage? In *Proceedings of the 6th International Conference on Electronic Commerce* (547–555). New York: International Center for Electronic Commerce Publisher.

Hurst, J. W. (1977). *Law and social order in the United States*. Ithaca, NY: Cornell University Press.

Lessig, L. (1999). *Code and other laws of Cyberspace*. New York: Basic Books.

Lips, M. (2001). Designing electronic government around the world. Policy developments in the USA, Singapore, and Australia. In J. E. J. Prins (Ed.), *Designing e-government: On the crossroads of technological innovation and institutional change* (pp. 199-216). Boston: Kluwer Law International.

Lusoli, W. (2006). Of windows, triangles and loops: the political economy of the e-democracy discourse. *Comunicazione politica, 7*(1), 27-48.

McMichael, P. (2000). States and governance in the era of "globalization". In J. D. Schmidt, *Globalization and social change* (pp. 181-198). London: Routledge.

Miller, A. S. (1976). *The modern corporate state: Private governments and the American constitution*. Westport: Greenwood.

Mosco, V. (1998). Myth-ing links: Power and community on the information highway. *Information Society, 14*(1), 57-62.

Mosco, V. (2004). *The digital sublime. Myth, power, and cyberspace*. Cambridge, MA: MIT Press.

National Partnership for Reinventing Government (2000). *Access America: Reengineering through information technology*. Washington DC.

Needham, C. (2004). The citizen as consumer: E-government in the United Kingdom and the United States. In R.K. Gibson, A. Römmele, & S. Ward, (Eds.), *Electronic democracy* (pp. 43-69). London and New York: Routledge.

Netchaeva, I. (2002). E-government and e-democracy: A comparison of opportunities in the north and south. *Gazette. The International Journal for Communication Studies, 64*(5), 467-477.

Oecd (2005). *Oecd input to the United Nations Working Group on Internet Governance*. Paris.

Oecd (2005). *e-Government as a Tool for Transformation*. Paris.

Osborne, D., & Gaebler, T. (1992). *Reinventing government: How the entrepreneurial spirit is transforming the public sector*. New York: Penguin.

Garson, G. D. (2004). The promise of digital government. In A. Pavlichev, & D. Garson, *Digital government: Principles and best practices* (pp. 2-15). Hershey, PA: IGI Publishing.

Peters, G. B. (1996). *The future of governing: Four emerging models.* Lawrence: University Press of Kansas.

Peters, G. B., and Pierre, J. (1998). Governance without government? Rethinking public administration. *Journal of Public Administration Research and Theory, 8*(2), 223-243

Rahman, A. (2006). *Access to global information. A case of digital divide in Bangladesh.* Northern University Bangladesh: Library and Information Division.

Sanders, L. (1997). Against deliberation. *Political Theory, 5*(25), 347-377.

Schedler, K. & Proeller, I. (2000). *New public management.* Bern, Stuttgart, Wien: Haupt.

Schneider, V. (2003, January). The transformation of the state in the digital age. Paper presented at the workshop *"The transformation of statehood from a European perspective"*. Vienna: Austrian Academy of Sciences.

Stahl, B. C. (2005). The paradigm of e-commerce in e-government and e-democracy. In W. Huang & K. Siau. *Electronic government strategies.* Hershey, PA: IGI Publishing.

Street, J. (2001). Electronic democracy. In N. J., Smelser & P.B. Baltes (Eds.), *International encyclopedia of the social & behavioral sciences* (pp. 4397-4399). Amsterdam: Elsevier Science.

United Nations (2001). *Human Development Report.* New York.

United Nations (2005). *Global e-government readinesss report 2005. From e-government to e-inclusion.* New York.

United Nations (2008). *Global e-government survey. From e-governmnet to connected governance.* New York.

Wade, R. H. (2002). Bridging the digital divide: New route to development or new form of dependency? *Global Governance, 8*(4), 443-466.

White House (1997). *A framework for global electronic commerce.* Washington DC.

Wimmer, M., Traunmüller, R., & Lenk, K. (2001). Electronic business invading the public sector: Considerations on change and eesign. In *Proceedings of the 34th Hawaii International Conference on System Sciences.* Maui, HI: IEEE Press.

World Bank (2002). *The handbook of e-government. The information for development program.* New York: infoDev

ENDNOTES

[1]   According to this point of view also some American historical precedents seem to justify such futurible, as the local institutional tradition constitutes a good example of decision-making practices; for instance, the trial jury duties are aimed to capture what's best about American democracy, encouraging citizens' direct involvement in the administration of justice (Sanders, 1997).

[2]   Like other social institutions, technologies may facilitate or discourage courses of action, contributing to define a specific power structure. In this chapter the expression "Electronic Constitution" will be used with a wide meaning, which does not refer to a legal text, but to something that is being constituted. The concept represents the act or process of composing, setting up, or establishing. At the same time the term constitution may be referred to the product of such make-up, e. g. the way in which things are composed.

[3]   On the profound role that information technologies has played in the American history see also: A.D. Chandler and J.W. Cortadaeds (2000). A Nation Transformed by

Information: How Information Has Shaped the United States from Colonial Times to the Present. New York: Oxford University Press.

[4] Shneider clearly explains the pillars of the public sector reform program, underlining the ICTs role in pursuing the following goals: «1) reduction of public welfare programs, 2) change in administration and management techniques, 3) introduction of competition through deregulation and liberalization, and, last but not least, 4) the privatization of economic activities of the state» (2003, p.10).

[5] Such initiatives are easily distinguishable from those related to a participative model, which represents the only model that assures to citizens the possibility to be able «to act communicating» (Habermas 1986). In this case customers not only receive information from the State, but also produces it, by activating a deliberative processes based on listening, comparison between opinions, dialogic exchange, common position construction.

[6] Chadwick (2003a) presents a wider conception when he states that the (often unforgotten) democratic promises of the ICTs introduction converge in four areas: online consultations integrating civil societal groups with bureaucracies and parliaments, the internal democratization of the public sector itself, the involvement of users in the design and delivery of public services, and the diffusion of open-source collaboration in public organizations.

[7] Systematic program of e-government are based on a long tradition of investments in technology. US public organizations have been applying IT since digital computers first have been introduced in the early 1950's. Yet only in the nineties investments have shown huge dimension, so that by 2002, federal government spending for IT was

$45 billion annually (Forman, 2003).

[8] According to Oecd glossary of statistical terms (http://stats.oecd.org/glossary/index. htm), electronic commerce refers to commercial transactions occurring over open networks, such as the Internet, including both business-to-business and business-to-consumer transactions.

[9] Such transfers have been also produced by the fact that governments have bought hardware and software that originally has been developed for the private sector and apply it to their tasks. Yet the implication of the adoption of an "e-commerce paradigm" in governmental activities may lead to a sort of analogy between political and economic systems, denoting a more or less hidden political ideology namely capitalist liberalism. See B. C. Stahl (2005).

[10] Npm paradigm suggests to overcome the rigid boundaries between business and governmental areas. As Osborne and Gaebler (1993: 21) put it, new managerial model «underlies success for any institution in today's world—public, private, or nonprofit».

[11] In this vision, the creation of an "information gateway" aims at overcoming «the barriers of time and distance to perform the business of government and give people public information and services when and where they want them. It can swiftly transfer funds, answer questions, collect and validate data, and keep information flowing smoothly within and outside government» (NPR report in 2001: 200).

[12] Chadwick and May (2003b, p.17) remind a significant sentence pronounced by the former Vice-President Al Gore, denoting a particular model of interaction between States and citizens: «We need a federal government that treats its taxpayers as if they were customers and treats taxpayers dollars with the respect for the sweat and sacrifice that earned them» (National Performance

Review, 1993, introduction, paragraph 10).

13   The Bush secrecy policy contradicts many aspirations of the so-called "information revolution". One of its most relevant point has consisted in removing thousands of documents and tremendous amounts of data be summarily from agency web sites. See Gary D. Bass and Sean Moulton, "The Bush Administration's Secrecy Policy: A Call to Action to Protect Democratic Values", OMBwatch working paper, October 2002.

14   The lack of a State as the unique center of authority has not to lead to the conclusion that, in the words of Mauro Calise (2002: 313), «American statelessness means political vacuum». Corporations represented «the most successfull organizational device in the forging of the first new nation. In several respects, the American corporations become a functional equivalent of European State».

15   In the same perspective James W. Hurst (1977: 228) has recognized that in the American history «no institution bulked larger than the market for organizing and adjusting social relations» and also law «provided substantial autonomy for private will in market».

16   The creation of cross-organizational integration and coordination among public agencies and stakeholders (government, private sector, academic institutions, NGOs and civil society) has been considered the most important challenge of e-government by a recent United Nations report (2008: XV and 16). Also Oecd contributions have indicated a shift in e-government paradigm towards connected form of governance: «While initially the political and managerial focus was on developing e-services within each public institution, with limited consideration being given to cross-organizational coherence, the focus today has clearly shifted towards coordinated services offering one-stop shops

to citizens and businesses» (Oecd, 2007: 16).

17   The shift from government to governace is tributated to redefine politics by grounding power in networks rather than institutions and producing a blending of public-sector and private-sector resources (Peters 1996). According to such vision, new economic power rests in looser structures, systems with nodal points whose power derives not from their geographical supremacy but from networked interdependence and flexibility. Yet empirical analysis has often to contradict such reassuring view, by demonstrating the persistence of old inequalities.

18   Such rule-making process, with its complex notice and comment system, could work in many contexts. Several US public agencies, such as the Department of Agriculture, the Nuclear Regulatory Commission, and the Environmental Protection Agency, have already embraced this process. According to Gary Coglianese (2004: 1) rulemaking may be able «to create binding legal rules affecting virtually every aspect of society, including drinking water quality, airline operations, electricity distribution, and car design, among other important fields».

19   A report by the Aspen Institute deals with the "private governance" which are developing for the net regulation: «Private, multinational entities such as ICANN, IETF, and W3C, which are not directly accountable to the public, leverage technical expertise to play growing rulemaking, standard-setting, and advisory roles in shaping the changing architecture of cyberspace. Private corporations such as AOL and Microsoft, whose first mission is to maximize shareholder value, control borders on the Net (such as borders around namespaces) that concern not only their own customers but also millions of other users» (Kalir & Maxwell 2002: 11-12).

[20]   See also the contribution of M. Santaniello "Who Governs Cyberspace? Internet Governance and power structures in digital networks", contained in this volume, for a classification of the different areas of Internet regulation.

[21]   Information technologies may be used as an element a specific project for reproducing capitalism and economic integration on the global scale. New forms of governance involve the rules of "market rationality" that abstract from the population to which they are applied. They present a "governance of flows", with a probable contradiction with the traditional "governance of space" (national territory). See Philip McMichael (2000).

[22]   The democratization process through new media is considered to conform to a similar path in countries with very different tradition and history. For instance such vision is embraced by a report produced by the World Bank, under a very meaningful title: "The Handbook of E-government" (2002).

[23]   Distinctions between countries have more to do with capacity to both facilitate and shape development within national borders in a manner that manages the challenges and opportunities of a globalizing world. See F. Fukuyama, 2004.

[24]   Often developing countries have doubted about a significant role of new technologies programs in for reducing global gaps among different geo-political areas. In 2000 the Group of 77 (G-77), a coalition of 133 developing nations, complained that the huge income gap between rich and poor has been exacerbated by a North-South "digital divide". See Unesco Observatory, Newsletter No 45, March 30, 2000. Available at: http://www.apnic.net/mailing-lists/s-asia-it/archive/2000/04/msg00015.html, accessed 30 January 2008.

[25]   Language is the one of the most important standard of internet design. It is very difficult to overestimate geo-political implication of the dominance of english in the digital developments. The position of english as a universal language and lingua franca on interenet underlines the imperialism of english-speaking countries, tending to polarize the world into Internet users and Internet illiterates. Yet, also starting from the assumption that even for the most revolutionary global communication technologies geography and governmental coercion retain fundamental importance, Jack Goldsmith and Tim Wu (2006) recognize ample autonomy for non American countries to define the rule og the Internet. For instance they dedicates and entire charter of their volume to the case of China, that «is not only an extreme example of control; it is also an extreme example of how and why the Internet is becoming bordered by geography» (Id.: p. 90).

Chapter V
The European Administrative Space and E–Government Policies:
Between Integration and Competition

Francesco Amoretti
University of Salerno, Italy

Fortunato Musella
University of Naples Federico II, Italy

ABSTRACT

The challenge of convergence has becoming a core issue in the European agenda, as the existence of «widely accepted administrative standards» represents one of the most important preconditions to promote socio-political development and to reinforce the single Market. Indeed many initiatives have been launched by European institutions to ensure uniformity in terms of administrative action and structures, and several communications by the European Commission have considered the impact of new technologies in creating systems of integrated and interoperable administration in the Old Continent. In this chapter it will be investigated the role of communication and information technologies in the formation of an European administrative space, the process for which administrations become more similar and close to a common European model. The contribution will consider ICTs as a key element of Europe's economic competitiveness agenda as well as the interconnection between e-government programs and the social dimension of development. In addition to this, in the final part of the chapter it will be also analyzed the nature and implications of the process of uniformity produced by the new digital infrastructures, a peculiar mix of attractiveness and imposition.

INTRODUCTION

The on-going difficulties in defining and developing the European Union landscape seem to have been overcome due to the widespread Image/metaphor of the so-called 'Common Space'. Aimed at describing the integration process within the European context, such metaphor is easily found in official documents and field research as well. Used under slightly different definitions —*European constitutional space* (the realm of shared constitutional values between member States and the EU itself), *European judicial space* (the cooperation existing in courtrooms and within an 'area of Security, Freedom and Justice'), *European public space*, and *European space of research* (Hofmann, 2006)—such metaphor is enabling the public discourse and its related policies, with a performing ability in a moment when it seems so difficult to define the EU boundaries and identity.

Even if under different names, the "challenge of convergence" has becoming a core issue in the European agenda. From the end of 1990s the construction of a "common information area" based on ICTs has been considered the key element of significant community programmes (i.e., IDA [Interchange of Data between Administrations] and TEN-TELECOM [from 2002 renamed eTen]). Moreover the so-called "Lisbon strategy" ambitiously stating the European aspirations "to become the most competitive and dynamic economy based on knowledge in the world", represents a decisive step to joint the Member States in a single "information society" (European Council, 2000). More recently an initiative has been launched by the Commission (the so-called i2010) to ensure uniformity across the new technologies policies in Europe, so demonstrating the value of standardization as a political value: «policy makers need to ensure that the legislation impacting on converging sectors provides the legal certainty needed for stakeholders to innovate. The aim is to respond to technological changes in a way that promotes competition, consolidates the internal market and benefits users. A review of the main policy issues at stake indicates that the overall legal and regulatory framework is favourable for the further development of convergence» (European Commission, 2007: 4).

Above all the need of a homogenized setting concerns prospects of strengthening the single Market and exploiting the industrial opportunities offered by new technological instruments (i2010 High Level Group, 2006: 5). In a workshop on the theme of the "single information space" for environment in Europe, it is complained the absence of «*widely accepted standards*» (Coene and Gasser, 2007: 6), as EU Member States have often failed to establish a common framework allowing technology companies to replicate the case for example in the USA or Japan: «this perpetuates a fragmented European market and therefore generates numerous obstacles to European competitiveness, as companies simply cannot implement strategies or solutions on a European or global scale. Such fragmentation reveals the Member States' tendency to continue to think and act based on national instead of European considerations» (Eu task force report, 2006: II). On the other hand, the European Union as supranational entity has considered harmonization and coherence among regulatory architectures as the basis of its own existence – and success.

The creation of interoperable infrastructures mainly respond to economic objectives, as they are directed to enhance competition, improve the quality/price *ratio* and stimulate innovation and investments. Nevertheless implications of the new administrative space formation do not only regard market considerations. Efforts in such field have concerned the achievement of general interests involving social and political objectives and the new standards for administrative action. As it will been shown in the following chapter, the definition of a common administrative "architecture" affects fundamental domains for future European development such as citizens rights, political authority regulation and public-private relationships. If the lack of a cultural integration is among the main causes of the delay of the European experiment (Shore 2000), the strategic use of new technology face the challenge to keep together organizations at different levels of government and therefore can be easily interpreted as a mere process of institutional building[1].

REINVENTING THE ADMINISTRATIVE STATE

A delay in the Constitutional Treaty approval[2]—scheduled and wished for in 2009—confirmed the euro-sceptics diagnosis, while also strengthening the many rhetorical grounds fostering the European integration process. Such context helps to prioritize the future development of a "European Administrative Space", as a normative program, an accomplished fact, or at least as an hypothesis (Olsen, 2002b: p. 1). First outlined in 1992, only in 1999 this program became an attempt to clarify that definition term against the background of the EU expansion to Central and Eastern European countries (CEECs). Siedentopf and Speer (2003) recall the definition introduced by SIGMA, a EU-OECD joint initiative established in 1992 dealing with CEECs government and administration reform:

The E[uropean] A[dministrative] S[pace] represents an evolving process of increasing convergence between national administrative legal orders and administrative practices of member States. This convergence is influenced by several driving forces, such as economic pressures from individuals and firms, regular and continuous contacts between public officials of member States, and finally and especially, the jurisprudence of the European Court of Justice (p.13).

Despite the traditional flavour of this definition, which essentially restricted the action scope to a

link between the implementation of EU law and the Europeanization of national administrative law, it was also uncertain that the influences were strong enough to create a uniform model of public administration within the EU. From this point of view, internal reasons and national traditions were decisive for the implementation of administrative reforms in each member State.

Within a few years the field research took different paths[3], and mapping the European administrative space has thus became a major task. In analysing the various stages of the Administrative Space development, and the importance of the European experience, Herving Hofmann (2006) argues that the developments towards an integrated administration «goes beyond forms of cooperation for implementation of EU law by community institutions and Member States' agencies [... and that the European administrative space] contains aspects which affect the very nature of the EU's system of shared sovereignty as well as the conditions for its accountability and legitimacy» (p. 1). On one hand, such developments dealt a harsh blow to the State's sovereignty principle within the administrative structure[4], while on the other hand outlined a very different framework from those applied in the national administrative systems[5]. The same developments justified the introduction of such categories as the "Europeanization" in regard to the transformation of national institutions' structures and practices activated by the EU integration process[6] (Börzel & Risse, 2000). Even if this process is producing different influences in the different contexts[7] (Heritier, 2001), there is no doubts that the various national administrations are currently undergoing a transformation stage where they share some common directions (Riekmann et al., 2006)[8].

As it will shown in the next section, e-government development policies represent one of the most important stage for the europeanization of national public administrations and for the creation of a "European administrative space". By providing standardization[9], ICTs turned out to be a crucial lever toward a greater integration within the European administrative structures and the computer-based network became a mirror – and a promise – for a new administrative set-up. In this way technology seems to constitute an essential element for the construction of the European entity, offering a premise for «cooperation mechanisms between Member States administrations, relevant national and European Union initiatives, standardisation and market initiatives, as well as research activities» (European Commission 2003a: p. 14). Although the intensive use of new technologies is considered a technical issue as a way of neutralising it, the impact of e-government has not to be underestimated. Indeed, it can be easily argued that «what can be recognized from the many initiatives and strategies toward e-government is a huge demand for holistic approaches going far beyond present-day technological developments» (Wimmer, 2002: p. 94).

A common element of the "European e-Government platform" is the attention to the development of a administrative framework favourable to business, especially through the reduction of the administrative costs, i.e. the costs that the corporate sector must make in order to comply with the information obligations resulting from Government-imposed legislation and regulations (International working group on Administrative Burdens 2004). For this reason administrative reforms are included as a key element of Europe's competitiveness agenda, as they may provide user-centred services and cutting red tape (i.e. unnecessary administrative burdens), requiring that information is shared across departments and different level of government. Although the correlation between digitization of public services and a more competitive economy remains complex and elusive, and it appears impossible to quantify the returns of e-government in increasing economic efficiency, wider benefits have been recognized in the introduction of new technologies. For instance a recent report produced by Idabc e-government Observatory (2005a) identifies seven types of

interconnected benefits: improved quality of information and information supply, reduction of process time; reduction of administrative burdens, cost reduction, improved service level, increased efficiency, and increased customer satisfaction (p. 13). Yet such tangible points seems to find a unity in the broader objective of increasing economic competitiveness: «e-government can provide a major contribution to increasing economic competitiveness at local, regional, national and Community level. By streamlining bureaucratic procedures and increasing public sector efficiency, it plays a significant roles in raising productivity levels in economy as a whole» (Id: p. 13).

Another important chapter of the European strategy indicates the interconnection between e-government initiatives and the social dimension of development. If it seems that «differences in economic performances between industrialized countries are largely explained by the level of ICT investment» (European Commission 2005: p. 3), such result is fulfilled only through a policy convergence and a willingness to adapt regulatory frameworks in order to facilitate the mobility of citizens and businesses. Cross-border company registrations and the interoperability of European e-Procurement are examples of how e-government could respond to single market necessities. Some initiatives refers also to the creation of a Web portals designed as a single entry point for businesses, which enabled the interaction between financial actors and institutions regardless their position at local, national or national level. The final objective here is the formation of an "Online one-stop government" which requires that all public authorities be interconnected and that the customer (citizen, private enterprise or other public administration) be able to access public services via a single point, even if these services are provided by different public authorities or private service. Although the failure of the treaty approval, it is not difficult to perceive that e-government represent a pillar of the European economic and administrative constitution, due to its contribution to the policies for efficiency as well and for social cohesion: a point deeply underestimated in the academic debate on the future of European Union.

EUROPEAN E-GOVERNMENT BETWEEN MARKET-POLITICS

Only recently object of focused analysis, the e-government issue has gained a good success in the policy-making realm after the approval, in June of 2000, of the Action Plan called "eEurope 2002: An Information Society for All". A success quite questionable, though, since one the most significant reports (Alabau, 2004) goes so far to even doubt the existence of a development policy about the e-government in Europe. According to Alabau, a unified and coherent strategy has been displaced, with the introduction of the Action Plan, in favour of a collection of initiatives linked to different matters, such as the information society, an internal market establishment, a territorial development. Such fragmentation has also been mirrored on the organizational level, due to a weak coordination strategy among the various action plans. The e-government program in Europe involved several accountability centres and administrative units, as it is evident from the lack of a unified expense budget for such initiatives. Also peculiar is the method chosen to promote the e-government, mostly based on specific agreements with member States in accordance with an *open method of coordination*[10].

In any case, the e-government policy[11] held a central role for the EU development due to its meaningful potential for transformation. Despite a poor definition for a comprehensive project, the heavy use of new technologies and the development of a supra-national computer-based architecture has became, indeed, much valued tools aiming at EU major targets, particularly in the financial area. For example, the recent *i2010 E-government Action Plan* (2006a) clearly states

that those countries with a higher e-government development degree are also at the top level in the main economy indicators: «this strong link between national competitiveness, innovation strength and the quality of public administrations means that in the global economy a better government is a competitive must» (Ibidem: p.3). There are in fact several evidences supporting the positive impact of new technologies, in both the short and long term, on innovation and growth within the public sector[12].

Securing the competition in the European market required the establishment of an integrated service system able to overcome the EU fragmentation and differences, in regard to both the increasing number of member States and the local government accountability due to the subsidiarity principle introduction. With the EU undergoing an expansion process, thus becoming more and more diversified, it was necessary to implement effective public services systems covering the entire European territory and ensuring the full mobility of citizen and goods (Ibidem: p. 3).

A recent report presented by the European Commission to the European Parliament (European Commission, 2006b) was also focused on the interoperability as the administrative structure's main goal. The public administration modernization is thus finalized to the establishment of a common market and an effective interaction between citizens and companies: «the single market relies on modern and efficient public administrations which facilitate the mobility and seamless interaction of citizens and businesses» (Ibidem: p.2). The same approach applies to the administrative structures integration in the various government levels. The same document states that «to be affordable and effective, implementation of the infrastructure required for the delivery of pan-European eGovernment services will have to be guided by an overall conceptual architecture, based on standards» (Ibidem: 9). Therefore, the interoperability goal is described as intertwined with the strengthening of the economic competition and the overcome of any obstacle to the common market establishment (European Commission, 2006a). Defined as "*the key enabler for the delivery of e-government services across national and organisational boundaries*", such interoperability systems are regarded as the right tools to ensure the mobility of businesses and individuals[13], a larger interaction among the stake-holders, and an effective administrative cooperation[14] (European Commission, 2006b).

Such interoperability can only be actualized by intervening on three different elements, respectively related to the organizational, technical, and semantic dimensions:

– Organisational interoperability is about being able to identify those players and organisational processes involved in the delivery of a specific eGovernment service and achieving agreement among them on how to structure their interactions, i.e. defining their "business interfaces".

– Technical interoperability is about knitting together IT systems and software, defining and using open interfaces, standards and protocols in order to build reliable, effective and efficient information systems.

– Semantic interoperability is about ensuring that the meaning of the information exchanged is not lost in the process, that it is retained and understood by the people, applications and institutions involved (Ibidem: 6).

These three aspects recall the major problems facing the public administrations in managing shared activities and relationship with their citizens: a dual communication/integration problem in contemporary public administration concerning the achievement of internal integration at the administrative intra- and inter- agency level, as well as external integration and user-centric communication channels with the overall society (Goudos, Peristeras & Tarabanis, 2007: 1).

The public agencies, therefore, have to address a structural problem concerning their own communication and coordination assets. While historically Europe's main administrative model encouraged a functionality division, now we are witnessing a restructuring process aiming at creating public agency networks. These integration procedures affects directly both the different authority levels and the relationship among administrative units at the same government level[15]. The new technologies, acting between and within the public administration agencies, transform the "*government to government*" relationship, by creating the assumption for a system development defined by its polycentric character.

Coherence and interconnection criteria are as much organizational as they are technological. Currently, emphasis is given to the need for a single access point for citizens to all eGovernment services. This implies a high degree of coherence between the different parts of public organizations and as well as an interconnection and fluid inter operability between them (Burgelman & Clements, 2003).

The technical infrastructure is then conceived as the basis enabling the administration to present itself to the citizen as a unified actor: information and services are offered through a single computer-based outlet, thus preventing the need to contact several public agencies. By pulling down the functional walls, it becomes possible to create a one-stop government unit, reducing the operative expenses and producing a more efficient and flexible administration (Realini, 2004). This approach enables users to access the public services through a single entry point, so that «the key issue of presenting and structuring information and services in a one-stop Government is that the customer does not need specific knowledge of the functional fragmentation of the public sector»[16] (Wimmer, 2001: 2). Such viewpoint gives way to the need of activating the administrative agencies interaction and also of overcoming those cultural differences that could jeopardize the access to the same service in different countries. The actual problem here is to locate those common procedures needed to use and read the available data, and to share the logical structures. Therefore, the computer network is used as an infrastructure allowing the introduction of shared organizational and cognitive models.

Also thanks to the pan-European e-government, the service system restructuring seems more apt to cover the enterprise desires than the citizen needs, as is the case of a larger diffusion and complexity of electronic services devoted only to the business community (Centeno et al, 2005). Indeed, the most recent benchmarking report about online services provided in Europe (Capgemini, 2006) reveals that the advancement degree of business-oriented services vastly exceeds those directed to the citizens: the former are included in "two-way interaction" category in many countries, while the latter are stuck in a "one-way interaction" mode (Ibidem: 9). Beside this quality aspect, other differences are notable in the delivery variable: business-oriented and citizen-oriented services cover, respectively, two thirds and one third of the overall distribution. (Ibidem: 10). Thus the service availability for financial entities creates a *de facto* priority for the business sector[17].

Despite of a renewed interest in the demand for services and for citizen-customer needs[18], EU policies will continue to be above all technologically and commercially driven[19]. This nature could generate some critical remarks, because e-government policies are still scarcely *citizen oriented*. As Strejcek and Theil (2002) state, in a context characterized by the change of the balance of component values that shape e-government policies, implementing e-government solutions can cause a conflict with the European Convention of Human Rights (ECHR) which ought to be integrated into the policies. Yet such two components seemed to set conditions and constrains so that the

many solutions adopted by national and regional decision-makers will not provide consequences for the development of a European Administrative Space, which remains the main objective of e-government policies for the future.

CONSTITUTIONAL PERSPECTIVES AND FUTURE TRENDS

The centrality of the public administration discourse and a constant reference to the free market commanding have often underlined some holes in the European initiatives, and particularly a lacking of programs directly fostering a democratic participation (Trechsel et al, 2003). Despite some controversial issues, the affirmation of new principles such as administrative transparency and *responsiveness* are promoting institutional participatory practices that, when supported by the new media, configure a specific form of democratic praxis: the *administrative e-democracy* (Amoretti, 2006). This strategy could become a way to approach the democracy deficit without aiming at advisory or deliberative practices, but nevertheless providing an important experimentation ground for rebuilding the relationship between public and private actors (Bignami, 2004)[20].

The principles of *good administration*[21] become then a pillar of the associative life, sometimes as accessories to traditional sources of democratic legitimacy. Affirmed as fundamental rights, such principles directly permeating the world of public administration (Fortsakis, 2005)[22] and embracing the e-government policies as a strong engine for institutional propagation and strengthening. Despite some discrepancies in their overall outcome, and in terms of implementation in the various member states, such reforms have a common theme: the desire to make the relationship between citizens and government more direct by streamlining or eliminating the government layers and complexities (Ansell & Gingrich, 2003).

Therefore, we witnessed the development of a

true corpus of principles and standards – fairness and transparency, the people's right to be heard and the administration duty to voice the reasons behind its decisions – able to effectively transform the public administrations, giving also way to «a process of procedural harmonization through the constitutionalization of administrative law norms» (Harlow, 2005: 289).

However, promoting such rights is neither the only engine pushing the convergence of administrative system nor the sole venue where new technologies enable this kind of operations. In fact, the ICTs generate - in a more significant way, even if not always well affirmed - a new form of harmonisation that redefines the local actors power and enables a unified system reconfiguration at the European level[23]. By focusing on technical and infrastructural aspects, the e-government policy in Europe operates a twofold process of «standardisation in technology and harmonisation in legislation» (Idabc, 2004: 24).

This framework is an open acknowledgment of the brilliant intuition by Lawrence Lessig (1999), who explained how the definition of a computer-based infrastructure established a true code able to regulate, probably even better than the traditional normative mechanisms, the various actors' behaviour. The growing awareness of the new technologies potential in producing constraints and opportunities within a certain institutional context, could also explain why a supra-national body such as the European Union embraced the ICTs to create a new political and institutional structure. The new element breaking up with the past is that all this situation could partially unfold without resorting to traditional legislative tools. In other words, the so-called technology revolution appears to represents, within the European landscape, one of the possible path leading to the member State's administrative convergence toward a series of shared principles.

The outline of this European administrative model can be described through a pair of (only seemingly conflicting) principles. In regard to the centre-periphery dynamics, the EU chose the subsidiarity principle by promoting the local government and the creation of new networks to manage some administrative functions. Indeed, from the point of view of organizational and economic studies, it could been noticed that the new technology enables the setup of coordination systems that allow for greater decentralized operation at more widely dispersed locations, because of the decrease of hierarchical costs (Schneider, 2003:16). Moreover what applies to firms and market actors concerns also the political institutions. Indeed, the new position of the state leads to a situation where resources become less and less controlled by national governments and to a large degree dispersed among an increasing number of actors at the local level. At the same time the inclination toward standardization and interoperability[24] revealed also top-down approach for the inter-institutional and trans-national relationship. A process of *homogenisation* is driving by a combination of political and technological logics, and ICTs can be expected to reinforce this trend (Baldersheim, 2006: 5).

The e-government policy attempts to preserve a sensitive balance between two European aspirations: along with the autonomy of local actors—local computer-based networks, specific administrative tasks, financial resources availability—it also established the standard needed for a wider control and uniformity in the action of public administrations at any government level. Yet the most recent trends show the prominence of the supranational element. Standardization and interoperability limit the autonomy of national and local powers, and, at the same time, lay the foundation of a trans-national e-government platform for a new European administrative space.

CONCLUSIVE REMARKS

Despite some limitations (Olsen, 2002a), the Europeanization category seems particularly apt to analyse the young e-government policy as applied in the Old World. It can be argued that the interventions in such policy area are already "Europeized", that is, they are developing under the EU push, thus embracing its initiatives' spirit and goals. In this context, for example, the *e-Europe* program was crucial in enabling specific national *action plans* and in opening up the EU funding activation (Idabc, 2005b). Also, the e-government policies are increasingly considered as important tools for the Europeanization process. Thanks to the creation of an inter-exchange and cooperation network for supra-national administrations and the identification of the *best practices*, the public agencies are moving toward a general convergence and an action uniformity. As outlined by the Italian experience, at the same time we are witnessing the growing importance and extension of autonomy choices for some local and regional offices. These dynamics, however, risk to produce *top-down* policies centred on technology and a fragmented *bottom-up* policy focused on decentralisation practices: aiming at maintaining a difficult balance, so far the e-government policy appears in tune with how occurs in other policy areas [25]. In addition to this, if the strategic feature of this policy will be considered, it could easily understood the steady attention addressed by the European Union towards the diffusion of infrastructure - an apparently neutral ground on which the Union is authorised to intervene[26].

The overall outline includes also significant and new elements. If the term Europeanization represents a useful concept to understand the supra-national policies' impact on national processes, we must investigate these European standards, in particular the nature and implications of the organizational behaviour produced or imposed by the computer code and other technology options. Due to its more defined characteristics, the e-government is doubtless one of the most effective tools for establishing common administrative standards. Therefore we should reconsider what argued by Olsen (2002b). According to the Norwegian scholar, the "member States' preferences for administrative autonomy has to be balanced against the Union's need for effective and uniform implementation. [...] The European context suggests that administrative convergence is more likely to follow from attractiveness than from imposition. Convergence is also more likely to be an artefact of substantive policies than the result of a coherent European administrative policy".

If the "European administrative space" will still be accepted as an effective metaphor to describe the existence of a harmonization and homogenization process of public administrations in Europe, then we should mostly consider the national and sub-national outcomes to those pressures described earlier (Overeem et al, 2007) and to the original mixture of action plans that will take shape. Can we talk of *attractiveness* for the inter-operational platforms implementation, when their refusal leads to the actual exclusion from Europe's institutional and financial network? And what is exactly the nature of such technology-based *imposition*? If there is no question that the EU does not still show a sufficient executive power if compared with the national governments' experiences, it can been asked how can we define the supranational ability of limiting the member states' autonomy in the choice of their organizational and functional arrangements, and in influencing the behaviour of so many individual and collective actors.

Along with these questions, the above described processes and their major characteristics suggest the difficulty of acknowledging and defining a European specificity. Indeed, on one hand these trends meet the needs of the EU building-process, while on the other hand they embrace a thrust aiming beyond the European borders: the establishment of an integrated market (including capitals, goods, and services), the development of a network supporting global administrative standards, and the affirmation of oligopolistic corporations that establish computer-based codes and platforms[27].

Finally, there is a need for further studies and field research concerning the e-government policies and their significance for the creation of a "European Administrative Space". Such analysis will surely help in better understanding the Europeanization idea and those processes it will likely generate in the future.

REFERENCES

Alabau, A. (2004). *The European Union and its eGovernment Development Policy. Following the Lisbon Strategy Objectives*. University of Valencia: Valencia.

Ansell, C., & Gingrich, J. (2003). Reforming the Administrative State. In B.E. Cain, R.J. Dalton, & S.E. Scarrow (Eds.), *Democracy Transformed? Expanding Political opportunities in Advanced Industrial Democracies* (pp. 164-191). Oxford: Oxford University Press.

Amoretti, F. (2006). La Rivoluzione Digitale e i processi di costituzionalizzazion euopei. L'e-democracy tra ideologia e pratiche istituzionali. *Comunicazione Politica, 7*(1), 49-74.

Baldersheim, H. (2006, May*). The Future of the Periphery in Information Society. Or: Stein Rokkan Meets Manuel Castells*. Paper presented at the conference on "*Towards a New Nordic Regionalism?*", Sogn og Fjordane University College, Balestrand.

Baptista, M. (2005). e-Government and State Reform: Policy Dilemmas for Europe. *Electronic Journal of e-Government, 3*(4), 167-174.

Bignami, F. (2004). Three Generations of Participation Rights before the European Commission. *Law and Contemporary Problems*, 68, 61-83.

Börzel, T., & Risse, T. (2000). When Europe Hits Home. Europeanization and Domestic Change. *European Integration Online Papers, 4*(15). Retrieved June 20, 2007, from http://eiop.or.at/

Brewer, G. A., Neubauer, B. J., & Geiselhart, K. (2006). Designing and Implementing E-Government Systems. Critical Implications for Public Administration and Democracy. *Administration and Society, 38*(4), 472-499.

Burgelman, J. C., & Clements, B. (2003). A New Paradigm for eGovernment Services. *The European Science and Technology Observatory*. Retrieved March 25, 2008, from http://www.jrc.es/home/report/english/articles/vol78/ICT1E786.htm

Capgemini (2006). *Online Availability of Public Services: How Europe is Progressing*. Brussels.

Cassese, S. (2006a, November). Four Features of the European Administrative Space. Paper presented at the Connex thematic conference "*Towards a European Administrative Space*", Birbeck College, London.

Cassese, S. (2006b). *Oltre lo Stato*. Roma-Bari: Laterza.

Centeno, C., van Bavel, R., & Burgelman, J. C. (2005). A Prospective View of e-Government in the European Union. *The Electronic Journal of e-Government, 3*(2), 59-66.

Coene, Y., & Gasser, R. (2007). *Joint Operability Workshop Report "Towards a single information space for Environment in Europe"*, Frascati.

DG Infortmation Society and Media, eGovernment unit (2007). *EU: Study on Interoperability at Local and Regional Level*. Brussels.

Eu task force report (2006), *Fostering the competitiveness of Europe's ICT industry.* Retrieved April 15, 2006, from ec.europa.eu/enterprise/ict/policy/doc/icttf_report.pdf

European Commission (2000). *eEurope 2002. Action Plan.* Brussels.

European Commission (2003a). *Linking up Europe: the importance of interoperability of e-Government services.* Brussels.

European Commission (2003b). *The Role of eGovernment for the Europe's Future.* Brussels.

European Commission (2005*). i2010 – A European Information Society for growth and employment.* Brussels.

European Commission (2006a). *i2010 – eGovernment Action Plan: Accelerating eGovernment in Europe for the Benefit of All.* Brussels.

European Commission (2006b). *Interoperability for Pan-European eGovernment Services.* Brussels.

European Commission (2007*). i2010 – Annual Information Society Report.* Brussels.

Fortsakis, T. (2005). Principles Governing Good Administration. *European Law Review, 11*(2), 207-217.

Goudos, S. K., Peristeras, V., & Tarabanis, K. (2007). *Mapping Citizen Profiles to Public Administration Services Using Ontology Implementations of the Governance Enterprise Architecture (GEA) models.* Proceedings of the 40th Hawaii International Conference on System Sciences, Maui (Usa): IEEE Press.

Harlow, C. (2005). Law and public administration: convergence and symbiosis. *International Review of Administrative Sciences, 7*(2), 279-294.

Heritier, A. (2001). Differential Europe: National Administrative Responses to Community Policy. In M.G. Cowles, J. Caporaso & T. Risse (eds).

Transforming Europe. Europeanization and Domestic Change (pp. 257-294). Ithaca-New York: Cornell University Press.

Hofmann, H. C. (2006, November). Mapping the European Administrative Space. Paper presented at the Connex thematic conference *"Towards a European Administrative Space"*, Birbeck College, London.

Idabc (2004). *European interoperability framework for Pan-European e-government services.* Brussels.

Idabc (2005a). *The impact of eGovernment on competitiveness, growth and jobs.* Brussels.

Idabc (2005b). *eGovernment on the Member States of the European Union.* Brussels.

Idabc (2006). Bringing government closer to the people. *Synergy. The Idabc quarterly*, Brussels.

International working group on Administrative Burdens (2004). *The Standard Cost Model; a framework for defining and quantifying administrative burdens for businesses.* Retrieved May 15, 2008, from www.compliancecosts.com

i2010 High Level Group (2006). *The Challenge of Convergence.* Brussels.

Kahler, M. (2002). The State of the State in World Politics. In I. Katznelson and H. Milner (Eds), *Political Science: the State of the Discipline.* New York-London: Norton.

Lisbon European Council (2000) *Presidency Conclusions.*

Retrieved August 13, 2007, from http://www.europarl.europa.eu/summits/lis1_en.htm.

Lessig, L. (1999). *Code and other laws of cyberspace.* New York: Basic Books.

McMicheal, P. (2004). *Development and social change. A global perspective.* London: Sage.

Olsen, J. P. (2002a). The Many Faces of Europe-anization. *Arena Working Papers*, *1*(2).

Olsen, J. P. (2002b). Toward an Administrative European Space?. *Arena Working Papers*, *2*(26).

Overeem, A., Witters, J., & Peristeras, V. (2007, January). *An interoperability Framework for Pan-European E-Government Services (PEGS)*. Proceedings of the 40th Hawaii International Conference on System Sciences, Maui (Usa): IEEE Press

Palan, R.P., Abbott, J. & Deans, P. (1996). *State Strategies in the Global Political Economy*. London: Pinter.

Piana, D. (2006). *Costruire la democrazia. Ai confini dello spazio pubblico europeo*. Novara: Liviana.

Radaelli, C. (2000a). Policy Transfert in The European Union: Institutional Isomorphism as a Source of Legitimacy. *Governance, 13*(1): 25-43.

Radaelli, C. (2000b). Whither Europeanization? Concept Stretching and Substantive Change. *European Integration Online Papers*, *4*(8).

Retrieved June 20, 2007, from http://eiop.or.at/

Realini, A. F. (2004). *G2G E-government: the big challenge for Europe*. Unpublished master thesis, University of Zurich, Zurich.

Riekmann Puntscher, S., Mokre, M., & Latzer, M. (2006). *The State of Europe: Transformation of Statehood from a European Perspective*. Chicago: Chicago University Press.

Schneider, V. (2003, January). The Transformation of the State in the Digital Age. Paper presented at the workshop *"The transformation of statehood from a European perspective"*, Austrian Academy of Sciences, Vienna.

Shore, C. (2000). *Building Europe. The Cultural Politics of European Integration*. Routledge: London.

Siedentopf, H., & Speer, B. (2003). The European administrative space from a German administrative science perspective. *International Review of Administrative Science*, *69*(1): 9-28.

Strejcek, G., & Michael, T. (2002). Technology Push, Legislation Pull? E-Government in the EU. *Decision Support System*, 34: 305-313.

Trechsel, A., Kies, R., Mendez, F., & Schmitter, P. (2003). *Evaluation of the use of new technologies in order to facilitate democracy in Europe, E-democratizing the parliaments and parties of Europe*. European University Institute and University of Genoa.

United Nations (2001). *Globalization and the State*. Washington.

van Ark, B., & Inklaar, R. (2005). *Catching up or Getting Stuck? Europe's Troubles to Exploit ICT's Productivity Potential*. GGDC, University of Groningen, September.

Vink, M. (2002, November). *What is Europeanization? and other Questions on a New Research Agenda*. Paper presented at Second YEN Research Meeting on Europeanization. Milan: Bocconi University.

Wimmer, M. A. (2001, Octomber-November) *European Development toward Online One-Stop Government: The "E-Gov" Project*. Proceedings of the ICEC2001 Conference, Vienna.

Wimmer, M. A. (2002). A European perspective toward online one-stop government: The eGOV project. *Electronic Commerce Research and Applications*, *1*: 92-103.

ENDNOTES

[1]   For the analysis of how new technologies of communication may produce institutional isomorphism and on an reinterpretation of

such concept see also F. Amoretti and F. Musella, *Institutional Isomorphism and New Technologies*, in Mehdi Khosrow-Pour (ed.), *Encyclopedia of Information Science and Technology*, Hershey, PA: Idea Group Inc, 2008.

2   This compromise was outlined in June 2007 by the European Council.

3   As confirmed also at the recent Connex thematic conference "Towards a European Administrative Space", held at Birbeck College, University of London, 16-18 November 2006.

4   With the European Central Bank establishment the EU made an exception to the principle, previously considered to be untouchable, protecting the autonomy and independence of member States' administrative systems.

5   Cfr. S. Cassese (2006a). According to this author, there are four differentiation points: 1. While domestic administrations depend on one centre - the President or the cabinet - the European administration does not provide a single centre of power; 2. While domestic administrations have exclusive implementation power, the European administration is not the only EU implementing authority; 3. While the domestic administrative law is binomial (there are relations between two poles, the executive and a private party), European administrative law is trinomial (there are relations among the European Commission, national administrations and private actors, and each of them may play multiple roles); 4. While domestic administrative laws are usually a privileged branch of law, full of executive prerogatives, in the European administrative law the administration does not generally enjoy any special right and privilege.

6   The term "Europeanization" has also acquired new meanings due also to cultural influences related to EU building process; therefore the Europeanization is being ad-dressed as a «a process of (a) construction, (b) diffusion and (c) institutionalization of formal and informal rules, procedures, policy paradigms, styles, ways of doing things and shared beliefs and norms which are first defined and consolidated in the making of EU decisions and then incorporated in the logic of domestic discourses, identities, political structures and public policies» (Radaelli, 2000a: 4).

7   Particularly some authors have underlined that, despite the convergence produced by the Europeanization process on the policy outcomes, this event does not exclude entirely a possibile divergence among member States about the path to gain such outcomes (Borzel & Risse, 2000). Due to its unique integration of a "negative" adaptation to the EU market and a "positive" role of political regulatons, the Europeanization has thus became a «a fashionable but contested concept» (Olsen 2002a: 1).

8   Such events unfolding in Europe are mirroring some more global dynamics. Due to an increasing market integration, there is an actual rethinking on the overall role of the State. One of the first reports focusing on the equation between each country's participation in the global market and national structures strenghtening, was the UN 2001 paper entitled *"Globalization and the State"*. As a result, this institutional transformation could be considered a key factor for a «successful participation in the world market» (McMichael, 2004: 116).

9   The standard definition is an important part of the current globalization process. For instance, Sabino Cassese (2006b) highlights the growing presence of global regulation entities, called international or intergovernament organizations, mostly establish in the last 25 years and covering different areas, from environmental issues to financial matters. We are witnessing new trend

of establishing behaviour standards based on the new technologies diffusion. Now the computer code is in charge of deciding the norms to be applied at a global level to cooperation, harmonization and standardization procedures.

[10] Since its inception, one of the e-Government major goals was to point the public administrations toward «new innovative ways of working, including partnership with the private sector» (Lisbon European Council, 2000). The e-government was one the major tools enabling the EU to proceed with administrative restructuring of the member States: «Although the EU has no competence over the government and public administration organisation of its Member States – and therefore does not exert any binding "policy" in this field – the European Commission has a "proto-policy", i.e. an "agenda" promoted through other areas of competence, such as the internal market, programmes including eEurope 2005, i2010, Idabc, and, to a lesser extent, research/development programmes such as IST and eTEN. This e-Government agenda of the European Commission simultaneously derives from – and feeds into – what we have called the transnational e-Government agenda» (Baptista, 2005: 172)

[11] Three years after the Lisbon resolution, the e-government has been defined as «the use of information and communication technology in public administration combined with organisational changes and new skills in order to improve public services and democratic processes and to strenghten the support to public policies» (European Commission, 2003b: 4).

[12] According to a recent study on the influence of new technologies use on Europe's productivity grade, such levels are quite lage, about half of the United States' (van Ark and Inklaar, 2005). Similarly, the technology

investments had an annual growth of 0.9% in Europe from 1995 to 1999 (1,7% in the US), while the productivity growth rate was a plus 0.5% (0.9% in the US).

[13] Several intiatives addressed the mobility issue within the EU, aiming at harmonise the administrative procedures and to provide citizens with cognitive tools to get engaged in cross-national activities. For example, the multi-language Web portal *Your Europe* was promoted by the European Commission to provide different information for business and individuals (Idabc, 2006). Businesses could find information on issues such as company registration, public procurement, accounting regulations, taxation laws, market information and regulations for funding opportunities. The citizen-oriented services provided instead practical information such as guides for moving to a new country, information on schooling and social security, employment search tips.

[14] An efficient administrative system and a fair application of law regulations are the basic premises for market expansion. As a result, the EU *good governance* principles stress the need for an environment able to ensure an open competition (Piana, 2006: 74).

[15] A recent interoperability study by the European (2007) detailed a best practice applied in Sweden that led to a effective process. Before the new technology implementation, two different authorities managed registration and taxation of commercial activites, while today a single Web portal (foretagsregistrering.se) enables both services delivery, thus providing a more effective integration between the two agencies.

[16] An effective example is the multi-language Web portal providing assistance to transnational activities exerted by individuals and businesses (http://www.europa.eu.int).

[17] A recent publication by the Idabc (European

eGovernment Services) provides data related to the accessibility variable, showing an unleveled development of e-government services for business as opposed to those devoted to individual citizens: «Priority in developing eServices is generally in favour of business with the result that companies in the EU-18 can access 74% of all services for businesses online whilst the comparable figure for the EU-10 is 55%. Citizens of the EU-18, on the other hand, can only access 37% of services on the Internet; those living in the ten new member States only 33%» (Idabc, 2006: 4).

[18] Cfr. Blakemore, Michael, Think Paper4: eGovernment strategy across Europe – a bricolage responding to societal challenges. Report prepared for the eGovernment unit, DG Information Sosiety and Media, European Commission: http://europa.eu.int/egovernment_research november 2006.

[19] Cfr. The ICT Policy Support Programme (part of the new Competitiveness and Innovation Programme), which devoted € 25 million to the implementation of the eGovernment Action Plan.

[20] One specific reform strategy to make agencies more responsive to citizen needs is the *Citizen's Charter* or *Service Charter*, where each government agency makes public declarations about service standards and improvement goals.

[21] The origins of the right to good administration can be traced back to some Council of Europe resolutions, as well as to the case law of the European Court of Justice. Before the adoption of the new Constitutional Treaty, the concept of good administration had been codified in two (not legally binding) documents: the *Charter of Fundamental Rights of the European Union*, which only has the ambiguous status of a solemn proclamation by three of the EU most important institutions, while this concept was further elaborated in the later *Ombudsman's Code of Good Administrative Behaviour.*

[22] The *European Charter of Fundamental Rights* states a citizen's fundamental right to good administration. If in the '90's the transparency principle was a tool used by individuals and associations to put pressure on the decision-making process, the most recent stage is focused on the inclusiveness of law-making and rule-making processes.

[23] According to some authors, this institutional isomorphism is being imposed by a coercive relationships, or by some kind of mimetic process; in any case it represents a powerful tool for integration and the basis for further developments in the EU building process (Radaelli, 2000b).

[24] According to the European Commission, the interoperability is «the ability of information and communication technology (ICT) systems and of the business processes they support to exchange data and to enable sharing of information and knowledge» (European Commission, 2003b).

[25] About the development of the integration of EU and national administrative principles and structures in many policy areas cfr. the contributions in *Law and Contemporary Problems*, 68, 2004.

[26] The Commission's desire to have a political role also in the strategies of e-Government policy can been noticed in many cases: indeed, the range of action has gradually extended from technological and infrastructural reorganisation of back-office issues, so that it includes also the politico-strategic running of front office issues.

[27] This action plan, according also to Lessig (1999: 59), could increase the importance of private actors in the (more or less concealed) management of the public administration: when the computer code decides what people can or cannot do, then the privatization of the law codes appears to be less remote.

Chapter VI
The EU and the Information Society:
From E-Knowledge to E-Inclusion, In Search of Global Leadership

Clementina Casula
University of Cagliari, Italy

ABSTRACT

The rhetoric used worldwide by policymakers in promoting the uptake of Information and Communication Technologies (ICT) emphasizes the advantages deriving for all citizens from the advent of the Information Society (IS). Among the democratic features of the IS particularly praised are despatialisation processes, leading to a sort of "death of distance" mainly benefitting the inhabitants of territories traditionally located in peripheral and backward areas, as well as the enlarged global market. However, research shows that the uptake of ICT varies territorially, mainly following wealth distribution, among other variables. This consideration would corroborate the view of those reading the rhetoric over IS as a facade covering the restructuring of capitalist economy at the global level and arguing that the uptake of ICT, based on an unequal model of development, further strengthens rather than reduces the territorial and socio-economic divides between centres and peripheries. The chapter confronts those two readings of the main rationale behind policymaking for the development of an IS by looking at the case of the European Union (EU). The argument is that, although global economic competition in the ICT sector seems to be the mainspring that led the EU to promote policies for the IS, social concerns are emerging as the flagship of the policy, increasingly tuned with other policies within a wider European developmental strategy, which may start up a new field on which to compete for global leadership.

INTRODUCTION: THE DOUBLE VISION OF INFORMATION SOCIETY

Information and Communication Technologies (ICT) are today considered as one of the crucial policy fields in which to invest in order to achieve economic competitiveness within a globalized market. This belief is strongly upheld in governmental records of different countries (Brazil, 2001; Commission, 1999; US, 2001) or in policy recommendations of international organizations (OECD, 1986, 2003; WB, 2002) supporting the expansion of the ICT sector, both at the national and international level. Although the measures proposed by those documents mostly have a techno-economic character (as in the case of infrastructure or e-business development), they are usually inserted within a wider discourse on the Information Society (IS). This evocative policy-frame generally implies the desirability of shifting from the age of industrial modern democracy to a new era brought about by digital revolution.[1] The 'Information Age' is in fact usually presented as offering unique opportunities for regenerating, restructuring, or reinventing democratic societies in the direction of a more participatory, accountable and efficient structure and a better provision of education, health care, business and administration services. Thus in information societies access and use of ICT become considered as part of citizenship rights to be granted by governments committed in the removal of 'digital divides' or 'digital inequalities'. A fundamental role in the production of IS democratic features is played by the despatialisation processes that would follow ICT uptake, leading to a sort of 'death of distance' mainly benefitting—in the medium to long term—the inhabitants of territories located in peripheral and backward areas.[2]

To allow communications to work their magic, poor countries will need sound regulations, open markets, and, above all, widely available education. Where these are available, countries with good communications will be indistinguishable. They will all have access to services of world class quality. They will be able to join a world club of traders, electronically linked, and to operate as though geography has no meaning. This equality of access will be one of the great prizes of the death of distance. (Cairncross, 2001, p.16)

However, research in the field describes the diffusion of ICT as featured with inequalities in terms of access and use, mainly related to wealth, social status, gender, age, and ethnicity (DiMaggio & Hargittai, 2001; Martin & Robinson, 2004; Norris, 2001). The persistence of those inequalities is advanced by those authors viewing the IS not as a democratic and open society, but as a new form of capitalism, oriented towards accumulating economic, political and cultural capital and based on structural inequalities (Schiller, 2000; Webster, 2002). Disguised behind the illusive image of an IS, 'transnational informational capitalism' (or 'digital capitalism') would enhance competition by promoting greater concentration of capital and centralization of production, further exacerbating existing social and territorial inequalities (Fuchs 2008; Dawson & Bellamy Foster, 1998).

This view relates current governments' policymaking on the IS to the neoliberal turn waving since the mid 1980s in Western democracies (Dyer-Witheford, 2002; Stewart *et al.*, 2006). A particularly evident case of this alliance is considered to be the United States (US), where the trend towards 'hyper-commercialism' in ICT policies[3] prompted by globalization processes appears to be little resisted by governments – of both sides – seen as increasingly prone to the dictates of the market.[4] From this view governmental policies, speaking of IS but promoting a commercial use of ICT, are seen not only as not favouring democracy but as representing a frontal assault to its principles (McChesney, 2000).

Based on the assumptions of market economics, the internet has come to be defined mostly as a commercial realm with clear boundaries to be protected and finite resources to be distributed. In this same light, internet policy primarily serves to facilitate competition and provide arbitration among conflicting interests. From this perspective, technology appears neutral, acultural and ahistorical. Cyberspace is separate from its very content, as well as from social, cultural and political realms. (Stewart et al., 2006, p.732)

This chapter aims to confront these opposite views on policymaking on the IS by looking at the case of the European Union (EU). First, the diffusion of ICT in the EU is considered, both at the national and regional level, showing the presence of digital territorial inequalities. Second, the evolution of EU cohesion policy—aimed at the reduction of inequalities within the European territory—is briefly reviewed to highlight the double-edged nature of its objective, matching competitiveness and cohesion. Finally, the more recent development of EU policy on the IS is appraised to question which of the two above-mentioned views it seems to match better.

DIGITAL TERRITORIAL INEQUALITIES IN THE EU

The spatial structure of the EU is strongly polarized, in that it concentrates most of its activities in the central regions (the so-called 'blue banana' area) accounting for only 14% of its territory but holding nearly a third of its population and almost half of its GDP (Commission, 2001b, p.30). Territorial inequalities are evident also in ICT uptake in terms of infrastructure, access and production, and mainly follow a North–South and (since the last enlargement) East–West divide. As in all fields of innovation, the picture is in a state of flux, influenced by the features of each phase of ICT diffusion in different countries.

The data from the 2004 Eurobarometer survey show an overall increase of the use of ICT in the EU(15): only from 2001 to 2003 the percentage of population using a mobile phone, a computer, internet increased by nearly ten percentage points.[5] However, in the same lapse of time, the relative position of each Member State in relation to the EU(15) average has not changed and sees among the greater users of the internet Sweden and Denmark (nearly 72%), and among the least users Portugal and Greece (nearly 22%).[6] The internet is mostly used at home, for other reasons than work (75% of users) and nearly each day for the 50% of men and 38% of women (Commission, 2004a, p.12). The reasons for which the internet is mostly used include personal contact, to keep in touch with friends and family (32%) and to obtain information on free services (27%). The information and services searched for the most are those concerning tourism (97%), health (33%), culture (32%), and administration (29%) (Commission, 2004a, pp. 20, 22). The main reasons advanced by more than half of the EU(15) population (57%) to explain why they do not to use the internet include the fact that they do not have a computer at home (38%) and that they are not interested in it (27%) (Commission 2004a, p. 14). With respect to the survey carried out in 2001, a greater number of the non-users appears to be skeptical on the possibility that the internet could bring about changes in their daily lives or make them feel more integrated within society (Commission, 2004a, pp.17–18).

The presence of digital territorial divides is reinforced for the EU(27) area. Differences in the uptake of ICT (use of computers and internet and broadband uptake), although smaller than those in the GDP per capita, seem strongly influenced by market forces: telecommunication investments generally follow population distribution, and thus mostly concentrate in the core of Europe (although a number of 'gateway cities' are emerging in more peripheral areas), where most e-commerce activities are also performed (ESPON, 2006, pp.68, 69). Nevertheless, territorial coverage of ICT infrastructure and services in the EU is progressing much faster than the development of transport infrastructure (ESPON, 2006, p.68) and many of the new Member States[7] have now outperformed several southern (such as Portugal, Greece, Italy) or eastern countries (such as Bulgaria and Romania) (Eurostat, 2008). In this respect, national policies seem to play an important role in terms of enhancing or delaying the development of the ICT sector and the endorsement of a culture favourable to their use: in Nordic countries governments have rapidly embraced ICT policies while those of southern and Eastern Europe do not seem to sufficiently invest in the development of the sector (ESPON 2006, p.69).

However, the national level is by no means the only factor influencing the territorial distribution of ICT use. Despite the lack of availability of standardised statistical data for EU regions, several studies and EU funded projects show the presence of digital inequalities at the sub-national level related to the industrial features of the area considered: ICT are more diffused in urban rather than rural areas, and industrial rather than agricultural ones (ESPON, 2006, p.70; Milievic & Gareis, 2003, p.7). Among the non-users,[8] residents from less developed regions seem less likely to use ICT also when access barriers have been overcome, especially in the case of online commercial services (ESPON 2006; Milievic & Gareis, 2003, pp.10–11). Also ICT industry tends to be concentrated territorially in areas located in the richest core of the EU, in a geographic arc that extends from the south of the UK and continues through the Benelux countries, the north of Italy, south of Germany to the Île de France region (MPRA, 2008, p.11). However, the

past decade has also witnessed the emergence of regional clusters in peripheral areas (such as the Madrid region, the south of Scotland, Ireland, the south of Finland, the western regions of Sweden, and some regions of the new Member States) in the growing subsector of computing services (MPRA, 2008, p.11). In sum, ICT uptake in the EU follows a pattern of unequal territorial distribution, mainly concentrating infrastructure and activities in EU's core areas. The pattern, although conditioned by ongoing changes in the sector, seems to be 'embedded' within a wider complex picture featuring development gaps between European territories as well as different national approaches to ICT policies. This explains why EU policymaking for the IS is increasingly searching to integrate actions taken on the field of ICT at the national and regional level, as well as in other EU policies, according to a wider European development strategy.

TERRITORIAL INEQUALITIES AND COHESION POLICY, LINKING COMPETITIVENESS WITH SOLIDARITY

The EU's commitment to the reduction of territorial inequalities dates back to the Treaty of Rome (1957), declaring among its objectives the 'harmonious development' of the European territory, for which the growth of national economies should have looked at reducing disparities among different regions. However, until 1988 European funds for the development of backward areas were limited in size and administrated by national governments. Neo-marxist scholars, who interpreted the process of European integration "as a means of institutionalizing the domination of the geographic core or of large-scale capital concerns versus more peripheral areas and smaller-scale enterprises" (Holland, 1980, pp.99–100; Overturf,

1986, p.126), saw those funds as a side-payment to those areas doomed to underdevelopment.

However, in 1988 the Delors Commission launched a reformed regional policy aiming "to make of cohesion the social counterpart to the dominant economic European project of the creation of a frontier-free market" (Hooghe, 1996a, p.5). The term 'cohesion',[9] chosen as the overall objective of the policy took into account a double-edged purpose (Hooghe, 1996b, p.123): on the one hand there was the value of *solidarity*, exhorting to help regions territorially disadvantaged or economically and socially deprived compared to EU averages to catch up and equally enjoy the benefits of EU membership; on the other hand there was the goal of *competitiveness*, prompting to stimulate and differentiate local growth in order to let also backward regions compete in the global market, thus making the most of the potentials of the EU's economic integration. Article 2 of the Maastricht Treaty (1993) recognized socio-economic cohesion as one of the pillars of the Community structure, while Cohesion policy became "the only redistributive policy of importance in an almost exclusively regulatory project of European integration", second only to the Common Agricultural Policy for the allocation of financial resources (Hooghe, 1996b, p.115).

While the impact of cohesion policy in terms of enhancing the process of economic convergence is still widely debated from a quantitative point of view,[10] less disputed are qualitative judgments on its promotion of administrative and institutional learning. Authors have in fact praised the positive outcomes obtained by the policy in promoting innovation and efficiency within regional and national administrations (through the diffusion of planning methodologies, monitoring and evaluations techniques and transparent management procedures) as well as in endorsing democratic practices (from the direct role recognized to local actors and regional institutions in the definition of development processes, to the adoption of participatory decision-making procedures, and integrated sustainable development strategies) (Viesti & Prota, 2004, pp.131-132).

From the end of the 1990s the Commission widened the definition cohesion objective to include a new dimension, that of 'territorial cohesion',[11] in front of the persistent polarised development of its territory. The argument to justify the introduction of the objective was again double-edged: on the one hand territorial cohesion was said to be *a matter of fairness* towards peripheral regions, for which "people should not be disadvantaged by wherever they happen to live or work in the Union"; on the other hand, it was presented from an *efficiency logic* as also in the interests of the central regions, since congestion problems and social consequences of disparities can be seen as a "suboptimal allocation of resources", which in the long term could "affect the overall competitiveness of the EU economy" (Commission 2004b: 27–28). In this regard, a European policy on IS was identified as a perfect ally of cohesion policy in its difficult task of matching the objectives of a more competitive economy and a more inclusive society to promote the EU's harmonious development:

New information and communication technologies (ICT) have an important part to play in realizing the vision of a more competitive and inclusive Europe. By bringing people together regardless of the distance between them, the information society brings new social and economic opportunities to Europe's regions. It can result in a more balanced development between urban and rural areas. For European citizens, this means that everyone is far less dependent on their location. (Commission, 2006, p.4)

A COMPETITIVE KNOWLEDGE-BASED ECONOMY *AND* AN INCLUSIVE INFORMATION SOCIETY?

The tension between economic competitiveness and social cohesion in the definition of the EU's main developmental objectives is present also in the case of policy making for the IS. The 'Delors White Paper' (Commission, 1993) defined the priorities to create a common European information space, which was considered as a fundamental step to achieve socio-economic cohesion within the Union, as well as to compete in the global economy. The consideration of knowledge as a public good left the State a crucial role in granting access to telecommunications services virtually to all citizens (Calenda, 2007, p.70). However, a different view on the matter was to be more influential in the setting-up of the process mainstreaming EU policy on the IS. In 1994 the High-level group on the Information Society presented to the European Council their report:[12] starting from the identification of the global market as the main driving force of the new industrial revolution based on information, it called the EU to take urgent action for the wiping out of monopolistic and anticompetitive environments of the sector, not to waste its openings to the advantage of its global competitors (Commission, 1994).

The neoliberal approach at the basis of the 'Bangemann report' raised several objections, especially over the fact that liberalised markets alone could allow peripheral territories to catch up with central areas, a fact which underestimated the polarisation of the spatial distribution of development in the European territory (Calenda, 2007, p.70; Ducatel *et al.*, 2000). However, the report represented a important step in the definition of a common approach to the IS, coordinating all measures on ICT (especially those on cohesion policy) at different territorial levels. Efforts towards rationalisation were strengthened in the following years, and after that the 'Lisbon strategy' (2000) identified as the EU's developmental objective that of becoming "the most competitive and dynamic knowledge-based economy in the world by 2010" (European Council 2000). Crucial to this ambitious objective was the development of the ICT sector, identified as one of the main factors behind the greater productivity and economic growth of the US and the Asian countries since the mid-1990s (MPRA 2008).

The reference of first EU action plans to the 'social dimension' of IS appears to be mainly evocative and symbolic (as in the "IS for all" slogan used in the title), while action mainly focuses on the techno-economic dimension of the strategy, which should have 'naturally' brought to all citizens the new opportunities offered by the digital revolution. The Commission's communication launching *e-Europe* (1999) begins with acknowledging the several steps already taken by the EU to promote the IS (in terms of liberalisation of telecommunications, establishment of a clear legal framework for e-commerce, the support for ICT industry and R&D measures), but also reckoning that "given the rapid pace of technological change and of the markets" a political initiative was needed to push policies ahead more rapidly (Commission, 1999). Although the initiative includes among its key objectives the concern over the fact that the process be socially inclusive (with special attention to disabled and youth), most of the ten priorities translating the initiative are focused on the completion of infrastructure and services and of the new economy (cheaper internet access, accelerating e-commerce, smart cards for secure electronic access, risk capital for high-tech SMEs, intelligent transport, healthcare and government online) (Commission, 1999; Martin, 2005, p.7). Those steps are followed by the next communication (*eEurope 2002: Impact and Priorities*), mainly focussing on internet connectivity in Europe: the key objectives identified are a cheaper, faster and secure internet, investing in people and skills and stimulating the use of the internet (Commission, 2001a).

Similarly, the following action plan *eEurope 2005* (2002) declares among its aims that of redressing the perceived imbalance between the techno-economic and the social component of the EU's policy for the IS (Commission, 2002). However, in order to do so, it mostly focuses on technological change, namely the establishment of a European-wide broadband infrastructure (Martin, 2005, p.7). The expectation is that the shift to a secure and widespread broadband infrastructure should bring increased economic productivity and improved quality and accessibility of services (in the field of e-government, e-learning, e-health, etc.) for the largest possible number of European citizens (Commission, 2002). Measures for e-inclusion are considered as a 'horizontal concern' for all fields, but this concern is mainly expressed in terms of ensuring accessibility to excluded groups and disadvantaged regions (Commission, 2002): "[h]owever, if the goal of these European programmes is the attainment of an Information Society rather than an Information Technology Society, then much more attention must be paid to the role of social factors within the policy agenda" (Martin, 2005, p.8).

A further important step towards greater attention for the social dimension within EU policy for the IS was made by the Commission with the 2005 Communication *i2010*, defined after discussions with stakeholders drawing a balance over previous action plans. One main innovation regards policy objectives: a better balance is found—in the definition of the three priorities set to be achieved by MS by 2010—between the technological, economic and social dimension of the IS. The first priority aims to establish a Single European Information Space through the increase in the speed of broadband services, the development and interoperability of multimedia and the enhancement of a more secure internet (Commission, 2005, pp.4–6). The second priority refers to the boosting of innovation and investment in ICT research, to close the gap with competitors, through the increase of community

funds (also suggested to MSs), the launching of measures encouraging private investment in the field, the definition of e-commerce measures to remove technological, organisational and legal barriers to ICT adoption (especially for SME), a better coordination with the objectives of EU policies on research and cohesion (Commission, 2005, pp.7–9). The third priority aims to promote inclusion, better public services and quality of life, also in regions lagging behind, by issuing policy guidance on e-accessibility and coverage of broadband, adopting an action plan on e-government and strategic orientations on ICT-enabled public services and measures improving citizens' quality of life (Commission, 2005, pp.9–11). In this regard, a more direct synergy with the objectives of cohesion policy is sought after, also in the terminology used:

ICT are becoming more widely used and are benefiting more people. But today over half of the EU population either does not reap these benefits in full or is effectively cut off from them. Reinforcing social, economic and territorial cohesion by making ICT products and services more accessible, including in regions lagging behind, is an economic, social, ethical and political imperative. (Commission, 2005, p.9, original emphasis)

To further strengthen the 'i2010's societal agenda', the Commission declares its intention to propose by 2008 a European Initiative specifically focussed on e-Inclusion, addressing issues such as equal opportunities, ICT skills and regional divides (Commission, 2005, p.11). A year later EU Ministers, meeting in Riga on the occasion of the Ministerial Conference titled "ICT for an inclusive society", unanimously approved a declaration providing strategic guidance on the topic. The Riga declaration specifies that e-inclusion has to be conceived both as an end in itself (*a more inclusive IS*) and as means to achieve a wider end (*a more inclusive European society*):

eInclusion" means both inclusive ICT and the use of ICT to achieve wider inclusion objectives. It focuses on participation of all individuals and communicates in all aspects of the information society. eInclusion policy, therefore, aims at reducing gaps in ICT usage and promoting the use of ICT to overcome exclusion, and improved economic performance, employment opportunities, quality of life, social participation and cohesion. (Council of Ministers, 2006, p.1)

Among the priorities identified by the Declaration are that of addressing the needs of older workers and elderly people, reducing geographical digital divides, enhancing e-accessibility and inclusive e-Government, promoting digital literacy and motivation towards ICT use, supporting cultural diversity and gender balance in the IS (Council of Ministers, 2006, pp.2–5).

A second innovation of the *i2010* action plan regards the policy method adopted, which follows an integrated approach coordinating its specific objectives with those of related policies in a cooperative effort undertaken by MSs to converge towards a common European 'institutional space'. Following the Open Method of Coordination[13], a first synergy is sought with the wider objectives of the Lisbon strategy[14], at its turn revised and refocused (Kok, 2004). Conversely, the i2010 objectives are included within the Integrated Guidelines used to monitor the implementation of the Lisbon strategy by MSs through the National Reform Programmes (NRPs). A second synergy is looked for with EU policies on audio-visual media (in order to facilitate the digital convergence of MSs and a common response to the challenges of the IS) and on R&D (to improve investments and integrated solutions in the most innovative sectors) (Commission, 2005). Last but not least, another strategic field of integration is that with the new regional policy (2007-13) that, as seen, has added to its objective of socio-economic cohesion, a territorial dimension particularly suitable to address the issue of geographical digital divides[15].

CONCLUSION

Today the uptake of ICT is considered as offering countries unprecedented opportunities in different social spheres: from economic growth, to modernisation of the public sector, to social and political participation. Because of the overarching nature of ICT applications, governmental measures for the development of the sector are usually included under the heading of 'Information Society', a concept evocative of a wide-ranging social change but often ambiguously defined.

The chapter has confronted the EU's model of Information Society with two opposite interpretations of the notion. The first one argues that the development of IS allows, through the uptake of ICT, the achievement of greater labour productivity and growth (necessary to compete in the globalized market) while reducing socio-economic and territorial inequalities. The second one sees IS as a facade covering the restructuring of capitalist economy and holds that the uptake of ICT, unchecked by States increasingly prone to the dictates of the global market, further reinforces patterns of social exclusion, structural unemployment, and the 'centre-periphery' models typical of capitalist development.

Those opposite views were contrasted with a reflection on existing digital territorial inequalities in the EU as well as on its wider developmental model, trying to match competitiveness with cohesion. As also shown, ICT uptake, like other socio-economic activities, tends to follow a polarised model of development oriented following a North–South and (since 2004) East–West divide. Differences are more pronounced at the regional level, although the picture is in a state of flux and there are exceptions to this general trend. Also national policies in the sector seem to have a direct influence in this process.

The fact that 'distance (still) matters' does not necessarily lead to confirm the view that IS policy reveals a strategic design of capitalist economies to exploit territories lagging behind. Scholars have explained the process of formation of centres and peripheries as derived from the tendency of economic activities to agglomerate in specific areas: once established, development paths tend to reinforce over time, if not contrasted with appropriate spatial strategies (Krugman, 1991; North, 1990). Thus also in the case of ICT uptake "we might expect to see changes in economic geography of the world economy, but not necessarily changes towards the 'integrated equilibrium' view of the death of distance" (Venables, 2001, p.3).

The acknowledgment of the potential role to be played by public policies in the achievement of a more balanced spatial distribution of development led us to consider the strategies pursued by the EU in its policymaking on the IS. The analysis showed that initially the mainspring to promote a European IS policy was economic competition in the global market and that action plans mostly focused on the techno-economic dimension of ICT uptake within a picture of liberalised free markets. However, growing awareness on the specificity of the 'social dimension' of the IS is evident in the following action plans (as in the case of e-Inclusion) trying to integrate their measures with those of other developmental EU policies oriented to the wider European strategy. It might be precisely this new attentiveness on the social agenda of the Information Society that will allow the EU increasingly to play a leadership role *vis à vis* its competitors in the global market.[16]

REFERENCES

Barney, D. (2004) *The Network society*. Cambridge: Polity Press.

Barro, R. J., & Sala-i-Martin, X. (1991). Convergence across states and regions. *Brookings Papers on Economic Activity, 1*.

Boldrin, M., & Canova, F. (2001). Inequality and convergence in Europe's regions: Reconsidering European Regional Policies. *Economic Policy*, April.

Brazil (2001). *Information society in Brazil: Green book*. Brasilia: Ministry of Science and Technology.

Calenda, D. (2007). La Società dell'Informazione nelle regioni del Mezzogiorno. Sfide, visioni, comportamenti. *Polis*, XXI, April, (pp. 65-94).

Castells, M. (1996). The rise of the network society. In M. Castells, *The information age: Economy, society and culture, 1*. Malden, Oxford: Blackwell.

Castells, M. (2000). End of Millenniun. In M. Castells, *The Information Age: Economy, Society and Culture, 3*. Oxford: Blackwell.

Commission of the European Communities (1993). *Growth, competitiveness, employment: The challenges and ways forward into the 21st century - White Paper*. Parts A and B. COM (93) 700 final/A and B, 5 December 1993, Bulletin of the European Communities, Supplement 6/93.

Commission of the European Communities (1994). *Report on Europe and the global information society: Recommendations of the high-level group on the information society to the Corfu European Council*. Bulletin of the European Union, Supplement No. 2/94.

Commission of the European Communities (1999). *eEurope – An information society for all*. Communication from the Commission of 8 December 1999, COM(1999)687 final.

Commission of the European Communities (2001a). *eEurope 2002 – Impact and priorities*. Communication from the Commission to the Spring European Council in Stockholm, 23-24 March 2001, COM(2001)140 final.

Commission of the European Communities (2001b). *Unity, solidarity, diversity for Europe, its people and its territory - Second Report on*

Economic and Social Cohesion. Luxembourg: OOPEC.

Commission of the European Communities (2002). *eEurope 2005 action plan – An information society for all.* Communication from the Commission of 28 May 2002, COM(2002) 263 final.

Commission of the European Communities (2004a). *Internet – Eurobarometre spécial 194,* Vague 59.2 – European Opinion Research Group EEIG.

Commission of the European Communities (2004b). *A new partnership for cohesion: Convergence, competitiveness, cooperation. Third Report on Economic and Social Cohesion.* Luxembourg: OOPEC.

Commission of the European Communities (2005). *i2010 – A European information society for growth and employment.* Communication from the Commission of 1 June 2005, COM(2005)229 final.

Commission of the European Communities (2006). *Information society and the regions: Linking European policies.* DG Information Society and Media, Luxembourg, OOPEC.

Council of Ministers (2006). *Ministerial declaration on the occasion of the Ministerial Conference* "ICT for an inclusive society". 11 June 2006, Riga (Latvia).

Dawson, M., & Bellamy Foster, J. (1998). Virtual capitalism. In R. W. McChesney, E. Meiksins Wood, & J. Bellamy Foster (Ed.), *Capitalism and the information age.* New York: Monthly Review Press.

DiMaggio, P., & Hargittai, E. (2001). *From the 'Digital Divide' to 'Digital Inequality': Studying Internet use as penetration increases.* Princeton University, Center for Arts and Cultural Policy Studies, Working Paper, 15. Retrieved September 23, 2007 from: https://www.princeton.edu/artspol/work-pap15.html

Ducatel, K., Webster, J., & Herrmann, W. (Ed.) (2000). *The information society in Europe, work and life in an age of globalization.* New York-Oxford: Rowman & Littlefield.

Dyer-Witheford, N. (2002). E-capital and the many-headed hydra. In G. Elmer (Ed.), *Critical perspectives on the Internet,* (pp. 129-64). Lanham, MD: Rowman and Littlefield.

ESPON (2006a). *Territory matters for competitiveness and cohesion – Facets of regional diversity and potentials in Europe – ESPON Synthesis Report III.* Results by autumn 2006. Retrieved May 13, 2008 from: http://www.espon.lu

Fuchs, C. (2008). *Internet and society: Social theory in the information age.* New York: Routledge.

Holland, S. (1980). *Uncommon market: Capital, class and power in the European community.* New York: St. Martin's Press

Hooghe, L. (1996a). Introduction: Reconciling EU-wide policy and national diversity. In L. Hooghe (Ed.), *Cohesion policy and European integration. Building multi-level governance,* Oxford: Clarendon Press.

Hooghe, L. (1996b). Building a Europe with the regions. The changing role of the European commission. In L. Hooghe (Ed.), *Cohesion policy and European integration. Building multi-level governance.* Oxford: Clarendon Press.

Japan (2004). *Information and communication in Japan - White Paper.* Ministry of Public Management, Home Affairs, Posts and Telecommunications. Retrieved May 11, 2008 from: http://www.soumu.go.jp/joho_tsusin/eng/whitepaper.html

Krugman, P. (1991). *Geography and trade.* Leuven: Leuven University Press.

Leonardi, R. (2005). *Cohesion policy in the European Union: The building of Europe.* London: Palgrave Macmillan.

Kim, J. (1998). Universal service and Internet commercialization: Chasing two rabbits at the same time. *Telecommunication Policy, 22*(4-5), 281-288.

Lusoli, W. (2006). Of windows, triangles and loops: The political economy of the e-democracy discourse. *Comunicazione Politica, 7*(1), 27-48.

Machlup, F. (1962). *The production and distribution of knowledge in the United States.* Princeton: Princeton University Press.

Martin, B. (2005). Information society revisited: From vision to reality. *Journal of Information Science, 31*(1), 4-12.

Martin, S., & Robinson, J. P. (2004). The income digital divide: An international perspective. *IT&Society, 1*(7). Retrieved April 3, 2008 from: http://www.itandsociety.org

Masuda, Y. (1981). *Information society as post-industrial society.* Bethesda, MD: World Future Society.

McChesney, R. (2000). *Rich media, poor democracy: Communication politics in dubious times.* New York: New Press.

Midelfart-Knarvik, K. H., & Overman, H. G. (2002). Delocation and European integration. Is structural spending justified? *Economic policy,* October.

Milievic, I., & K. Gareis (2003). *Disparities in ICT take-up and usage between EU regions.* Presentation at NESIS workshop in Milan. Retrieved September 23, 2007 from: http://www.biser-eu.com

MPRA (Munich Personal RePEc Archive) (2008). *Mapping the ICT in EU regions: Location, employment, factors of attractiveness and economic impact* (S. Barrios, M. Mas, E., Navajas, J. Quesada. Institute for Prospective Technological Studies, Joint Research Centre, European Commission, 31 January 2008. MPRA Paper No. 6998, posted 04.

February 2008 / 10:08. Retrieved March 3, 2008 from http://mpra.ub.uni-muenchen.de/6998

Norris, P. (2001). *Digital divide: Civic engagement, information poverty and the Internet worldwide.* Cambridge: Cambridge University Press.

North, D. C. (1990). *Institutions, institutional change and economic performance.* Cambridge: University of Cambridge.

NTIA (National Telecommunications and Information Administration) (2002). *A nation online: How Americans are expanding their use of the Internet.* Retrieved September 23, 2007 from: http://www.ntia.doc.gov

OECD (Organization for Economic Cooperation and Development) (1986). *Trends in the Information economy.* Paris: OECD Publications.

OECD (2003). *Policy brief: The e-government imperative: main findings* (OECD Observer, March 2003). Paris: OECD Publications.

Overturf, S. F. (1986). *The Economic Principles of European Integration.* New York: Praeger.

Radaelli, C. M. (2003). *The open method of coordination: A new governance architecture for the European Union?* Preliminary Report, SWEPS (Swedish Institute for European Policy Studies). Retrieved May 24, 2008 from: http://www.epin.org/pdf/RadaelliSIEPS.pdf

Schiller, D. (2000). *Digital capitalism.* Cambridge, MA: MIT Press.

SIBIS (2003). *New e-Europe indicator handbook.* European Commission. Retrieved September 23, 2007 from: http://www.sibis-eu.org/

Stewart, C. M., Gil-Egui, G. Y. T., & Pileggi, M. I. (2006). Framing the digital divide: A comparison of US and EU policy approaches. *New Media & Society, 8*(5), 731-751.

US (United States of America) (2001). *Leadership for the new millennium, delivering on digital progress and prosperity, third annual report of the electronic commerce working group.* Department of Commerce. Retrieved September 23, 2007 from: http://www.lib.umich.edu/govdocs/stsci.html

Van Dijk, J. (2006). *The network society.* London: Sage.

Venables, A. J. (2001). *Geography and international inequalities: The impact of new technologies.* Retrieved April 12, 2008 from: http://siteresources.worldbank.org

Viesti, G., & Prota, F. (2004). *Le politiche regionali dell'Unione Europea.* Bologna: Il Mulino.

Webster, F. (2002). *Theories of the information society.* London: Routledge.

WB (World Bank) (2002). *The e-government handbook for developing countries.* Infodev, Centre for Democracy and Technology, November. Retrieved April 12, 2008 from: http://www.cdt.org/egov/handbook/2002-11-14egovhandbook.pdf

ENDNOTES

[1]   The concept of Information Society has appeared in the literature since the 1960s (Beniger, 1986; Machlup, 1962; Masuda, 1981) but gained momentum from the mid-1990s, when the diffusion of ICT – and especially the Internet – extended to most sectors of private and public life (Barney, 2003; Castells, 1996, 2000; Van Dijk, 2006).

[2]   "Anytime, Anywhere, by Anything and Anyone" is the slogan of the u-Japan initiative launched in 2004 by the Koizumi government to make Japan by 2010 an 'ubiquitous society', where every citizen may, at a low price, benefit from the use of a interconnected ICT infrastructure (Japan, 2004).

[3]   Of the several signs of his tendency commentators notice that since 1997 the commercial domain ('.com') has replaced the education one ('.edu') as the largest one queried in the web (Kim, 1998).

[4]   For a discussion on this view see Lusoli (2006), picturing the techno-political encounter between democratic crisis and ICT celebrated by political discourse on e-democracy in the US and the United Kingdom (among other countries), and on this volume Musella (2008), relating new public–private relations and ICT policy in the US with its historical 'stateless' traditions.

[5]   From 63% to 70% for the use of mobile phones, from 44% to 52% in the use of computer, 34% to 44% for the use of the internet (Commission, 2001; 2004, p.5).

[6]   Similar to the internet use is that for the computer, which sees again Sweden and Denmark (78%) at one end and Greece and Portugal (28% and 30%) on the other; for the mobile phone, there is Sweden (86%), Finland (83%), Italy and Luxembourg (both 81%) and on the other side Portugal (59%) and France (56%) (Commission, 2004, p.6). Portugal (80%), Greece (78%) together with Italy (81%) are the countries showing the highest number of citizens who have not benefitted from education in informatics. Overall, the use of the internet appears correlated with age, level of education, occupational condition, and gender (Commission, 2004, pp.8,24).

[7]   For the indicator measuring the percentage of households having access to internet Latvia, Lituania or Poland registered respectively 3%, 4% and 11% in 2002 and 51%, 44% and 41% in 2007. The latter data are higher than those registered in 2007 for Italy (43%), Portugal (40%), Greece (25%), Romania (22%) and Bulgaria (19%) (Eurostat, 2008).

[8]   For the year 2005 the percentage of Europeans that did not regularly use the internet

is 57%. Of the Europeans using the internet the percentage among people over 65 is 10% (against 68% of those aged 16–24); among people with a low level of education it is 24% (against 73% of those with a high level of education); among unemployed persons it is 32% (against 54% of the employed) (Commission, 2006).

[9] "Chosen among a variety of alternatives considered by the Commission ('structural policy', 'redistribution', 'solidarity/equity mechanism', 'convergence', 'regional and social development', 'social dimension', etc.). 'Cohesion' – a term of French derivation – resulted in the most politically imaginative". See Hooghe (1996b), p.123, n.9.

[10] Among the large amount of research on the issue, see Barro & Sala-i-Martin (1991) and Leonardi (2005), endorsing the 'convergence thesis', and Boldrin & Canova (2001) or Midelfart-Knarvik & Overman (2002), supporting a 'non-convergence' one.

[11] The objective of territorial cohesion was introduced in the Constitutional Treaty establishing a European Constitution among the main objectives promoted by the Union (Title I, Common Provisions, Art.3.3) and among its shared competences (Title XVIII, Economic Social and Territorial Cohesion; Protocol n.28) and, under the heading of 'territorial cooperation', among the three priority objectives of EU regional policy for 2007-13.

[12] Also commonly called the 'Bangemann report' after Martin Bangemann, president of the High-Level group and Commissioner for Industrial Affairs, Information and Telecommunication Technologies in the Santer Commission (1995–99).

[13] The OMC refers to a methodology using mechanisms of 'soft law' (such as the exchange of good practices, monitoring of indicators, benchmarking) to promote voluntary convergence towards common European policies. In offering policymakers with a common vocabulary and "a legitimising project" it enables them "to deal with new tasks in policy areas that are either politically sensitive or in any case not amenable to the classic Community method. The result is that practices that up until a few years ago would have been simply labelled 'soft law', new policy instruments, and benchmarking are now presented as 'applications' if not 'prototypes' of 'the' method (Radaelli, 2003, p.7).

[14] This is evident from the subtitle of the action plan: "A European Information Society for growth and employment".

[15] Of the nearly 308 billion Euros destined to EU regional policy for the 2007-13 period (nearly one third of total EU spending), about 7 billion euros is planned to be invested in ICT, with half allocated to infrastructures and half to services for citizens and enterprises. Another significant element of this synergy is that the management of ICT measures within cohesion policy (managed through the 'Community method') should grant the Commission, although indirectly, a greater influence in the management of IS policies (using the OMC).

[16] As stated in the consideration added at the end of the Riga Declaration, Ministers agree that by implementing the declaration "Europe creates the opportunity for global leadership in eInclusion" (Council of Ministers, 2006, p.7).

Chapter VII
World Wide Weber:
Formalise, Normalise, Rationalise: E–Government for Welfare State – Perspectives from South Africa

Nicolas Pejout
Centre d'Etude d'Afrique Noire (CEAN), France

ABSTRACT

Many of African States are focusing on ICTs and developing e-government infrastructures in order to fasten and improve their "formalisation strategy". This philosophy drives the South African State in its impressive efforts to deploy an efficient and pervasive e-government architecture for its citizens to enjoy accurate public services and for this young democracy to be "useful" to them. By focusing on the South African case, people will be able to understand the role of ICTs as tools to register, formalise and normalise, supporting the final objective of Weberian rationalisation. The author will consider the historical process of this strategy, across different political regimes (from Apartheid to democracy). He will see how it is deployed within a young democracy, aiming at producing a balance between two poles: a formal existence of citizens for them to enjoy a "delivery democracy" in which they are to be transparent; an informal existence of citizens for them to live freely in their private and intimate sphere. In this tension, South Africa, given its history, is paradigmatic and can shed light on many other countries, beyond Africa.

Numerous governments, particularly those of developing countries, have to deal with challenging economic, socio-economic and political *realities.* More challenging is to deal with *unrealities*, i.e. realities that do exist but that governments can't manage because they don't know about them. These realities are real but informal: one typical example is moonlight work. They all do exist but have no official, formal, legal-administrative and statistical existence. They are "parallel" to the official-formal world of public action and stay "underground", in the shadow of public policies.

This problem of "informality" is particularly experienced by governments in developing countries. They face tremendous difficulties in terms of public action upon realities that they don't and can't know of. That is due to a lack of measuring resources and public management capacities. Various examples are: the absence of a satisfactory statistical machinery, the ineffectiveness of a formal civil status (for instance, the registry of birth), the inefficiency of tax rolls…

For governments to act upon realities, they need to know them and therefore to reveal and measure them. In other words, they need to formalise them so as to be able to control them. Governments have to normalise human activities, i.e. to put them into norms, into measurable and controllable frameworks. This explains, for instance, the importance of statistical machineries into the construction of nation-states (Desrosières, 1993).

This quest of formalisation has been growing with the strengthening of nation-states throughout time. It has been using various tools, the last generation of which being Information and Communication Technologies (ICTs). Governments consider these technologies as powerful instruments to formalize and normalize realities. The use of ICTs to rationalize reality and therefore public action is set along the deployment of electronic government (e-government). The ultimate objective is to make a society (individuals as well as groups) highly visible—some might say transparent—to the power in place. By formatting knowledge for

the State, e-government is supporting a move towards genuine rationalisation: ICTs enable an extreme degree of accuracy and sophistication (data mining) so that everything and everyone can be labelled, measured, compartmentalised.

Obviously, such power of knowledge, based on the knowledge of power (Foucault, 1997), can threaten democracy: full transparency of individuals to the State is impossible, due to the absolute necessity of protecting the private sphere. Nevertheless, the development of the welfare State requires the administration to know most of personal data, so as to provide relevant services, for instance well-measured pensions or health care (Gilliom, 2001). This is all the more true when the welfare State is getting ICT-intensive, making the most of e-government to provide e-services. For such provision with efficiency and cost-recovery, the State needs to be scientific, somehow omniscient. That is why transparency of the society to the State is necessary (Lyon, 2003), but to a certain extent beyond which democracy is at risk.

Most of governments in African countries are confronted with informal realities, particularly in hard socio-economic contexts. They don't have enough resources—financial, human, …—to know of realities that they nevertheless need to tackle with. That is why some African States are focusing on ICTs and developing e-government infrastructures in order to fasten and improve their "formalisation strategy": by getting to know their society better, they can act upon it better (Cheneau-Loquay, 2005). South Africa is one of these and certainly the most advanced on the continent in that regard. The South African State is indeed deploying a remarkable e-government architecture for its citizens to enjoy accurate public services (Péjout, 2004; Péjout, 2007). The challenge is not just that of administrative efficiency but is also highly political: this young democracy needs to be "useful" to its citizens.

This chapter will highlight the role of ICTs as tools to register, formalise and normalize realities. We will first show how formality is key to the e-welfare State. We will see how "the World Wide Web meets Weber": ICTs are ideal the ideal tools to implement the Weberian administrative organization (Weber, 1978). This ICT-based administration must serve an e-welfare State that has to guarantee the effectiveness of the social contract. We will then consider the historical process of the South African e-strategy, from Apartheid to democracy. We will see how it is deployed within a young democracy, aiming at producing a balance between two poles: on the one hand, a formal existence of citizens—even a transparency to the State—so that they can enjoy the deliveries of democracy; on the other, an informal existence of citizens so that they can live freely in their private and intimate sphere. This tension is magnified by the intrusion of e-government. South Africa, given its history, is paradigmatic and can shed light on many other countries, beyond Africa.

FORMALITY IS KEY TO E-WELFARE STATE

ICT are powerful tools for e-government to be deployed in the most comprehensive way. The condition for this scenario to happen and for e-government to mean something for everyone in everyday life is: what can e-government deal with? Indeed, e-government functionalities can only be put into place if they get the clearest picture of the reality to be managed or to be transformed.

According to Lautier (2004), the word "informal" was first proposed in 1971 by Hart (1973) to describe the complementary revenue that is necessary to face wage stagnation, inflation, insufficient kin solidarity and limited access to credit opportunities in developing countries. The term was then popularized by the International Labour Office (ILO) in 1972 in its report "Incomes and Equality – A Strategy for Increasing Productive Employment in Kenya". Whereas Hart was tak-

ing "informality" for a set of practices, the ILO is using it to depict a situation, the "informal sector", the "informal economy". As Lautier (2004) shows it, searching for a definite list of criteria to define informality is a vain activity. He rather focuses his analysis on the economic informality, that of economic activities that are characterised by the following: law infringement, relatively small-scale, under-employment, poverty, survival strategies…

Because we agree with Lautier (2004) on the impossibility of a comprehensive definition of informality, we shall push for a narrower understanding of this notion by focusing on the *administrative informality*. It gathers all human activities and products that do not exist in the eye of the State, that are not part of any statistical apparatus and thus can not be acted upon because they simply don't exist for the administrative machinery.

The construction of e-government requires the constitution of vast and complete databases. An "ignorant" e-government, that does not satisfy its appetite for information and data, can not run efficiently. Furthermore, these data must be usable by the technical architecture in place and thus must be standardised, formatted into the frameworks of the administrative system that collects, manages, produces and diffuses these information. In this regard, any e-government strategy must develop "policies of formalisation" (Lautier, 2004). Using the notion of formalisation refers to four objectives: making some realities official in the eye of the administration; formatting realities into specific statistical and administrative frameworks; informing authorities to produce knowledge; controlling these now well-known realities.

Aiming at satisfying its statistical appetite in order to ensure the exhaustivity of its knowledge and of its control, the South African State is implementing several ICT-based initiatives. Citizens, their identities and their activities must be visible to the State. Some might say they must be "transparent" to the Leviathan. Let's note that whereas the total population of South Africa was about 45,4 million in 2002 (StatSA, 2002), only 28 million South Africans had an identity card in May 2003 (Buthelezi, 2003). Three million unidentified children were still out of the social security system in 2004 (Mapisa-Nqakula, 2004).

The existence of people in the eye of the administration means that they can enjoy their "second-generation" socio-economic rights and access the facilities of the Welfare State. For instance, the *Vital Registration Programme* is using an online data collection and compilation system in order to register births with no delay in hospitals and clinics. The *Child Support Grant* is thus correctly paid to eligible households. The ID card is truly an enabling documentation. This formalisation of people's identity is also a condition of their citizenship : thus, according to the Home Affairs Minister, Mapisa-Kqakula, the attribution of ID documents to San communities in 2004 enables them to live "their full participation as citizens" (Van Der Berg, 2004). Geographical data are also crucial: the South African Post Office (SAPO) and the firms Spatial Technologies and MapInfo Corporation are involved in the ICT-based cleaning up of all street addresses of the country (Minnaar, 2004b).

Another example is the possibility for South African women to check their marital status online by typing in their ID number in a specific area of the Web site of the Department of Home Affairs. In July-August 2004, about 200 women discovered they were registered as married but … to men they didn't know. The "*Check Your Marital Status Campaign*" enabled 5 000 women to check their status (Burrows, 2004a). This campaign has enabled the prosecution of five Home Affairs officials who facilitated 1 500 fraudulent marriages between South Africans and illegal immigrants and 200 faked registrations of birth and unlawfully nullified marriages (Engelbrecht, 2007a).

This policy of administrative formalisation is essential to the tight administration of the South African population. In this regard, formalisation is

feeding "normalisation" as understood by Michel Foucault (1997) when he looks at the "society of normalisation" and the generalisation of disciplinary projects mixing knowledge and power. Here comes in the notion of "normalisation" as proposed by Guillaume (1978, p.8): normalisation is a compromise between order and disorder that can "lock up individuals, assign them to specific places and dictate their representation in the past, the present and the future". For instance, the officers of the *Johannesburg Metropolitan Police Department* can now get the personal data of any driver by entering into their cell phone the car registration number, its serial number, the engine number or even a spare part's number; the information is sent by SMS to the main server (CPSI, 2003, p.58).

The ICT-based formalisation applies not only to identities but also to activities, with the perspective of (economic) transactions between the State and citizens. The tax objective is also a driving force behind this move. 7 million South Africans have no bank account. The private sector, and banks at the forefront through the *Financial Services Black Economic Empowerment Charter*, is developing numerous projects to get citizens registered in the economic transactions. In that regard, they back up the State effort to formalise people's identities. In June 2004, the Unemployment Insurance Fund launched a web-based process to officialise the employment of domestic workers. The campaign was led by the Department of Labour (DL) and more than 600 000 workers got registered, who thus could claim their unemployment insurance (SAPA, 2004c). In a more comprehensive manner, aiming at formalising its knowledge about the working population of South Africa, the DL is computerising its 125 *labour centres* and 529 *satellite offices* and installing an ERP-like system called *Lesedi* that ought to accelerate the resolution of work-related conflicts and the response time to these. The installation is a core piece of the Public-Private Partnership that ties the Department and Siemens Business Services. The State also wants to introduce 20 mobiles offices units targeting Limpopo, Eastern Cape, KwaZulu-Natal and Northern Cape provinces.

All these initiatives recall that the constitution of data sets is closely "associated with the construction of the State, its unification, its administration" (Desrosières, 1993, p.16; Desrosières, 1997). South Africa provides us with an example of the British tradition of "political arithmetic" that aims at registering, referencing, codifying people and their activities. This enterprise reveals the tight connection of "knowledge instruments" with "discipline and constraint instruments" (Lascoumes & Le Galès, 2004, p.27). The best example thereof is the Apartheid State apparatus.

THE APARTHEID REGIME: THE "MANIA OF MEASUREMENT" FOR TOTALITARIAN POWER

The Statistical Frenzy at the Core of Apartheid State-Building

For totalitarian purposes, the Apartheid regime has generated an incredible statistical apparatus, answering to the huge data appetite of the segregationist administration.

The use of ICTs was aiming at three objectives: developing a comprehensive *knowledge* of the population; reinforcing the *surveillance* capacity of the State; consequently, ensuring its ability to *control* people. The construction of the Apartheid regime did not absolutely follow a grand plan but was rather a collection of discrete administrative and political acts (Bonner & al., 1993; Fauvelle-Aymar, 2006; Posel, 2000).

In this scenario, the "power of number" was crucial for the nascent administration to build "totalising modes of racialised knowledge" (Posel, 2000, p.116). The obsession of numbering, counting and classifying was remarkable within the Department of Native Affairs (DNA): the formalisation and categorisation of people was aiming

at normalising them because the "feasibility of apartheid came to rest on the pervasive presence of the State in every facet of life" (Evans, 1997, p.1). To put it shortly, this Department provided the DNA to the Apartheid State! The government and particularly the DNA were thus characterised by "an explosion of 'scientific' empirical investigation into the lives of Africans[1] in the urban and rural areas" (Evans, 1997, p.11).

The better the State knows the population and its members and activities, the easier the control thereof can be. The "mania of measurement" is supporting the conquest of omniscience by the Apartheid state which aims at developing "'modern' modes of political rationality" (Posel, 2000, p.116). The quest for numbers is fuelled by an administrative scientism and a racialist technocracy. The State becomes a super-calculator, a computer in its original meaning i.e. compiling data to process them for specific purposes.

However, this mania did begin before the start of the apartheid regime with the *Census Act* (1910) which enabled the first national census to take place in 1911; then, the *Statistics Act* (1914) established the Office of Census and Statistics. In 1937, the *Native Laws Amendment Act* forced the local authorities to conduct a bi-annual census of the Black people. These actions pushed for a "scientific statecraft" (Posel, 2000, p.122) which was all the more necessary for the apartheid regime that is was deeply based on a geographical approach of reality: the influx control policy, that was inaugurated in 1923 with the *Natives (Urban Areas) Act*, was to realise the following equation: *n* African people authorized to work in urban areas = *n* working African people = *n* African people lodging in townships. This transcription of reality into State mathematics enabled the administration to calculate the "surplus" of Black population to be rejected from "White spots" (Posel, 2000, p.122). Though the Apartheid regime does not inaugurate the birth of statistics, it does implement it with a remarkable systematisation. For François-Xavier Fauvelle-Aymar (2006, p.33), the Apartheid is a "delirious ideal of organising the world and of indexing the human diversity". It is

an engineering of the population that is crossed by a delirious ideal, that of rationalising diversity, that of summing it up to a series of social objects that are differentiated, homogeneous, singular, consisting as many as separate races, ethnies or tribes, in an unchangeable hierarchy. (Fauvelle-Aymar, 2006, p.102)

This "political arithmetic" (Sadie, 1950) is powerful because the "Apartheid statecraft represented a hankering for 'total order'" (Posel, 2000, p.127). It is fully implemented with the *Population Registration Act* (1950) and the National Population Register. To depict this "statistical frenzy", Posel (2000, p.134) and Breckenridge (2002a) use the term "hubris", qualifying an administration that is obsessed with its measurement tools.

In the 1960s, the cognitive system of the administration is getting more and more complex because the State is determined to settle a new order of "social engineering". The extension and the sophistication of the measurement system foster the creation of the Department of Planning in 1965. It complements the Department of Bantu Administration and Development whose statistical routine was involving

monthly counts of the numbers of work-seekers' permits issued on national basis; the number of service contracts registered (showing the area of origin of each of the registered workers); the number of registered vacancies in urban areas and in bantustans and on farms; annual counts of the numbers of 'Bantu' seeking assistance from special 'aid centres' which assisted with the work of the labour bureaux; the compilation of an 'occupation register'...in respect of each national unit... containing certain particulars of the Bantu concerned'; the creation of a separate data base of 'existing and anticipated employment opportunities in the homelands for Bantu with advanced

qualifications'; annual counts of the number of workers recruited by various recruiting organisations and the sites of recruitment; annual counts of the numbers of curfew proclamations issued; the number of 'new Bantu residential regulations' promulgated as well as amendments to the existing regulations; annual counts of the number of bodies designated as urban local authorities and the number of 'promulgations and redefinitions of Bantu residential areas'; annual counts of the numbers of permits issued in terms of the Group Areas Act for 'recreation and health services', such as cinema and hospitals; numbers of applications to 'conduct' church services for Bantu in white residential areas'; amounts of money accruing in respect of the Bantu Services Levy Act; annual income from the sale of 'Bantu Beer' by local authorities and employers; numbers of inspections undertaken of a whole range of different types of sites on which particular types of developments were being considered (each type of site being enumerated separately); 'numbers of cases in which comments... from inspectors were furnished on group areas planning'; numbers of townships planned and developed, per ethnic group; numbers of sub-economic houses constructed; numbers of families 'removed' from urban townships, on an ethnic basis; numbers of families removed from 'black spots' and white rural areas, per ethnic group; separate counts of numbers of 'Bantu traders, industrialists and professionals' resettled; numbers of children, placed in various welfare institutions; annual counts of numbers of reference books issued to males and females; numbers of duplicate reference books issued to males and to females; numbers of identification documents issued to 'foreign Bantu'; comparisons of 'numbers of Bantu whose identity numbers were known with numbers of existing records of fingerprints'; numbers of fingerprints 'classified and searched to determine whether the fingerprints of the persons concerned are not already on record'; numbers of sets of fingerprints 'added to existing record'; numbers of births, marriages and deaths registered each year; annual counts of the numbers of 'Bantu males' on the National Population Register; annual counts of the numbers of 'Bantu females' on the Population Register; numbers of inquiries into 'the tax particulars and movements of Bantu' on the Population Register; numbers of beneficiaries of social pensions, maintenance grant beneficiaries and pneumoconiosis grant beneficiaries; along with records of monies spent on salaries, capital equipment, land and other projects. (Posel, 2000, pp. 132-133)[2]

In that regard, the Apartheid apparatus is an "encyclopedian State". But the requirements that this statistical mania involves are far too demanding for the capacities of the administration. This has motivated the mechanisation of the system in order to satisfy the data appetite of the State (Breckenridge, 2002a; Breckenridge, 2002b). From the early 1970's, ICTs – at this time, computers – are considered as useful auxiliaries of the oppressive system (Chokshi & al., 1995). In that sense, the most rational instruments of obsessional statistical arithmetic are serving a folly made of violence, racism and authoritarianism. For Dubow (1995, p.248), "when discussing apartheid, it is important not to draw false distinctions between rationality and irrationality, sanity and madness: there was method in the madness of apartheid, just as there was madness in its method".

The Use of "Control Electronics" to Boost the State Knowledge

Formalising reality is an obsession of the Apartheid regime. This formalisation is aiming at strengthening the efficiency of the totalitarian power structure. It is heavily based on the use of ICTs, depicted at this time as "control electronics". At this time, the South African State has to rely on the intervention of foreign firms to get appropriate ICT products and services. The IT sector is the most advanced of the Apartheid

business (First & al., 1972, p.106). In 1977, only the United States and Great Britain are spending more money than South Africa on the acquisition of ICT equipment (KSG, 2004).

In the 1970s, the country is totally dependent on foreign imports for ICTs. In 1980, 70% of computers are of American origin (KSG, 2004). Some U.S. firms involved in this business are Unisys, Hewlett Packard, IBM, Westinghouse (Slob, 1990). Knight (1986) lists 14 U.S. companies as main IT providers for the State. For example, in 1986, the State is the first buyer of computers, absorbing 42% of the total IT sales (KSG, 2004). The United Nations military embargo that was established in 1963 did not prevent the State from acquiring IT equipments for authoritarian purposes.

The prominent role that was played by telecom and ICT infrastructures is shown by the prosecution of firms like IBM, ICL and Fujitsu in the United States. Because they turned down the invitation to appear before the Truth and Reconciliation Commission (TRC), the South African non-governmental organisation Khulumani Support Group (KSG) ledged a complaint against these firms in the State of New York in 2002 but lost the case. Indeed, ICL equips the police, local authorities and the defence administration with hundreds of computers. It particularly provides IT equipment to the Bantu Reference Bureau in 1967 and to the Department of Plural Affairs (DPA) that is using a computer network to coordinate its 14 Bantu Administration Boards and store fingerprints and personal data of 16 million South Africans in 1978. The DPA digitalises the National Population Register (NPR) that is aiming at "managing" 25 million South Africans. An impressive database is constructed upon several criteria: name, sex, age, home address, identity picture, civil status, driving license, work address, fingerprints... The government can thus rationalise the influx control and its surveillance on political opponents. On that specific issue, the National Intelligence Service (NIS, succeeding to the Bureau of State Security in 1970) is using an advanced ICT infrastructure to document the activities of anti-Apartheid movements (Slob, 1990, p.26). The South African police is using Series 2900, 190s and 2960s ICL computers, notably in the implementation of the *pass laws* system. In the mid-1960s, ICL automatizes the system that is aiming at running the flows of Black people across the country. IBM, on its side, implements a system targeting at non-Black people (Crush, 1992). IBM started its activities in South Africa in 1952 when it provides the first supercalculator that is used in the ICT-based management of population databases. The Department of Interior is using IBM computers (notably 370/158 Series) to store the *Book of Life* files of 7 million people classified as "non-Black" (Coloured, Asians, Whites). The Department of the Prime Minister, Department of Statistics and Department of Prisons are also using IBM products, as well as the South African firm Infoplan that was the main provider of the national defence force. In the same time, Phillips is promoting its *Access Control System* that stores the identity and movements of people entering and leaving a public place. AEG-Daimler-Benz is proposing the same product.

This digital automatisation must improve the efficiency of the national identification system. This system is the basis of the formalisation process: to normalise the population under a totalitarian order, the State needs to know all people's personal data. In the 1950s and 1960s, the national identification system is based on three main pillars that are heavily reliant on ICTs. The first component is the Reference Book (*Bewysboeks*) nicknamed Dompas (the dumb pass), which is replaced in 1986 by the current green bare-coded I.D. book. According to Breckenridge (2002a, p.17), the bureau in charge of these passes (*Bewysburo*) has quickly been used by the regime as a key piece of the surveillance process. The second pillar is the national fingerprints registry. The last pillar is the attribution of an identity number (*persoonsnommer*) to each individual

In order to implement the cognitive and repressive process, the creation of fingerprint databases has been accelerated by the South African authorities. The first fingerprint was collected in 1900 by Sir Edward Henry in the Natal colony in order to establish the first fingerprint bureau in Pietermaritzburg. This administrative arrangement was concretising the "panoptic fantasy" of the government (Breckenridge, 2002a, p.2). In 1972, the Department of Home Affairs computerises the NPR. In 1979, the *National Research Institute for Mathematical Sciences* improves the system. Until 1986, all personal data and fingerprints of the Black population are stored by the Plural Affairs Department. In 1987, the Department of Home Affairs uses the Model 370/158 mainframe IBM computers to store the NPR and holds 18,9 million fingerprints (Slob, 1990, p.24-25).

This historical process shows that the ICT-based formalisation has served the totalitarian normalisation led by the Apartheid regime. The results that were obtained have an important impact on the nature of the post-Apartheid Welfare State.

E-GOVERNMENT IN DEMOCRACY: THE USE OF ICTS TO ENHANCE THE WELFARE STATE – AT WHAT PRICE?

The post-Apartheid South Africa is deeply ambivalent: since 1994, the country has experienced both radical changes and continuities with the former regime (Guillaume & al., 2004). As far as e-government is concerned and especially its formalisation and normalisation component, the ambivalence is clear: the sophisticated infrastructure deployed by the apartheid government is an opportunity for the Welfare State to be more efficient in the provision of public services but it can also be a burden for the population if this infrastructure is used on a non-democratic way.

The Formalisation of the South African Population

The formalisation of the population remains a priority for the government, especially in its push for the development of a Welfare State. We will focus on one specific project that is being implemented for that purpose: the Home Affairs National Identification System (HANIS).

This project is the major building block of the DHA renewal strategy to rationalise the e-Welfare State. The DHA IT "turn-around strategy" could reach a total amount of 2 billion Rands and will be implemented from January to December 2008. It is implemented by a public-private DHA "Turnaround Action Team". One project is the "track and trace" system for IDs. It must allow the DHA – and the public – to track each ID book transaction from application to collection by means of a website, SMS and a call centre. The system should start being implemented in February 2008 (Engelbrecht, 2007b).

In this context, HANIS aims at providing a centralised database detailing the profile and activities of the South Africa population. In 2003, the then DHA Director-General, Barry Gilder, describes the system in these terms:

as I understand the dream, it was that any client of the Department would be able to go into a Home Affairs office and have their fingerprints, photograph, signature and application taken on the spot electronically, checked immediately against the system and entered in real time onto the system. (Gilder, 2003)

HANIS was conceived by the DHA in 1993 and approved by the Presidency in January 1996 for an initial total amount of 930 million Rands[3]. In February 1999, the tender was awarded to the Marpless consortium to implement the first steps of the system. The consortium is combining the Japanese conglomerate Marubeni Inc. and the South African firm Plessey in a joint venture.

In January 2000, the project was officially launched.

By officialising the existence of people in the eye of the State, HANIS is expected to strengthen the fight against crime and terrorism, to reduce fraud (notably in the payment of pensions and social grants[4]) and to improve the performances of the South African Welfare State. To fulfil these objectives, HANIS is made of three components. The first one is the Automated Fingerprint Identification System (AFIS) which is a fingerprint recognition technology developed by the firm NEC. About 40 million fingerprints must be digitalised: it is the back record conversion. New fingerprints (for instance, those of new born or people getting their first identity card) are directly stored in the HANIS database. In September 2007, almost all of the fingerprints that were stored at the DHA had been migrated to a digital format: the AFIS contained about 30 million sets of fingerprints, the target being 31 million out of the country's 45 million citizens. The remainder are either children younger than 16, who are not fingerprinted, or the estimated 6% of adults who do not have an identity document and whom the department is trying to reach (Engelbrecht, 2007c).

To secure the data, the DHA has spent 207 million Rands in 2005-2007 creating the HANIS disaster recovery system at a secure location which backs up all our records every fifteen minutes on a continual basis (Mapisa-Nqakula, 2007).

The second component of HANIS is the South African Multi-Application Identification Card (SAMID): a chip-based "smart card" must gather all personal data and facilitate the provision of everyday public services. This Secure Electronic National Identity Card (SENID) will combine fingerprints, an identity picture, an identity number and an electronic chip gathering the following information: unemployment insurance number, health data (blood group, allergies, the list of the 10 last medical treatments and prescriptions), social transfers (family and housing benefits), number and amount of pensions, driving license and car registration numbers, tax information (Brümmer, 2002a). In Foucault's terms (1995, p.163-165), this card will be a tool of the "political anatomy of the details" that characterizes the formalisation project of the State. The cost of the SAMID component is estimated to be around 2,5 billion Rands (Brümmer, 2002b). It is not yet decided whether the card will be free of charge. The first cards will be distributed to retired people in order for pensions to be paid more rapidly and waiting queues to be shorter (Department of Home Affairs, 2001, p.48). The other reason why the focus is put on retired people is the amount of fraud in the payment of pensions that would approximate one billion rand (Buthelezi, 1997).

The last component of the HANIS project is the integration of HANIS in the National Population Register (NPR) which generates birth and death certificates. The firm Unisys is expected to integrate the different elements of the system (identity image capture, identity checks in the database, deployment of the telecommunication infrastructures). Beyond this, all DHA databases must be integrated in the HANIS system: the Electronic Document Management System (EDMS[5]), the Movement Control System (MCS), the Visa System, the Refugee Database and the Illegal Foreigners Database (Gigaba, 2004).

As in the movie *Matrix*, one talks of HANIS *reloaded* because the database must now integrate all individuals that are interacting with the DHA: citizens, foreign residents, foreign visitors (tourists for instance), refugees[6], documented and undocumented immigrants, the total number of people concerned reaching about 300 million individuals. A priority will be given to the equipment of the 57 official points of entry into the country and particularly to the Lindela retention and repatriation centre that is dealing with undocumented people (Lambinon, 2003). According the government, the formalisation of flows in and out of the country must rationalise the management of the migrant population through a better interconnection of all entry points, in order to avoid a situation whereby

a migrant candidate can jump from one entry point to the other in order to get a go-ahead. For Barry Gilder (2003), this situation is not totally without consequence on the levels of criminality.

Because the HANIS is not totally operational yet, it is too early to evaluate its achievements or shortages but one can already says that it is the first priority of the DHA in its formalisation-normalisation enterprise since it represents 22% of the IT expenditure of the DHA for the 2007/2008 budget period. However, prospects for an operational HANIS are still blur: the supplier contract for HANIS expired in June 2006 and by June 2007 no action had been taken whether to reopen it for tender or to extend it with the current supplier (Vecchiatto, 2007a). In July 2007, a contract was signed for equipment to be delivered in March 2008. Also, the HANIS maintenance contract with the Marpless consortium has been extended by a further three years. The introduction of the first smart cards could start in April 2008 (Engelbrecht, 2007c; Senne & Engelbrecht, 2007) and the government intends to introduce "e-passports" (containing a chip and, probably, a RFID) in April 2008 (Engelbrecht, 2007c).

The HANIS project is part of a bigger move illustrated by the fact that, in October 2007, the South African firm GijimaAst-led consortium won a 2 billion rand tender to implement an integrated citizen-centric documentation system called "Who am I online (I am I said)" which gives a "single view of the citizen" and visitors. The tender process started in March 2006. The consortium is expected to integrate birth, death and marriage certificates and identity documents with the existing AFIS (Engelbrecht, 2007d).

The Risk of a Non-Democratic Normalisation

The fact that the Welfare State needs to know as much as possible about the population as a whole for efficient policy making and about individuals for appropriate public service delivery is not surprising. However, more contentious is the way this formalisation of reality is being implemented: what data for whom? In other words, because the State is a huge machinery, a specific matrix must be designed so that all data can not be accessed by all parts of the administrative apparatus.

This is the heart of the ambivalence of e-Welfare State: on the one hand, citizens can get the most of the formalisation of their existence by getting more visible to the State and thus having access to appropriate public service; on the other, they must accept to be transparent to the administration and get their lives scrutinized by the Leviathan. To put it Bourdieu's terms (1993, p.221-223), the "left hand" of the State can serve the population only if the "right hand" of the State knows and controls it.

Who knows what? That is the critical question in the formalisation and normalisation of reality. Two issues are at stake: the publicisation of private life and the commercialisation of personal data. They are both core pieces of the rise of the e-Welfare State that can be depicted as a collection of *Little Sisters* rather than a single *Big Brother*.

The construction of a seamless administrative apparatus implies that most of (if not all) parts of the administration can communicate with each other and notably exchange information about individuals that they have to deal with. The intimate life of everyone must be accessed and "networked" in the State databases. The protection of personal data can not be but relative: social life (i.e., to live within a society and get the most of it) and the Welfare State make absolute private life impossible (Davies, 1997; Lyon, 2002; Lyon, 2003). The individual can not be invisible; otherwise, he/she would lose any sociality. Moreover, Lyon (2002) recalls that the everyday private life is positively affected by State control because it is the condition for citizens to enjoy the benefits of a Welfare State.

Nevertheless, the publicisation of private life that results from the formalisation / normalisation dynamics is damaging when it's implemented unbeknown to the citizens. It is especially the case when citizens know that the State collects data about them but don't know where these data go to in the end: which administration? Judiciary? Social? Etc… What connections are being established between the different databases? The risk is that of an "administrative panopticism—the urgent desire to complete and centralise the state's knowledge of its citizens". Breckenridge even sees here the "informational legacy of Apartheid" (Breckenridge, 2002b, p.3 & 13; Brümmer, 2002a).Van Tonder (2003) and the NGO Bridges.org (2002) point at the lack of any suitable institutional arrangement which could protect citizens against the misuse of their personal data for political or other purposes. Indeed, there might be a necessity of an independent agency that could set limits to the formalisation/surveillance process. This agency could implement the dispositions that are contained in the *Promotion of Access to Information Act* (PAIA, n°2, 2000) and the *Electronic Communications and Transactions Act* (n°25, 2002).

In 2002, the South African Law Reform Commission started drafting a comprehensive national Data Privacy Act (De Kock, 2005). In October 2005, the South African Law Reform Commission published a Discussion Paper (n°109) on "Privacy and Data Protection" that invites the government to push for the introduction of legislation regulating the collection, storage and processing of personal information by the public and private sector. To that end, the Discussion Paper contains a draft Protection of Personal Information Bill that must improve the South African data protection system which is based on the Bill of Rights and developed by the PAIA Act.

One improvement must deal with the "compartmentalisation" of data within the administrative apparatus: in a networked Welfare State, who can access what? In the United States, Loïc Wacquant (2004, p.49) diagnoses the rising of such a State through the implementation of various instantaneous identification and surveillance instruments, what he calls an "infrastructural power" – the capacity for the State to "penetrate" within the population. The regulation of data access is particularly important with the crime situation in South Africa: ICTs are powerful tools in the fight against crime but they can foster the undue mixing up of civilian and criminal data. The DHA and the South African Police Service (SAPS) databases are built upon the same architecture, in order to facilitate their joint exploitation. In 1997, the then Minister of Home Affairs, Mangosuthu Buthelezi, was declaring that the government was considering the integration of both databases (Buthelezi, 1997). In 2004, some high-rank police officials were also considering this option. The Court Process Project is sometimes using both systems so that the Departments of Justice, of Correctional Services and of Social Development can cross their information for court decisions to be taken quicker. This might illustrate what Wacquant (2004, p.19) calls the "criminalisation of social insecurity".

Beyond the data access within the Welfare State administration, another issue is crucial: the commercialisation of data outside the administration. Thus, the HANIS card is expected to have a financial functionality dedicated to banks, insurance and medical aid companies. The DHA intends to sale the HANIS-related data to the private sector, particularly for data mining initiatives. The fight against fraud is the main argument that must justify this commercialisation (Buthelezi, 2002). For Van Tonder (2003), this privatisation is a misuse of biometrics. In June 2004, according to the South African press, the Independent Electoral Commission (IEC) had used the national election roll to provide data to the South African Post Office (SAPO) and to the firm Intimate Data that was contracted by the SAPO in February 2004 to update, clean up and maintain its database of postal addresses. The IEC and the SAPO refuted this allegation (IEC, 2004).

The development of the e-Welfare State can not go without that of Little Sisters: this ambiva-

lence of e-government is well illustrated in the South Africa case. We know that knowledge is power. In the e-Welfare State, information, data and knowledge are the building blocks on which both social services and surveillance systems can operate. Both need to deal with a social reality that is formalised and, *in fine*, normalised. In the e-Welfare State, standardised knowledge is power. And this standardisation is based on ICTs. The collection of *Little Sisters* (Armatte, 2002; Castells, 20000; Lyon, 2003) is essential to the existence of the social Leviathan : ICTs are improving the "reason of State" for knowledge production that is aiming at both social and control purposes. Citizens are compelled to "produce truth" about their lives (Foucault, 1997, p.22-23). In South Africa, the legacy that was left by the Apartheid regime is an opportunity for the Welfare State to make the most of e-government; but it is also a burden that must fit into the set of democratic requirements.

REFERENCES

Armatte, E. (2002). Informatique et libertés: de big brother à little sisters ? *Terminal, n°88*, 11-21.

Bonner, P., Delius, P., & Posel, D. (1993). The shaping of apartheid: Contradiction, continuity and popular struggle. In P. Bonner, P. Delius, & D. Posel (Eds.), *Apartheid's genesis 1935-1962* (pp. 1-41). Johannesburg: Wits University Press.

Bourdieu, P. (1993). La démission de l'Etat. In P. Bourdieu (Eds.), *La misère du monde* (pp. 219-228). Paris: Seuil.

Breckenridge, K. (2002a, May). *From hubris to chaos: The making of the bewysburo and the end of documentary government.* Paper presented at the Wits Interdisciplinary Research Seminar, WISER, University of the Witwatersrand, Johannesburg.

Breckenridge, K. (2002b, October). *Biometric government in the new South Africa.* Paper presented at the The State We Are In Seminar Series, WISER, University of the Witwatersrand, Johannesburg.

Brdiges.org (2002). *Government efficiency vs. citizens' rights: The debate about electronic public records comes to developing countries.* Cape Town: Bridges.org.

Brümmer, S. (2002a). From dompas to smart card. Johannesburg: *Mail & Guardian*, 22 February.

Brümmer, S. (2002b). Buthelezi and Masetlha at it again. Johannesburg: *Mail & Guardian*, 22 March.

Buthelezi, M. (1997). *Department of Home Affairs budget vote 1997/1998.* Cape Town: National Assembly – Republic of South Africa, 17 April.

Buthelezi, M. (2002). *Parliamentary media briefing.* Cape Town, GCIS – Republic of South Africa, 11 February.

Buthelezi, M. (2003). *Home Affairs budget speech.* Cape Town, National Assembly – Republic of South Africa, 19 May.

Castells, M. (2000). *The information age: Economy, society and culture, vol.1 The rise of the network society*, Oxford: Blackwell.

Center for public service innovation. (2003). *Government unplugged – mobile and wireless technologies in the public service*, Pretoria: Center for public service innovation.

Cheneau-Loquay, A. (2005). Comment les nouvelles technologies de l'information et de la communication sont-elles compatibles avec l'économie informelle en Afrique. *Annuaire Français de Relations Internationales*, 5, 345-375.

Chokshi, M., Carter, C., Gupta, D., Martin, T., & Robert, A. (1995). *Computers and the apartheid regime in South Africa.* Retrieved October 28, 2003, from http://www-cs-students.stanford.edu/~cale/cs201/

Crush, J. (1992). Power and surveillance on the South African goldmines. *Journal of Southern African Studies*, *18*(4), 825-844.

Davies, S. G. (1997). Re-engineering the right to privacy : How privacy has been transformed from a right to a commodity. In P. E. Agre & M. Rotenberg (Eds.), *Technology and privacy: The new landscape* (pp. 143-165). Cambridge MA: MIT Press.

De Kock, E. (2005). Data protection in South Africa. *De Rebus*, December.

Department of Home Affairs (2001). *Strategic Plan 2002/2003 to 2004/2005*, Document n°2/6/8/P, Pretoria : Department of Home Affairs.

Desrosières, A. (1993). *La politique des grands nombres: Histoire de la raison statistique.* Paris: La Découverte.

Desrosières, A. (1997). Du singulier au général: L'argument statistique entre la science et l'Etat. In B. Conein & L. Thévenot (Eds.), *Cognition et information en société* (pp. 267-282). Paris: Ecole des Hautes Etudes en Sciences Sociales.

Dubow, S. (1995). *Scientific racism in modern South Africa.* Cambridge: Cambridge University Press.

Engelbrecht, L. (2007a). Five in court for marriage fraud. Johannesburg: *ITWeb*, 10 December.

Engelbrecht, L. (2007b). Blank cheque for Home Affairs information technology. Johannesburg: *ITWeb*, 5 September.

Engelbrecht, L. (2007c). HANIS gets Rands 30m 'refresh. Johannesburg: *ITWeb*, 6 September.

Engelbrecht, L. (2007d). Getting Home Affairs cup-ready. Johannesburg: *ITWeb*, 5 November.

Evans, I. (1997). *Bureaucracy and race: Native administration in South Africa*. Berkeley, CA: University of California Press.

Sadie, J. L. (1950). The political arithmetic of the South African population. *Journal of Racial Affairs, 1*(4), 3-8.

Fauvelle-Aymar, F. X. (2006). *Histoire de l'Afrique du Sud*. Paris: Seuil.

First, R., Steele, J., & Gurney, C. (1973). *The South African connection: Western investment in apartheid*. Harmondsworth: Penguin.

Foucault, M. (1995). *Surveiller et punir: Naissance de la prison*, Paris: Gallimard.

Foucault, M. (1997). Cours du 14 janvier 1976. In F. Ewald, A. Fontana & M. Bertani Ed.), *Il faut défendre la société* (pp. 21-36). Paris: Gallimard-Seuil.

Gigaba, M. (2004). Home Affairs department budget vote 2004/2005. Cape Town: National Assembly – Republic of South Africa, 11 June.

Gilder, B. (2003). Media Briefing. Johannesburg: Department of Home Affairs – Republic of South Africa, 5 November.

Gilliom, J. (2001). *Overseers of the poor: Surveillance, resistance and the limits of privacy*. Chicago, IL: University of Chicago Press.

Guilaume, M. (1978). *Eloge du désordre*. Paris: Gallimard.

Guillaume, P., Péjout, N., & Wa Kabwe Segatti, A. (2004). *L'Afrique du Sud dix ans après: Transition accomplie?*. Johannesburg – Paris: Institut Français d'Afrique du Sud – Karthala.

Hart, K., (1973). Informal income opportunities and urban employment in Ghana. *The Journal of Modern African Studies, 11*(1), 61-89.

International Labour Organisation. (1972). *Incomes and equality: A strategy for increasing productive employment in Kenya*, Geneva: International Labour Organisation.

Independent Electoral Commission. (2004). Independent Electoral Commission on reports about giving out details from voters' roll. Pretoria: Independent Electoral Commission, 15 June.

Khulumani Support Group. (2004). Complaint. South District Court, New York State, 15 December.

Knight, R. (1986). *US computers in South Africa*. Retrieved October 26, 2003, from http://richard-knight.homestead.com/files/uscomputers.htm

Lambinon, I. (2003). *Briefing at the Home Affairs portoflio committee*. Cape Town, National Assembly – Republic of South Africa, 18 March.

Lascoumes, P., & Le Gales, P. (Ed.). (2004). *Gouverner par les instruments*. Paris: Presses de Sciences Po.

Lautier, B. (2004). *L'économie informelle dans le tiers monde*. Paris: La Découverte.

Lyon, D. (2002). Everyday surveillance : Personal data and social classifications. *Information, Communication & Society, 5*(2), 242-257.

Lyon, D. (2003). Introduction. In D. Lyon (Ed.), *Surveillance as social sorting: Privacy, risk and digital discrimination* (pp. 1-9). London: Routledge.

Mapisa-Nqakula, N. (2004). *Home Affairs department budget vote 2004/2005*. Cape Town: National Assembly – Republic of South Africa, 11 June.

Mapisa-Nqakula, N. (2007). DHA budget speech for budget vote 2007. Cape Town: National Assembly – Republic of South Africa, 7 June.

Martin, D.-C. (1998). Le poids du nom: Culture populaire et constructions identitaires chez les 'métis' du Cap. *Critique Internationale*, n°1, 73-100.

Minnaar, C.-L. (2004). Putting maps to work. Johannesburg: *ITWeb*, 12 March.

Péjout, N. (2004). Big brother en Afrique du Sud? Gouvernement électronique et contrôle panoptique sous et après l'apartheid. In P. Guillaume, N. Péjout & A. Wa Kabwe Segatti (Ed.), *L'Afrique du Sud dix ans après: Transition accomplie?* (pp. 79-103). Johannesburg – Paris: Institut Français d'Afrique du Sud – Karthala.

Péjout, N. (2007). *Contrôle et contestation. Sociologie des politiques et modes d'appropriation des technologies de l'information et de la communication (TIC) en Afrique du Sud post-apartheid.* Unpublished doctoral dissertation, Ecole des Hautes Etudes en Sciences Sociales, Paris.

Posel, D. (2000). A mania for measurement: Statistics and statecraft in the transition to apartheid. In S. Dubow (Ed.), *Science and society in Southern Africa* (pp. 116-142). Manchester: Manchester University Press.

Senne, D., & Engelbrecht, L. (2007). Home Affairs admits ID inefficiencies. Johannesburg: *ITWeb*, 30 August.

Slob, G. (1990). *Computerizing apartheid: Export of computer hardware to South Africa.* Amesterdam: Holland Committee on Southern Africa.

South Africa Press Agency, 2004. More than half a million domestics registered with unemployment insurance fund (UIF). Pretoria: South Africa Press Agency, 2 June.

Statistics South Africa. (2002). *South African Statistics.* Pretoria: Statistics South Africa.

Van Der Berg, R. J. (2004). First ID for !Xhu and !Khwe communities, *Bua News*, Pretoria: Government Communication and Information System – Republic of South Africa, 9 June.

Van Tonder, K. (2003). Biometric identifiers and the right to privacy. *De Rebus*, 28 (8), August, Retrieved October 3, 2004, from http://www. derebus.org.za/archives/2003Aug/articles/Biometric.htm

Vecchiatto, P. (2007a). Languishing HANIS needs attention. Johannesburg: *ITWeb*, 7 June.

Vecchiatto, P. (2007b). SA trials smart ID cards. Johannesburg: *ITWeb*, 2 November.

Wacquant, L. J. D. (2004). *Punir les pauvres: Le nouveau gouvernement de l'insécurité sociale*, Marseille: Agone.

Weber, M. (1978). *Economy and society: An outline of interpretative sociology*, Berkeley: University of California Press.

ENDNOTES

[1] For the Apartheid regime, the term "African" designates the Black population (Constant-Martin, 1998, p.75).

[2] This list is compiled by Posel (2000) from : Republic of South Africa. (1971). *Report of the Department of Bantu Administration and Development for the Period 1 January 1971 to 31 December 1971.* Pretoria: Department of Bantu Administration and Development, RP 41/73.

[3] 1 rand = 0.098 euro (15/01/2008).

[4] The annual fraud would amount to about 2 billion Rands (DHA, 2001, p.47). Government lost more than 1 billion Rands to social grant fraud in 2006. The private sector is also suffering from ID-related fraud: for instance, in 2006, insurance lost in excess of 3 billion Rands due to their inability to verify information (Senne & Engelbrecht, 2007). In the United States, Gilliom (2001) analyses how ICTs are fuelling a "welfare surveillance": the Client Registry and Information System – Enhanced (CRIS-E) refines the meticulous control operated by the welfare administration upon beneficiary families.

5   Under its Electronic Data Management System, the DHA has scanned a total of 57 million records including birth, marriage and death records (Mapisa-Nqakula, 2007).

6   The Department of Home Affairs was planning to roll out a refugee online verification system in January 2008 (Vecchiatto, 2007b).

Chapter VIII
ICT Challenges and Opportunities for Institutionalizing Democracy in Ghana:
An Integrative Review of the Literature

Joseph Ofori-Dankwa
Saginaw Valley State University, USA

Connie Ofori-Dankwa
University of Michigan, USA

ABSTRACT

Several African countries have begun to introduce and implement Information and Communication Technology (ICT) policies. In the context of such developing countries, it is important to assess the nature of research focus on the ongoing ICT revolution and its potential to stimulate institutionalization of democracy in Africa. This chapter reviews and integrates literature by scholars focusing on ICT in Africa in general and more specifically on Ghana. The authors incorporate several key points in their discussion. First, they provide a summary of ICT trends and policies in Ghana and their emphasis on helping to institutionalize democracy and its related free market system. Next, they provide a description of some of the major challenges to institutionalizing democracy that scholars writing about ICT in Ghana have identified. In addition, the authors discuss several opportunities for enhancing democracy that scholars writing about ICT in Ghana have highlighted. Finally, they make a few general recommendations for mitigating the potential problems and enhancing the opportunities of the ICT revolution for Ghana as well as the entire African continent.

INTRODUCTION

The term "Information and Communication Technology" (ICT) refers to emerging technology revolving around the increasing availability and use of the Internet, personal and organization-wide computer systems that are faster, more powerful, smaller, and less expensive, and increasingly more accurate global wireless and satellite systems. The result of the ICT revolution is an increasing level of connectivity at a rapidly evolving pace, and the heightened possibility of fully-realized national and global connectivity.

There is a corresponding increase in research by Western scholars of the political and democratic governance implications of the ongoing digital revolution (Evans & Yen, 2005; Evans & Yen, 2006; Amoretti, 2006). Different researchers have emphasized different approaches such as depicting the potential e-governance process in Europe as having different stages (Layne & Lee, 2001) and typologies (Amoretti, 2006; Schelin, 2003), or involving the technology enactment process (Fountain, 2001). Nevertheless, these scholars all point to ICT as providing important means for political information, and for broad and mass participation in the governmental decision-making process (Amoretti, 2006).

A noted caveat of ICT is its provocation of a technological divide between more- and less-developed countries (Ifinedo, 2005). If unattended, this technological divide will continue to grow wider as time goes on (Singer et al, 2005). Evans & Yen (2006) notes that e-government in the United States is rapidly expanding and that the total information technology budget will exceed $48 billion in 2002. Referencing a 2003 United Nations Crossroads study, Evans & Yen (2006) compares and ranks global regions in terms of e-government readiness—Africa placed last. A similar study arrives at the same conclusion (Ifinedo, 2005). In essence, African countries have been identified as places where the digital divide

is a potentially major and serious socio-economic threat (Wright, 2004).

However, several African countries have begun to introduce and implement ICT and its related policies (Haruna, 2003). Further, to the extent that ICT provides new ways of increased citizenry participation in governance, several major international organizations such as the World Bank, the Organization for Economic Cooperation and Development, and the United Nations have begun initiating reforms to help harness the potentially positive aspects of these new technologies for nation building and the democratization process in developing countries (Amoretti, 2007).

In the context of such developing countries, there has also been sustained research attention to the potential challenges and opportunities of technology for economic development with a wide range of publication outlets such as the *Journal of Technology Transfer* (established in 1975) and *Information Technology for Development* (established in 1990). Focusing more specifically on ICT, the online journal *International Journal of Education and Development Using ICT* was established in 2003 and has begun to play an equally pivotal role. It is heartening to note the substantial number of studies looking at different aspects of ICT revolution and its implications for developing countries. Indeed scholars focusing specifically on developing economies have both conceptually and empirically begun to examine ICT implications for increasing the democratization process (e.g. Narayan & Nerurkar, 2006; Mensah, 2005)

We have one important research question driving this essay: What are African scholars identifying as critical challenges and opportunities for ICT to enhance the democratization process in African countries? In answering this research inquiry, this chapter focuses specifically on Ghana, West Africa as a case study. This particular country is our focus because of the pioneering roles it has historically established in Africa. Ghana was at the forefront of the Sub-Saharan independence drive from Great Britain in 1957. In the early 1980s, Ghana was again one of the first African countries to undertake the drive towards the free market liberalization of the national economy. It also represents the emerging institutionalization of democracy through the electoral process that is taking place in Africa. Perhaps even more importantly, Ghana is at the forefront of the ICT revolution in Africa and was one of the first countries on the continent to begin to develop and implement a broad national ICT strategy.

Specifically, in this chapter we seek to do the following: first, we briefly highlight ICT policy initiatives and trends in Ghana and their emphasis on helping to institutionalize democracy and its related free market system. Second, we discuss our research methodology in carrying out an integrative review of the literature. Then, we synthesize and describe five major ICT-related challenges associated with institutionalizing democracy that several scholars have identified. Finally, we describe five major democracy-enhancing opportunities that ICT offers Ghana.

GHANA'S ICT POLICY INITIATIVE: A BRIEF HISTORICAL OVERVIEW

In Africa, there has been a more liberal policy to improve the infrastructure of its existing ICT system and to attract investors. Wright (2004) describes the increasing trend of privatization and liberalization of the telecommunication industry in several countries in Africa. A leading advocate of African ICT is the African Information Society Initiative (AISI) that was set up by the Economic Commission for Africa (ECA) which advocates the development and implementation of national information and communication infrastructural (NICI) plans. This is part of a trend noted by Juma (2005) where developing countries are beginning to work together to meet their technological needs.

By the end of the year 2000, thirteen countries in Africa had NICI plans in place, while ten were actively designing them, including Ghana. While Ghana has always been at the forefront of the ICT revolution in Africa, the ICT capabilities of firms in Ghana are significantly weaker than those of newly industrialized countries in Asia and Latin America, thus revealing a major development constraint (Lall, Navaratti, Teitel & Wiggnaraja 1994). Indeed, Zachary (2003) describes an emerging drive to make Accra, Ghana's capital, the information technology hub of Africa with the included acknowledgement that although Ghana is at the forefront of ICT compared to other African nations, it is still notably far behind more developed countries.

In August 1994, Ghana was one of the first countries to liberalize basic telecommunication services, resulting in a highly competitive telecommunications market. The National Communications Authority Act of 1996 set up an independent regulatory body that oversees communication by wire, cable, radio, television, satellite and similar technologies. In addition, the National Communications Authority (NCA) provides licenses for the operation of communication services, and for the application of radio frequencies (www.ict.gov.gh).

Several authors have noted the need for an effective telecommunication policy that will complement and assist the economic development of the country (e.g. Alhassan, 2004; Lall, Navaratti, Teitel & Wiggnaraja, 1994). For example, Lall, et al (1994) highlights the need for appropriate technological policies if Ghana's economic development initiatives such as the structural adjustment program are to be successful. Ghana's ICT process was undertaken under the auspices of the African Information Society Initiative. The process, termed "ICT for Accelerated Development," identified the key role that ICT can play in furthering national economic development if these technologies are effectively utilized (www.ict.gov.gh). There is the recognition that ICT could be a potentially new source for "the creation of quality jobs, wealth generation, income distribution and poverty alleviation, as well as for rapid economy development, prosperity and a source for facilitating global competitiveness" (www.ict.gov.gh). However, there is also the recognition that unlike the industrially advanced countries where existing institutional and infrastructural levels facilitate the movement into information and knowledge based economies (IKE), developing countries such as Ghana need to do much more in order to surmount their existing double void contexts and to develop IKE.

In August of 2002, the National ICT Policy and Plan Development Committee in Ghana highlighted a three-phased process based on the AISI/ECA methodology. In essence, the first phase detailed the nature of the challenge to be faced; the second phase identified what needed to be done, and the third described how the goal was to be achieved. The committee sought to be highly approachable and consultative, and met with several different constituents such as women's organizations, professional associations, and government ministries (www.ict.gov.gh).

In 2005, the National Telecommunication Policy was presented. It had the vision of creating a "true Information Society" which would provide "the greatest opportunity for economic growth, social participation, and personal expression." The policy sought to integrate Ghana with the new emerging economic order that uses IKE to increase "investment, development of human capacity, and increased governance leading to wealth creation and national prosperity" (www.ict.gov.gh). The policy indicates that this is to be achieved through the development of a highly competitive and liberalized telecommunication market structure. The Ministry of Communication is primarily responsible for the stipulation of government policy, and assessment as to the extent to which a policy has been effectively implemented. This reaffirms the role of the National Communications Authority as the regulatory body of the telecommunications sector and the implementer of policy. Regulatory issues include distributing licenses and ensuring that Ghanaians interactions with all competing service providers be done in a "transparent, open and non-discriminatory" way. In addition, the NCA seeks to ensure that telecommunication providers offer adequate quality services and protect consumers' private information. In line with the competitive and open market principle, the NCA will make sure that no single organization has "significant market power," (www.ict.gov.gh) which is generally defined as controlling at least 40% of a relevant market segment.

Thus, it is clear that Ghana has undertaken a serious effort to develop and implement an ICT policy. It is also clear that one of the primary goals of its ICT policy is to further the economic development of the country and to increase the level of citizen participation in the democratic process. However, some scholars (e.g. Jaeger, 2005; Yildiz, 2007) caution that with respect to the democratization process, ICT can be a double-edged sword with the potential to present increased levels of both challenges and opportunities. For example, it has the potential to improve the deliberative democratization process by helping to provide reflection about significant issues and by increasing active participation. On the other hand, however, it may impede the deliberative democratic process through aiding in group polarization (Jaeger, 2005). With this caveat in mind, a critical concern centers on understanding the potential that ICT initiatives in Ghana would have to either diminish or enhance the democratic process.

METHODOLOGY OF THIS STUDY

In order to identify the major challenges and opportunities that ICT might generate for the democratic process, the following methods were used:

Initially, a broad list of studies in which scholars discussed the implications of ICT in developing countries (particularly Ghana) was compiled. Given the substantial number of studies relevant to understanding ICT and the democratization process it was necessary to limit the number of studies used. Consequently, this integrative discussion should be seen as an illustrative and not necessarily exhaustive review of scholars focusing on ICT and the democratization process. In order to ensure that we also reflect the more practical implications of ICT, our review incorporates information from current African business magazines such as *Africa Business*. Finally, we identified several central themes associated with the potential challenges and opportunities that these scholars identified and several ways in which these challenges can be minimized and the potential opportunities increased. Once again, given both time and space constraints, we limited our study to identifying only five of the most pertinent challenges and five of the most significant opportunities that researchers have identified.

Defining the Phrase "Institutionalizing Democracy"

An important issue of our study centers on what we mean by "institutionalizing democracy" and what definitional approach should be used. For example, some scholars suggest that the democracy institutionalization of the ICT process be reflected in terms of different stages (e.g. Layne & Lee, 2001). Other scholars suggest the need to develop typologies (e.g. Amoretti, 2006; Schelin, 2003). For example, Amoretti (2006) identifies four different types of e-democracy governance models: consultative, participative, deliberative and administrative. Other scholars such as Fountain (2001) highlight the need to think of ICT and its implementation as a "technology enactment process" with two distinct steps—its adaptation to suit the environment in which it is implemented,

and simultaneously, the change the environment undergoes as a result of the adoption and implementation process. Focusing more specifically on the context of developing economies, Narayan & Nerurkar (2006) also proposes a three-phase model of ICT: enabling, awareness, and diffusion. Irrespective of the relative emphasis that these different scholars place, the common themes associated with their different models is that they are primarily democratic in their assumptions, and that they all emphasize distinct but interrelated designs that ensure increased participation by the general public in the governance of a country (Yildiz, 2007).

In general, the potential for ICT to increase the democratization process within a nation centers on two distinct variables—the government and individual citizenry (Yildiz, 2007; Fountain, 2001; Gil-Garcia & Pardo, 2005). For the purposes of this study, we will use very broad and generalized definitions. First, we will define democracy enhancement in terms of *e-government*, (sometimes referred to as "digital government" or "virtual state") which seeks to reflect both national, regional, and district level implications (Fountain, 2001). While some scholars such as Yildiz (2007) suggest that the definition of e-government may be loose in part because it does not fully take into account potentially different meanings arising from the nature of the "regulatory environment, dominance of a group of actors in a given situation, [and] different priorities in government strategies," there is a relatively broad consensus in viewing e-government as the intensive or generalized use of information technology in government for the provision of public services, the improvement of managerial effectiveness, and the promotion of democratic values and mechanisms through providing citizens with easy access to political information (Gil-Garcia & Pardo, 2005; Gil-Garcia & Luna-Reyes, 2003).

From the perspective of the individual citizen, the potential for ICT to enhance the democratic process is sometimes referred to as *e-participa-*

tion (Amoretti, 2006). Again, we will use a very general definition in this study, reflecting both organizational and individual citizenry. Thus, e-participation centers on the effective contribution of ICT to the democratization process, and the extent to which it allows for communication between the citizenry and its elected representatives and government officials (Nugent, 2001).

Both processes of e-government and e-participation capture several key ICT democracy-enhancing imperatives. First, ICT can ensure a greater and faster flow of information from government to the citizenry. Such information makes it possible for informed decision-making to take place. In addition, ICT can facilitate a similar link between citizens, making it easier for them to mobilize and organize around specific issues. However, ICT makes it difficult for the government to control the nature and dissemination of relevant political and economic information. For example, the use of the Internet, with its increasingly widening scope and availability, makes it difficult for governments to tightly control sensitive information.

POTENTIAL PROBLEMS AND CHALLENGES ICT POSES FOR DEMOCRACY INSTITUTIONALIZING IN GHANA

As we have noted above, ICT has the potential to service citizens efficiently. However, there are potential obstacles for effective implementation (Evans & Yen, 2005). In reviewing the literature, we identified several barriers that ICT implementation in Ghana faces, especially in relation to democracy enhancement. Specifically, scholars highlight five major e-government and e-participative challenges in regards to human capital, technological infrastructure, governmental cognitive mindset, cultural norms, and financial resources.

Human Capital

Scholars point to the importance of human capital for the effective development of e-governance (e.g. Gil-Garcia & Pardo, 2005; Srivastava & Teo, 2007; Hinson, 2005). Gil-Garcia & Pardo (2005) offers an extensive review and identify several challenges that e-government initiatives face. The list that their study generated includes lack of skilled technology-related human resources, and organizational and managerial issues associated with top management attitudes as top concerns.

In the Ghanaian context, it is clear that there is a substantial lack of requisite human capital and skilled ICT personnel (e.g. Zachary, 2003; Binns et al, 2005). For example, Zachary (2002) points out the reality of very limited manpower in IT-related areas in Ghana. In 2002, Accra had fewer than 50 experienced computer programmers. Binns, Porter, Nel & Kyei (2005) also point to a lack of qualified staff as a major constraint in the drive to use ICT to decentralize and modernize local governments.

In spite of this, there is a significant drive to increase ICT-related skills and expertise in Ghana, and several public and private institutions have been set up to achieve this end (Mangesi, 2007). In a comprehensive review of the role of ICT in education within Ghana, Mangesi (2007) describes several initiatives and ongoing programs. These ICT initiatives and projects include: expanding the deployment of ICT in both secondary and tertiary institutions in Ghana, the production of a television show on ICT that promotes distance learning, the establishment of a high speed ICT infrastructure at the Kwame Nkrumah University of Science and Technology, awards for teachers who excel in the utilization and teaching of ICT, and the establishment of several ICT-focused youth clubs.

In addition, Ghana Telecom University is a publicly funded tertiary institution focusing specifically on ICT. The Ghana Telecom University was inaugurated in August of 2006 and within a year attracted over 500 tuition paying students. The university is affiliated with the Kwame Nkrumah University of Science and Technology and DePaul University in the U.S., and is planning to set up more centers of telecommunication and information nationwide (Mangesi, 2007).

Plans are also underway to introduce online distance education in Ghana. The Ghana Telecommunications Training Center also provides several ICT-related workshops and seminars and trains more than 2,500 people a year mostly from the Ghanaian telecommunications sector (Mangesi, 2007). Despite these initiatives and programs, it is clear that a lack of sufficient human capital still remains a major ICT challenge in Ghana.

Technological Infrastructure

Srivastava & Teo (2007) points to the importance of existing technological infrastructure for the effective development of e-governance. Indeed the success of e-government initiatives with respect to the democratization process relies on ICT having the appropriate infrastructure to provide high-quality information that is relevant to the citizenry (Gil-Garcia & Pardo, 2005).

An important ICT challenge in the Ghanaian context stems from the inadequacy of current information, communication and technological infrastructure (Hinson, 2005; Adjei, 2004; Adjei & Ayernor, 2005). In a survey Hinson (2005) identifies several ICT challenges that firms doing business in Ghana often face. Some of the major problems noted include frequent breakdowns in network connections, and equally frequent power outages. In addition, online payment methods are still unavailable. The study goes on to indicate that ISP providers in Ghana are generally not reliable (as of yet), and that Ghanaian businesses are quite vulnerable to potential virus attacks.

Another ICT challenge, especially associated with the increasing technological divide between developed and developing economies, comes from the increasing prevalence of "legacy information

systems" (Ebbers & van Dijk, 2007). Indeed as the ICT divide widens, so does the potential that inherited language, platforms and techniques contained in folder generations of hardware and software will in several instances not work across platforms, and thus hinder use.

Scholars identify a third technological challenge stemming from the current lack of modern information and communication technologies. Adjei & Ayernor (2005) indicates that the retention of medical records in hospitals in Ghana is poor and not well managed. Adjei (2004) also describes substantial limitations of the medical record tracking system for Ridge Hospital in Accra, and proposes a design for an automated computerized system. Their study concludes that Ghanaian hospitals could well benefit from the increasing use of modern technologies.

Finally, a major ICT challenge stems from threats to the democratization process that are associated with the potential for government systems and firm data systems to be compromised (Hinson, 2005). Indeed with reports of increasing Internet fraud, identity theft scams and hackers breaking into the data systems of such technologically advanced sites as major multi-national corporations that spend millions of dollars on Internet security and firewalls, and the Pentagon in the United States, the threat of compromised security for developing countries such as Ghana is very real. In fact, the reason that potential damage is so substantial is in part because countries such as Ghana do not have the requisite resources or the well-trained personnel to prevent the occurrence of security breaches.

Governmental Cognitive Mindset

Several scholars note the major cognitive challenge associated with the bureaucratic and centralized mindset that government institutions often possess, and the mindset of citizenry empowerment that is associated with the decentralized institutional frameworks that they are supposed to institute (Mensah, 2005; Owusu, 2005; Haruna, 2003). Indeed, Haruna (2003) notes that a key focus of reform in the Ghana Public Service has been decentralization. While Owusu (2005) uses two districts in Ghana's Central Region as a case study to depict how Ghana is carrying out the decentralization program through the strengthening of district capitals, Haruna (2003) notes that institutional reforms by their very nature are slow and difficult to achieve.

There is a highly bureaucratic mindset at the national and regional/district level of government (Mensah, 2005; Abdulai, 2005). This prevalence can be attributed to the failure of numerous developmental plans, and thus there is no reason to presuppose that the implementation of an ICT plan will not encounter such a major barrier. Indeed, Abdulai (2005) highlights possible reasons why Ghana's numerous development plans have failed, and suggests that it is "because the traditional planning system, which inspired their development, was defective. The plans were highly centralized in nature, biased towards top-down decision making, conceived by technocrats who dwelt mainly on their perspectives in the planning process, eliciting little or no participation of people from local or grassroots levels, and were consumed by implementation problems" (p. 31). Furthermore, a key assumption is that decentralization is an important aspect of local government development. Yet as Binns, Porter, Nel & Kyei (2005) insists, there is a major problem associated with a reluctance to give genuine autonomy.

In essence, a major challenge arises because the implementation of ICT initiatives, especially at the national and regional governmental level, will run into this major mindset divide. By definition, ICT implies a highly decentralized, fast-paced, and continually evolving mindset. At every moment new advances of ICT are being developed and brought to market. These advances are in terms of basic ICT use and also in terms of its increasing applications.

Another ICT-related challenge is the extent to which it is deployed and utilized in the governmental ministries and public sector organizations. A survey conducted by the National ICT Policy and Plan Committee indicates that as of 2003 the majority of government ministries and public sector organizations spent less than 10% of their total budget on ICT-related items such as acquisition of hardware, software, ICT training and the maintenance of ICT systems (www.ict.gov.gh). Furthermore, over 90% of the government ministries and public sector organizations that were surveyed were not involved in e-commerce activities such as trading goods and services on the Internet, and utilizing the Internet to remain in contact with vendors and suppliers (www.ict.gov.gh).

Finally, there are a host of ICT-related challenges that are specifically related to the decentralization process in Ghana (Mensah, 2005). Such reforms experience major organizational and managerial issues associated with top management attitudes (Gil-Garcia & Pardo, 2005). Gil-Garcia & Pardo (2005) also identifies multiple and potentially conflicting goals, resistance to change and the improvement of existing structures, turf conflicts, legal and regulatory restrictions (including one year budgets), and institutional and environmental problems such as political, economic, and cultural constraints. Abdulai (2005) points to a lack of data, indiscipline, waste of resources, poverty, lack of continuity, and difficulty mobilizing financial and human resources as leading problems hindering the success of these development plans.

Cultural Norms

We have identified several challenges that scholars suggest are associated with the governmental bureaucratic mindset—a mindset that is at variance with the empowerment imperative underlying most governmental reforms. At the e-participative and firm/individual level of analyses, scholars also point to the existence of several Ghanaian cultural challenges to ICT development and implementation (Hinson, 2005; Parent et al, 2005; Haruna, 1999; Abdulai, 2005; Guseh & Oritsejafor, 2005).

Hinson (2005) identifies a cultural reluctance to change old ways of doing business. The Ghanaian tradition of bargaining is one such example; to the extent that it cannot be carried out effectively and consistently online, it provides a constraint for business. Furthermore, ICT also assumes a certain level of trust by the citizenry. However, Parent et al (2005) points out that ICT often does not build trust; rather trust appears to be based on pre-existing levels, which suggests a limit of ICT because it provides for little (if any) face-to-face contact. To this end, the lack of client trust that Hinson (2005) identifies as one of the several problems that firms doing business in Ghana face, is important.

Another ICT-related challenge that is of a cultural nature stems from the gender divide (Donkor, 2002; Elijah & Ogunlade, 2006; Kwapong, 2007). For example, women form only about 18% of the middle-to-top public sector executives in Ghana (Haruna, 1999). In particular, women are underrepresented in education involving science, mathematics, and technology (Haruna, 2003; Donkor, 2002). Therefore, there is an important need for educational policies to help bridge the gender divide by minimizing the gender differential, especially when considering the advent of the technological revolution (Elijah & Ogunlade, 2006; Kwapong, 2007). Some programs have already been established to achieve this end. In particular, the Girls Education Unit (GEU) was set up in 1997. Yet, more should be done to encourage and educate women about ICT.

Financial Resources

Finance is another area in which ICT-related challenges are encountered (Hinson, 2005; Binns et al, 2005; Vesley, 2003). Information communication technology is not inexpensive or readily available, especially in resource-constrained economies such as Ghana. Thus for the broad mass of Ghanaian citizenry, ICT (and even more particularly, the ready access of the majority of the citizenry to personal computers and the Internet) is still far off.

The financial difficulties that contribute to the broad mass of Ghanaians having little or no access to ICT characterizes the rural-urban divide that most developing countries in Africa exhibit (Bertolini et al, 2001; Kwapong, 2007). This divide is important especially because the socio-economic level of the citizenry is associated with the extent to which ICT can be effectively utilized and ICT policies implemented. For example, considering such socio-economic indicators as income, education, and age, Kwapong (2007) found that the level of household income and the education of rural women influenced citizens' choice of information delivery technology, and also their willingness to pay for that technology.

An important manifestation of the rural-urban divide with respect to ICT is encountered when examining the concentration of phone lines in Ghana. Between 50 - 70% of the telephone lines are located within the capital city of Accra, even though more than 70% of Ghanaians live in rural areas, although they have substantially lower levels of income (Kwapong, 2007). In 1999, the teledensity of the largest cities in Ghana were 5.43 compared to 0.25 for the remainder of the country. The disparity between the rural and urban areas is likely to grow even wider with the implementation of new technology. The Ghanaian government is cognizant of this problem; thus in 2005, the National Telecommunication Policy sought to provide "universal access" to ICT and to set up the Government Investment Fund for Telecommunications (GIFTEL). GIFTEL generates contributions from telecommunication licenses and uses these funds to establish projects that will help bridge the rural-urban technological divide. While this is clearly a step in the right direction, the rural-urban divide is undoubtedly an important ICT-related issue, especially as it pertains to ensuring democracy assumes, in part, broad general access of information to and by all Ghanaian citizens.

The challenge that finance poses to ICT in Ghana is also manifest in several other areas. Take for example university libraries in Ghana. Badu's (2004) surveys of twenty-one academic librarians in Ghana identify lack of information technology and lack of funding as major concerns. In another instance, Hinson (2005) surveyed several small- and medium-sized firms who were members of Ghanaian export firms and identifies several problems faced in the adoption of the Internet and other forms of ICT. A major problem that was recognized is the substantial expense faced in the development and maintenance of a website. The Hinson (2005) study also highlights the relatively high costs of broadband connection, training employees for the use of the Internet, and telephone bills generated by ICT as major problems faced.

Finally, the challenge of finance for ICT is also manifest in the area of governmental reform. Binns, Porter, Nel & Kyei (2005) finds that a lack of adequate funding and financial capital is an important problem that African governments face as they try to implement decentralization plans, especially at the regional and district levels.

POTENTIAL ICT OPPORTUNITIES FOR DEMOCRACY INSTITUTIONALIZING IN GHANA

Despite the above challenges, there are numerous ICT-related opportunities that can serve to strengthen democratic institutions in Ghana. We will focus on five major opportunities: the ability to increase the availability of ICT, more comprehensive utilization of existing technologies in Ghana, more innovative explorations of global linkages, an increase in the utilization of

non-governmental organizations (NGOs) and government fostered dialogue using ICT to enhance democracy.

Increasing the Availability of Information Communication Technology

Several scholars indicate the increasing availability of ICT in Ghana (Zachary, 2002; 2003; Zelnick, 2000; Obeng, 2003). To the extent that ICT becomes more readily available to the broad mass of Ghanaian citizens, the democratization process is greatly enhanced. For example, in the past there were only a few ways of transmitting information, all of which were government controlled. In the 1970s, there was only one national television broadcasting station and one radio broadcasting station. In contrast, now in 2008 there are over 15 independent and privately-owned broadcasting stations. These radio stations are free to broadcast what they choose, and are in several instances highly critical of the government and elected officials. They often hold duly-elected representatives highly accountable for the general state of the economy and its resources. This trend is very helpful in the institutionalization of democratic principles

In addition, Zachary (2002) adds that the increasing availability of ICT in Ghana means that the technological divide is slowly narrowing. As an example, in the 1990s there were no Internet cafes in Ghana (and thus only severely limited ways of accessing the Internet), and the average wait time for the installation of a telephone line in 1998 was one year at the approximate cost of $1000 per line (www.uneca.org). In stark contrast, in 2002 there were over 600 Internet cafes in the Accra area, and today Internet cafés are located in almost every town in Ghana. Furthermore, mobile phones are becoming increasingly accessible to all Ghanaians, and thus can be seen in a sense as a technology-equalizing agent (Zelnick, 2000). The fact that they are currently in frequent use and that their purchase prices are dropping points to their potential for helping to bridge the technological divide. Estimates suggest that mobile telephone customers in Ghana represent about 20% of the population. Bertolini et al (2001) suggests that vendors in the Southern Volta region in Ghana utilize mainly cell phones to discuss fluctuating market prices, delivery and payment schedules, and the prices of competitors' goods.

It also appears that government agencies in Ghana are slowly increasing their use of ICT. Obeng (2003) describes the use of ICT to modernize the court system resulting in the resolving of a complex case in two years that would have taken over 15 years in the past. This automation has included training staff in the use of computers, scanners, the public address system, and machines that automatically translate court proceedings into text. The Ghana–India Kofi Annan Centre of Excellence in ICT is another good example of this increase in the utilization of ICT. It was founded in 2003 through a partnership of the governments of Ghana and India. Its state-of-the-art complex houses West Africa's first supercomputer. It is also clear that despite potential problems and challenges that have been identified, business firms and private sector organizations in Ghana are increasingly using more modern ICT (Hinson, 2005; Vesley, 2003).

More Comprehensive Utilization of Existing ICT in Ghana

Several authors suggest a more aggressive approach in the utilization of existing ICT in Ghana and indicate several possibilities that would arise as a result. Vesley (2003), for example, describes the technology fair set up by Ghana's African Information Technology Exhibition & Conference (AITEC) in Accra. Such technology fairs have great potential for increasing the awareness of the citizenry and businesses about ICT and its advantages.

Hess (2003) describes the rich historical and cultural heritage that Ghana possesses. This is captured by places such as the Kumasi Fort Museum, and "the treasure storehouses" of the Ghana National Cultural Center and the Manhyia Palace Museum. The artifacts, documents, and other displays that these museums house can be better preserved through the utilization of more modern technologies. For instance, Adams (2005) indicates the need for more effective management of chieftaincy records in Ghana. In addition, Kankpeyeng & DeCorse (2004) provides an extensive review and description of the ongoing destruction of valuable Ghanaian antiquities and archeological records. Akussah (2006) highlights the major deterioration of documents in the National Archives, with some 51% of existing documents in need of restorative treatment. ICT may be an important aspect of a plan to preserve documents in the National Archives in Ghana. Modern technology would aid, in this respect, in cataloguing and preserving valuable historical relics with online databases containing images of these documents and artifacts for public viewing. Several authors also indicate the need for increased utilization of ICT in Ghanaian universities (Adanu, 2006; Badu, 2004). For example, there is an important need for the planning and implementation of a library automation project in the University of Ghana (Adanu, 2006).

Moving to another sector of importance, Entsua-Mensah (2005) points to the possibility of substantially revitalizing the indigenous agricultural marketing system through the use of e-commerce practices. This is important because about 40% of Ghana's gross domestic product is agrarian-based, and about 60% of the population are located in rural areas. Such a revitalization relies on ICT being conceptualized as a process that would lead to a reduction of transaction costs, reduction of market uncertainty, the expansion of markets, and the overall implementation of market-oriented institutional changes (Bertolini, et al, 2001).

Extensive studies conducted by Sraku-Lartey indicate the tremendous potential of ICT in effective management of existing forests in Africa, and particularly in Ghana (2003; 2006a; 2006b). While Sraku-Lartey (2003) identifies the need for the development of a computerized management information system for the forestry sector in Ghana, Sraku-Lartey (2006a) also discusses efforts to improve skills of forestry information managers in Africa, and the potential role and advantages that such skilled personnel bring to the forestry management process. Sraku-Lartey (2006b) describes the Global Forest Information System (GFIS) spanning Ghana, Senegal, Gabon, Zimbabwe and Kenya, in which, despite several difficulties and challenges, an integrated forest data management system was developed, and personnel from the different countries were trained in its utilization.

Evans & Yen (2006) provides case studies from the United States of information technology applicability in a wide variety of areas such as distance learning, electronic monitoring of prisoners, fraud detection, electoral voting, jury selection, the social security administration, law enforcement, the emergency response system, military and defense, and space exploration. It is clear that Ghana would benefit from as broad and systematic an implementation of ICT as possible. Indeed, as Tagoe, Nyarko & Anuwa-Amarh (2005) indicates, to the extent that ICT is associated with good information management systems, record-keeping by firms in countries such as Ghana will be strongly associated with the reduced perception of business risk by foreign investors, and thus subsequently associated with the extent to which small-and medium-sized enterprises (SME) are able to acquire financing.

Some scholars (e.g. Mayur & Daviss, 1998; Jacobs & Herselman, 2005) describe new ICT that make it possible for more rapid and sustained development of economies without the draining and negative aspects of the industrialization process. As an example, Mayur & Daviss de-scribes "information kiosks" that can be set up in villages and that would contain a cell phone, radio, television, and computer, bringing important information to secluded villagers. This is broadly similar to the ICT–hub model that Jacobs & Herselman proposes. Another example is the ongoing project in Bangladesh by Nobel laureate Mohammad Yunus, founder of the now famous micro-lending Grameen Bank. This project involves placing a cell phone in each of the 65,000 villages in Bangladesh. An additional illustration of this point is the recent decision by Nigeria to provide a significant number of "$100 laptops" to a large segment of their school populations, and to teach them how to effectively use these for educational and civic communal purposes. This program, if effectively implemented, could serve as a major catalyst for advancing ICT use. This, in a way, would be similar to the increasing ubiquity of the cellular phone, a phenomenon briefly documented above. In addition, creative ways can and should be explored to use ICT to reduce youth unemployment in countries such as Ghana (Braimah & King, 2006). Finally, the increase and potential use of technologies fueled by renewable fuel sources such as solar energy raises all sorts of possibilities for increasing the availability of ICT in African countries.

There is however the disturbing issue that the existing ICT infrastructure already in Ghana is not being fully utilized. Roberts-Witt's (2000) article describes the discovery in the 1990s of fiber optic wiring that had been laid in Ghana but subsequently never used. Two businessmen from America, Greene and Johnson, uncovered these "dangling" wires and recognized their potential for increasing the communication and information process in Ghana. What is disturbing and sobering is the bureaucratic red tape that they encountered in attempting to get the fiber optics into operation. The Roberts-Witt (2000) article indicates that at the time of its publication the fiber optics project had still not been implemented.

More Extensive Exploration of Global ICT Linkages

Global linkages generate additional opportunities for increasing the utilization of ICT in Ghana. Roach (2002) describes the joint initiative between the Massachusetts Institute of Technology's Center for Advanced Educational Services, and the African Virtual University, where students in advanced computer programming courses in countries like Ghana, Kenya and Tanzania are taught Java programming based on MIT's course guidelines. Headquartered in Nairobi, Kenya, The African Virtual University was set up in July of 1997, and provides high-quality online education from top universities worldwide with over 30 learning centers across the continent. Roach (2001) mentions another project based on the use of computerization to explore Africa and African art. Adanu (2006) describes the recent planning and implementation of a library automation project at the University of Ghana with external funding provided by the Carnegie Corporation of New York.

Global ICT linkages are beginning to provide jobs and employment opportunities in Ghana (Brah, 2001; Zachary, 2002; Zachary, 2003). Brah (2001) describes how Aqsolutions, founded by Ghanaians living in the United States, is entering into the data entry outsourcing process using employees who are based in Ghana. This is representative of an emerging trend where developing countries tap into the expertise of their nationals living abroad (Juma, 2005). Zachary (2002) indicates the use of satellite links to bypass telephone constraints and enable data to be sent overseas instantaneously. In 2000, Dallas-based Affiliated Computer Services processed insurance forms for companies such as Liberty Mutual and HealthNet through data entry facilities located in Ghana with American supervisors paying occasional visits. In 2002, Data Management Internationale also set up a similar data entry system.

Several other studies describe in detail the volunteer program known as Geekcorp (e.g. Briggs, 2001; Calleja, 2002; DiNicolo, 2002). Geekcorp was founded in 2000 by Ethan Zuckerman, a "dot.com millionaire." It is a program that allows technicians in developed countries like the U.S. and Canada to spend two to three months introducing the Internet and other forms of ICT to small businesses in developing countries such as Ghana.

Non-Governmental Organizations as Technology-Bridging Institutions

An article by Closson, Mavima & Siabi-Mensah (2002) highlights a shift of paradigm from state-centeredness to a more decentralized emphasis on agencies such as non-governmental organizations (NGOs). Several scholars point to the potential use of NGOs in Europe to increase the utilization of ICT, and thus serve as a tool for democracy enhancement (Taylor & Burt, 2005; Bingham, 2005). For example, Taylor & Burt (2005) highlights the important role that voluntary sector organizations, through their independent gathering and dissemination of relevant information, can play in ensuring the effective delivery of e-governance services, and in ensuring higher levels of accountability and transparency. The article goes on to suggest that the capability of voluntary organizations is important for e-governance because they typically provide easy access to in-house information. By providing hot links to websites of other organizations that are the sources of relevant and related information, knowledge and expertise, they assist the citizenry in engaging in the democratic process, usually making sure that this can be done with ease and immediacy, and typically engaging traditionally marginalized sectors and communities of a nation in the democratization process. Bingham et al (2005) also suggests that voluntary organizations, through quasi-legislative and quasi-judicial governance processes such as deliberative democracy, e-democracy, public conversations, participatory budgeting, study circles, and collaborative decision-making, assist the common citizenry in actively participating in the democratic governance process.

In the Ghanaian context, NGOs can play similar roles as potentially powerful media sources for the promotion of public awareness and education regarding e-democracy issues and implications (Closson, et al, 2002).

RECOMMENDATIONS

We end this essay by identifying two generalized but significant recommendations that scholars writing specifically about ICT in Africa and Ghana, and those writing in regards to ICT in the European context, broach. These recommendations will help minimize the challenges and enhance the opportunities that we have identified in our above discussion. Specifically, we suggest that the successful implementation of ICT would be enhanced by developing planned and integrated approaches beforehand, and by governments making an effort to foster dialogue regarding ICT and the democratization process.

Government-Fostered Dialogue Utilizing ICT to Enhance Democracy

There are several ways in which government-sponsored dialogue could utilize ICT to enhance democracy. For example, e-government websites should be set up to present differing and alternative views of political issues. Jaeger's (2005) study is particularly illustrative as it uses the U.S. federal government's promotion of the "No Child Left Behind" initiative as a case study. In this particular example, the United States Department of Education's website in August of 2004 only highlighted the purported benefits of the initiative, and there was no acknowledgement of alternative views and thus no opportunity for

further dialogue. Jaeger proposes an alternative where a government website could highlight the various differing perspectives to issues being proposed, and thus encourage increased reflection and dialogue by citizenry.

These increased levels of dialogue could, for instance, provide access to expression mediums [websites, phones, short message services (SMS), online videos, online chats, and etcetera] that are specifically geared to the promotion of civil liberties and democracy. The implementation of these new technologies also allows for the promotion of citizenship. It provides a way to better acquire, collect, and circulate information about the nation and use it to the benefit of citizens.

The encouragement and development of blogging is another potential way public participation in governance can be increased. In addition, the continued introduction of computerization in the educational system, and a drive to provide the Ghanaian citizenry with access to low-cost computers and Internet connection, are all ways in which government-fostered dialogue can be enhanced.

In essence, for government-initiated ICT programs to be successful in enhancing democracy, it is important that it be carefully developed and dialogue fostered (Jaeger, 2005; Werlin, 1998; Werlin, 2003). Such dialogue must take into account the need to adopt a more "in-depth analysis of the political nature of the e-government development process and a deeper recognition of the complex political and institutional environment" (Yildiz, 2007, p. 647).

CONCLUSION

Guseh & Oritsejafor (2005), using Ghana and South Africa as case studies, suggests a strong link between democracy and economic growth. There is a broad underlying assumption that increased democracy in the different countries in Africa will be associated with positive market reforms (Bratton, Mattes & Gyimah-Boadi, 2005). Abrahamsen (2000) also suggests that the vicious cycle where less democratic governance inevitably leads to less individualistic, entrepreneurial activities, and thus less development, can be effectively combated with its converse, where increasing democratic governance will lead to greater entrepreneurship and economic development. Given the above, ICT-related initiatives and programs that encourage and promote increased levels of democracy in developing countries such as Ghana are important, yet oftentimes underestimated, tools that assist with national economic development.

Undoubtedly, there have been several challenges and problems associated with the development and effective implementation of an ICT policy in Ghana (Alhassan, 2005; Alhassan, 2007). In this discussion we have identified several potential challenges that the ICT revolution faces in Ghana with respect to the institutionalization of the democratic process. We have also described several potential opportunities that can be realized in response to the further development of ICT, including the further institutionalization of the democratization process. Indeed, syntheses of these studies suggest that the increasing utilization of ICT has the potential to either substantially improve or hinder the extent to which the democratic principles in a nation are institutionalized.

Zachary (2002) suggests that the main lesson garnered from his case study of Ghana is that "information technology is not the great leveler that enthusiasts champion, but it also is not as far out of reach as skeptics say" (p.72). Thus, to the extent that some of the ideas, suggestions and recommendations gleaned from ICT scholars can be taken into account in developing and implementing ICT programs in Ghana, there is the hope that Ghana could soon be considered the Silicon Valley of Africa and be on a comparable level of ICT competency and utilization as more-developed nations.

ACKNOWLEDGMENT

We wish to thank Daniel Ofori-Dankwa for several insightful ideas and also for his input and comments on the structuring of this chapter.

REFERENCES

Abdulai, A. I. (2005). Of visions, development plans and resource mobilization in S Africa: The case of Ghana vision 2020. *Africa Insight*, *35*(1), 28-35.

Abrahamsen, R. (2000). *Disciplining democracy: Development discourses and good governance in Africa*. London: Zed Books.

Adams, M. (2005). The management of chieftaincy records in Ghana: An overview. *African Journal of Library Archives and Information Science*, *15*, 1, 67-73

Adanu, T. S. A. (2006). Planning and implementation of the University of Ghana library automation project. *African Journal of Library Archives and Information Science*, *16*(2), 101-108.

Adjei, E. (2004). Retention of medical records in Ghanaian teaching hospitals: Some international perspectives. *African Journal of Library Archives & Information Science*, *14*(1), 37-52.

Adjei, E., & Ayernor E. T. (2005). Automated medical record tracking system for the Ridge hospital, Ghana Part 1: Systems development and design. *African Journal of Library Archives and Information Science*, *15*(1), 1-14.

Akussah, H. (2006). The state of document deterioration in the national archives of Ghana. *African Journal of Library Archives and Information Science*, *16*(1), 1-8.

Alhassan, A. (2004). *Development communication policy and economic fundamentalism in Ghana*. Finland: Tampare University Press.

Alhassan, A. (2005). Market valorization in broadcasting policy in Ghana: Abandoning the quest for media democratization. *Media, Culture & Society, 27*(2), 211-228.

Alhassan, A. (2007). Broken promises in Ghana's telecom sector. *Media Development,* 3, 45.

Amoretti, F. (2006). *The digital revolution and Europe's constitutional process. E-democracy between ideological ad institutional practices.* Paper presented at the VII Congresso Espanol De Ciencia Politica Y De La Administration, Grupo De Trabajo 9 Communicacion Politica.

Amoretti, F. (2007). International organizations ICTs policies: E-democracy and e-government for political development. *Review of Policy Research, 24*(4), 331-344.

Badu, E. E. (2004). Academic library development in Ghana: Top managers' perspectives. *African Journal of Archives and Information Science, 14*(2), 93-107.

Bertolini, R., Dawson Sakyi O., Anyimadu A., & Asem P. (2001). *Telecommunication use in Ghana: Research from the Southern Volta Region.* University of Ghana- Center for Development Research, Bonn University, Working Paper.

Bingham, L. B., Nabatchi, T., & O'Leary, R. (2005). The new governance: Practices and processes for stakeholder and citizen participation in the work of the government. *Public Administration Review,* 65(5), 547-558.

Binns, T., Kyei P., Nel, E., & Porter G. (2005). *Africa Insight, 35*(4), 21-31.

Brah, K. (2001). Ghana goes for IT lead. *African Business,* July/August, p.25.

Braimah, I., & King, R.S. (2006). Reducing the vulnerability of the youth in terms of employment in Ghana through the ICT sector. *International Journal of Education and Development using ICT, 2*(3), 23-32.

Bratton, M., Gyimah-Boadi, E., & Mattes, R. (2005). *Public opinion, democracy and market reforming Africa.* Cambridge, U.K.: Cambridge University Press.

Briggs, J. (2001). Geeks in do-good rampage. *Fortune, 144*(1), 182.

Calleja, D. (2002). Heart of geekness. *Canadian Business,* April 29, (pp. 65-71).

Closson, R. B., Mavima, P., & Siabi-Mensah, K. (2002). The shifting development paradigm from state centeredness to decentralization: What are the implications for adult education? *Convergence, 35*(1), 28-42.

DiNicolo, D. (2002). The African beam. *Canadian Business, April 29,* 73.

Donkor, M. (2002). Educating girls and women for the nation: Gender and educational reform in Ghana. *International Education, 32*(1), 72-85.

Ebbers, W. E., & van Dijk, J. A. M. G. (2007). Resistance and support to electronic government, building a model of innovation. *Government Information Quarterly, 24*(3), 554-575.

Elijah, O., & Ogunlade, I. (2006). Analysis of the uses of information technology for gender empowerment and sustainable poverty alleviation. *International Journal of Education and Development using ICT, 2*(3), 45-69.

Entsua-Mensah, C. (2005). Revitalizing the indigenous agricultural marketing system in Ghana through the e-commerce project: A performance appraisal. *IAALD Quarterly Bulletin, 50*(3), 141-147

Evans, D., & Yen, D. C. (2005). E-government: An analysis for implementation: Framework for understanding cultural and social impact. *Government Information Quarterly, 22,* 354-373.

Evans, D., & Yen, D. C. (2006). E-government: Evolving relationship of citizens and government, domestic and international development. *Govern-*

ment Information Quarterly, 23, 207-235.

Fountain, J. E. (2001). *Building the virtual state: Information technology and institutional change.* Washington DC: Brookings Institution Press.

Gil-Garcia, J. R., & Luna-Reyes, L. F. (2003). Towards a definition of electronic government: A comparative review. In A. Mendez-Vilas, et al. (Eds.), *Techno-legal aspects of information society and new economy: An overview.* Extremadura, Spain: Formatex Information Society Series.

Gil-Garcia, J. R., & Pardo, T. (2005). E-government success factors: Mapping practical tools to theoretical foundations. *Government Information Quarterly,* (pp. 187-216).

Guseh, J. S., & Oritsejafor, E. (2005). Democracy and economic growth in Africa: The cases of Ghana and South Africa. *Journal of Third World Studies, 22*(2), 121-137

Hinson, R. E. (2005). Internet adoption among Ghana's SME nontraditional exporters. *African Insight, 35*(1), 20-27.

Haruna, P. F. (1999). *An empirical analysis of motivation and leadership among career public administrators: The case of Ghana.* Unpublished doctoral dissertation, University of Akron, Ohio.

Haruna, P. F. (2003). Reforming Ghana's public service: Issues and experiences in comparative perspective. *Public Administration Review, 63*(3), 343-354.

Hess, J. B. (2003). Imaging Architecture 11, "treasure storehouses" and constructions of Asante regional hegemony. *Africa Today, 50*(1), 27-48.

Ifinedo, P. (2005). Measuring Africa's e-readiness in the global networked economy: A nine-country analysis. *International Journal of Education and Development using ICT, 1*(1), 53-71.

Jacobs, S. J., & Herselman, M. E. (2005). An ICT-hub model for rural communities. *International*

Journal of Education and Development using ICT, 1(3), 57-93.

Jaeger, P. T. (2005). Deliberative democracy and the conceptual foundations of electronic government.*Government Information Quarterly, 22,* 702-719.

Juma, C. (2005). The way to wealth. *New Scientist, 185*(21), 15-21.

Kankpeyeng, B. W., & DeCorse, C. R. (2004). Ghana's vanishing past: development, antiquities, and the destruction of the archaeological record. *African Archaeological Review, 21*(2), 89-128.

Kwapong, O. A. T. (2007). Problems of policy formulation and implementation: The case of ICT use in rural women's empowerment in Ghana. *International Journal of Education and Development using ICT, 3*(2), 1-21.

Lall, S., Navaratti, G. B., Teitel, S., & Wiggnaraja, G. (1994). *Ghana under structural adjustment.* New York: St Martin's Press

Layne, K., & Lee, J. (2001). Developing fully functional e-government: A four stage model. *Government Information Quarterly, 18,* 122-136.

Mangesi, K. (2007). *ICT in education in Ghana, Survey of ICT and Education in Africa: Ghana Country Report.* www.Infodev.org/ict4edu-Africa

Mayur, R., & Daviss, B. (1998). The technology of hope: Tools to empower the world's poorest peoples. *Futurist, 32*(7), 46-51.

Mensah, J. V. (2005). Problems of district medium-term development plan implementation in Ghana. *International Development Planning Review, 27*(2), 245-270.

Narayan, G., & Nerurkar, A. N. (2006). Value-proposition of e-governance services: Bridging rural-urban digital divide in developing countries. *International Journal of Education and Development using ICT, 2*(3), 33-44.

Nugent, J. D. (2001). If e-democracy is the answer, what's the question? *National Civic Review, 90*(3), 221- 233.

Obeng, K. W. (2003). Ghana pursues justice and development through computer training. *Choices, 12*(4), 20-21.

Owusu, G. (2005). The role of district capitals in regional development. *International Development Planning Review, 27*(1), 59- 89.

Germino, A. C., Parent, M., & Vandebeek, C. (2005). Building citizen trust through e-government. *Government Information Quarterly, 22*, 720-736.

Roach, R. (2001). Distance education course to explore African and African-American art. *Black Issues in Higher Education*, April 12, (p. 46).

Roach, R. (2002). African students take MIT course without leaving home campuses, *Black Issues in Higher Education*, May 9, *47*.

Roberts-Witt, S. (2000). Unused fiber serves as springboard for broadband networking in this West African nation. *Internet World,* September 1, (p. 38–39).

Schelin, S. H. (2003). *E-Government: An Overview. In G.D. Garson (ed) Public Information Technology: Policy and Management Issues,* (p. 120-137) Hershey, PA: Idea Group Publishing

Singer, P. A., Salamnca-Buentello, F., & Daar, A. (2005). Harnessing nanotechnology to improve global equity. *Issues in Science and Technology, 4*, 57-64.

Sraku-Lartey, M. (2003) The role of information in decision making in the forestry sector: Developing a computerized management information system (MIS) for forestry research activities in Ghana. *IAALD Quarterly Bulletin, 48*(1), 105-108.

Sraku-Lartey, M. (2006). Building capacity for sharing forestry information in Africa. *IAALD Quarterly Bulletin, 51*(3), 186-190.

Sraku-Lartey (2006). Developing the professional skills of information managers in the forestry sector in Africa. *IAALD Quarterly Bulletin, 51*(2), 75-78.

Srivastava, S., & Teo, T. S. H. (2007). What facilitates e-government development? A cross-country analysis, Electronic Government. *An International Journal, 4*(4), 365-378.

Tagoe, N., Nyarko E., & Anuwa-Amarh, E. (2005). Financial challenges facing urban SMEs under financial sector liberalization in Ghana. *Journal of Small Business Management, 43*(3), 331-343.

Taylor, J., & Burt, E. (2005). Voluntary organizations as e-democratic actors: political identity, legitimacy and accountability and the need for new research. *Policy and Politics, 33*(4), 601-616.

Vesely, M. (2003). New technology for an old continent. *African Business*, July, 20-21.

Website: http://www.ict.gov.gh

Website: http://uneca.org

Werlin, H. H. (1998.) *The mysteries of development: Studies using political elasticity theory*, Lanham, MD: University Press of America.

Werlin, H. H. (2003). Poor nations, rich nations: A theory of governance. *Public Administration Review, 63*(3), 329-342.

Wright, B. (2004). Telecoms around the continent. *African Business*, May, 16-17.

Yildiz, M (2007) E-government research: Reviewing the literature, limitations, and ways forward. *Government Information Quarterly, 24*(3), 646-665.

Warschauer, M. (2003). Demystifying the digital divide. *Scientific American, 289*(2), 42-47.

Zachary, P. G. (2002) Ghana's digital dilemma, *Technological Review, July/August,* (pp. 66-73).

Zachary, P. G. (2003). A program for Africa's computer people. *Issues in Science & Technology*, Spring, *79.*

Zelnick, N. (2000). Colonialism? Not again. *Internet World, 6*(18), 15.

Chapter IX
The Politics of the Governing the Information and Communications Technologies in East Asian Authoritarian States:
Case Study of China

Chin-fu Hung
National Cheng Kung University, Taiwan

ABSTRACT

China has vigorously implemented ICTs to foster ongoing informatization accompanying industrialization as a crucial pillar to drive its future economic development. The institutional and legal reforms involved were initiated and put into practice in order to meet the increasing demand for technological convergence and the negotiations for the expected entry into the World Trade Organization (WTO). The Chinese government has nevertheless long been torn by the ambivalence brought about by the Internet. It regards the Internet as an engine to drive economic growth on the one hand, and as a subversive challenge to undermine the ruling Communist Party on the other hand. As soon as ICTs were introduced and Web sites mushroomed, the Party was so determined to harness the new medium to assure the Internet's economic and scientific benefits. As a consequence, controls other than stifling ICTs would be critical for the CCP's agenda to achieve the century-long modernization process and in the meantime, consolidate its power.

Arguably information and communication technologies (ICTs) are nowadays integral part of the whole realm of human activities. Manuel Castells, for example, notes that we have seen that ICTs are (re)shaping the material basis of society at an accelerated pace. He argues that the new information-centered technological revolution is now fundamentally altering every aspect of our lives (Castells 2000). In other words, it seems we live in a world, that in the expression of Nicholas Negroponte, has become "digital." (Negroponte 1995) In another work, Castells goes on to describe work the specific role of the Internet as "the fabric of our lives," adding that,

...if information technology is the present-day equivalent of electricity in the industrial era, in our age the Internet could be likened to both the electrical grid and the electric engine because of its ability to distribute the power of information throughout the entire realm of human activity (Castells 2001: 1).

It is usually believed that "authoritarian systems are inherently fragile because of weak legitimacy, overreliance on coercion, overcentralization of decision making, and the predominance of personal power over institutional norms." (Nathan 2003: 6) Yet, Shanthi Kalathil and Taylor Boas reject the idea of "blind optimism" about the Internet, suggesting that the Internet may not necessarily transform authoritarian regime. The new media may instead be exploited by the authorities there as an effective tool to further strengthen their governing capability in the information age (Kalathil and Boas 2003). Similarly Nina Hachigian holds one-party states who embrace the Internet are not more likely to fail than those that attempt to constrain the medium (Hachigian 2001, 2002). This chapter does not, however, suggest that (new) technology absolutely determines social activity, nor does the realm of society and politics condition the entire course of technological change. Instead, it is likely to be a two-way interaction between the technology and sociopolitical development.

In the Chinese context, as China is gearing up to transform its economy from central planning into one of the world's key IT-driven economies, it provides a crucial test case for other like-minded regimes—Vietnam and North Korea, particularly—as to the ways in which governments may handle the threat or grasp the economic opportunities from cyberspace. As Hu Angang, a renowned Chinese scholar who is close to the central leaders in Beijing, enthusiastically holds, China, under economic globalization, ought to adopt the knowledge-driven strategy as its most significant national development approach in the twenty-first century. He explains that the application of ICTs can not only "bridge the divide between China and developed countries in terms of knowledge development, but also shrink the digital gap between hinterland and coastal China." (Hu 2002: 15)

The Chinese government has no doubt acted as a vital driving force for boosting Internet and e-commerce diffusion. In retrospect, from March 1993 the Chinese central government embarked upon a series of so-called "Golden Projects"[1]—including Golden Bridge, Golden Card, and Golden Customs—to give it information on and control over the rapid decentralization of decision-making that was taking place as a result of the move towards a market economy. The year 1993 can also be remembered as the formal start-up stage of China's informatization. (Lu 2002: 53) On the one hand, this was aimed at laying the infrastructure for the digitization of China's telecommunications network, on the other hand, the central government started indeed to utilize the infrastructure of the Internet to improve its own administrative control over provincial and local offices, enhancing its governing capacity and authority, as well as sociopolitical stability. In November 1998 the authorities further announced the "Government Online Project" whereby the end of 1999 and 2000, at least 60 and 80 percent respectively of

China's government offices and ministries were going online: all ministries and provincial authorities would establish their own Web sites for citizens to consult.[2] China even christened 1999 "The Government Online Year" (People's Daily, 3 January 1999) and 2000 the year of "Enterprise Online Project."[3]

To date, Internet access has been expanding rapidly and extensively chiefly due to direct support and promotion by the government. As we may examine in the recent semi-annual survey report on the development of China's Internet, released by the quasi-official China Internet Network

Information Center, also known as CNNIC, the estimated total number of Internet users by the end of December 2008 was more than 298 million, the world's largest Internet market that has suppassed the United States since June 2008.[4] The overall Internet penetration rate in China has also reached up to 22.6 percent.[5] Anyhow, such an amazing achievement within a relatively short period of time coincides with Dali Yang's observation that, although China is a latecomer to the Internet world, the government can act swiftly to play a key part in unleashing the Internet's economic potential (Yang 2001: 65). Table 1 illustrates this.

Table 1. Internet growth in China

	Computer Hosts	Internet Users	Domain Names (.cn)	Web Sites	International Bandwidth (Mbps)
Nov 1997	299,000	620,000	4,066	1,500	18.64
July 1998	542,000	11,750,000	9,415	3,700	84.64
Jan. 1999	747,000	2,100,000	18,396	5,300	143
July 1999	1,460,000	4,000,000	29,045	9,906	241
Jan. 2000	3,500,000	8,900,000	48,695	15,153	351
July 2000	6,500,000	16,900,000	99,734	27,289	1,234
Jan. 2001	8,920,000	22,500,000	122,099	265,405	2,799
July 2001	10,020,000	26,500,000	128,362	242,739	3,257
Jan. 2002	12,540,000	33,700,000	127,319	277,100	7,597.5
July 2002	16,130,000	45,800,000	126,146	293,213	10,576.5
Jan. 2003	20,830,000	59,100,000	179,544	371,600	9,380
July 2003	25,720,000	68,000,000	250,651	473,900	18,599
Jan. 2004	30,890,000	79,500,000	340,040	595,550	27,216
July 2004	36,300,000	87,000,000	382,216	626,600	53,941
Jan. 2005	41,600,000	94,000,000	432,077	668,900	74,429
July 2005	45,600,000	103,000,000	622,534	677,500	82,617
Jan. 2006	49,500,000	111,000,000	1,096,924	694,200	136,106
July 2006	54,500,000	123,000,000	1,190,617	788,400	214,175
Jan. 2007	59,400,000	137,000,000	1,803,393	843,000	256,696
July 2007	67,100,000	162,000,000	6,149,851	813,357	312,346
Jan. 2008	78,000,000	210,000,000	9,001,993	1,503,800	368,927
July 2008	84,700,000	253,000,000	11,931,000	1,919,000	493,729
Jan. 2009	N/A	298,000,000	16,826,198	2,878,000	640,287

Source: Zhongguo Hulianwangluo Fazhan Zhuangkuang Tongji Diaocha (Statistical Reports on the Internet Development in China), several years, China Internet Network Information Center, <http://www.cnnic.cn/index/0E/00/11/indexhtm>.

POLITICAL HISTORY OF THE INTERNET GOVERNING STRUCTURE

China has vigorously implemented ICTs to foster ongoing informatization accompanying industrialization as a crucial pillar to drive its future economic development. The unfettered perspective on the free use of the Internet has been largely challenged mainly because the Chinese authorities at all levels have aggressively regulated public use of the Internet, in particular through control over the political and dissent use of the Net. Many measures as well as legalizations have been enacted with the purpose to make the Internet behave like any other form of mass media under its firm control (Hung 2006: 142-5).

It is useful to understand how the Internet is being governed in China if we are to better comprehend and further assess the Internet's political impact upon Chinese society. To facilitate discussions, the regulatory regime of the Internet governance there is arguably divided into three stages since its inception from the late 1980s: the experimental and fragmented period, which was before 1994; the transitional regulatory period, which ranged between 1994 and 1998; and the current period since 1998 (Tan 1999: 265-70).

The establishment of the Economic Information Joint Committee in 1993 marked a milestone in the development and regulation of the Internet in China. It is primarily because it shifted focus from initially "formulating policies for the development of a national information infrastructure," (Triolo and Lovelock 1996: 28) to a more particular attention to the Internet medium. With the increasing development of the Internet over the following years, the Committee was later in 1996 developed into the State Council Steering Committees on National Information Structure (SCSCNII). The set-up of SCSCNII reflected several competing and rival bureaucracies, such as the Ministry of Posts and Telecommunications (MPT), the Ministry of Electronics Industry (MEI), the Ministry of Broadcasting, Film and Television (MBFT), the Ministry of Public Security, and the Xinhua News Agency, which was actively involved in formulating, and implementing Internet policy. Among them, the MPT enjoyed enormous commercial and political advantages over its rivals as a result of its historical status as Internet operator and regulator. The main competitor of the MPT came from the MEI, especially from late 1993 when the MEI created a separate but affiliated corporation called "*Ji Tong*" (the Auspicious Telecommunications Company). One of the main tasks *Ji Tong* commissioned was the so-called "Golden Projects," in which they promoted the wider linkage of financial institutions and government agencies with digitalized communications and information networks. Given there had not been any paralysis in the Internet development, some real regulatory problems arose mainly between MPT and MEI.[6] A Steering Committee was accordingly required to coordinate and oversee the Internet development.

The bureaucratic body of SCSCNII under the State Council had the following major responsibilities: (Cullen and Choy 1999: 116)

1. To formulate guiding principles, policies, rules and regulations in the developing process of national informatization;
2. To formulate the strategy for developing national informatization and its overall and stage-by-stage plans;
3. To organize and coordinate the construction of important information projects;
4. To be responsible for the coordination of and solutions for important issues arising from the computer networks and the Internet, and
5. To establish the standards for the technology and application related to informatization.

The new Ministry of Information Industry (MII, *Xinxi Chanye Bu*) was approved in March of 1998 by a decision of the Ninth National People's

Congress (NPC). This was against the backdrop of the Asian financial crisis (1997), political succession issues (1997-98), and the restructuring of government bureaucracy (1998). The MII was basically set up by merging the MPT with the MEI, while the MBFT was converted into a "general bureau" (*zong ju*) under the State Council. The major task for MII was, as officially announced, to administer the national manufacturing of IT products, national communication and software industries, facilitating the informatization of the national economy and social services.[7]

The set-up of such a super-ministry was in a sense to "...reduce jealousies between the MPT and MEI so that genuine competition within the telecommunications industry could finally be introduced." (Lynch 1999: 173) In this aspect, the institutional and legal reforms involved were initiated and put into practice in order to meet the increasing demand for technological convergence and the negotiations for the expected entry into the World Trade Organization (WTO). Above all, it implies that the authorities in Beijing intended to restore administrative control over the telecommunications sector from previous stages of devolution, which had resulted in fragmented governance and intensified pluralization in terms of efficient flow of information among several telecommunications service providers. As Philip Sohmen notes, "Although the SCSCNII is in theory a superior body to the MII, in practice it seems as if the MII is responsible for policies up to the highest level." (Sohmen 2001:18)

Recently, the *Decision of the First Session of the Eleventh National People's Congress (NPC) on the Plan for Restructuring the State Council* was adopted on March 15, 2008, involving the installation of "super ministries" to "streamline government department functions, and to form some 'bigger departments' to strengthen macro-economic regulation, [to] maintain national security of energy supply, [and to] integrate information development and industrialization." (Xinhua, 11 March 2008; Straits Times, 12 March 2008)

Among this round of government reshuffle, one of the five "super ministries" is the Ministry of Industry and Informatization (*Gongye he Xinxihua Bu*), which is established out of the merger of units including the Ministry of Information Industry, the National Development and Research Commission, the State Council Informatization Office, and the Commission of Science, Technology and Industry for National Defense. To some extent, several "super-ministries" are formed to curtail sinecures and excessive expenditures, and more importantly, to improve and/or enhance administrative efficacy. This is chiefly because different departments during Wen Jiabao's first five-year term as premier had at times issued instructions contradictory to the ones of the central governments in Beijing. Since central orders and authorities seem to have often been diluted and distorted by the localities over time, the initiative at the NPC is, in other words, intended to organically integrate the functions of smaller departments to resolve the problem of overlapping responsibilities and of powers not being matched by responsibilities on the one hand, and to strengthen Beijing's administrative macro-control on the other hand (Lam 2008: 2-4).

Apart from that, the intrinsic characteristics of the bureaucratic rivalry among the competing parties in the Internet industry since the 1990s have been to take respective organizational interests in consideration. Regulatory control of the Internet is complicated: the MII has been assigned to superintend the general development of the Internet but specific authority has been divided between the Internet Services Providers (ISPs) and Internet Content Providers (ICPs). While the MII oversees ISPs and their related infrastructure, the Internet Information Management Bureau under the State Council Information Office is in charge of supervising the ICPs. In addition, the State Administration of Radio, Film and Television (SARFT) and the Ministry of Culture are still enthusiastic in sharing the development and management of the Internet, as is the Netnews Bureau of China Internet Information Canter (Shoesmith and Hearn 2004: 101-14). Barry Naughton vividly suggests that the economic decisions in China are made by a "broad and diverse group of economic agents." (Naughton 1997: 1)

Simply put, the processes of the de-concentration prior to 1998 and concentration course from 1998 onwards simply exhibit the competitive tension between the forces, with each "respectively supporting either centralized control or the break-up of the monopoly." In other words, both ideas are based on self-interest of the respective group. (Chung 2002: 55) The rationale behind the de-concentration process is basically in line with the greater political economy setting: de-concentration of decision-making authority over economic policies. Accompanied by the ideological decompression of the post-Mao period, and the rapid diffusion of new communications technologies, these have all contributed to produce significant administrative fragmentation.

INTERNET CENSORSHIP AND CONTROL

What are the online contents deemed undesirable by the Chinese authorities? This is exemplified in the "Measures for Managing the Internet Information Services (*Hulianwang Xinxi Fuwu Guanli Banfa*), which holds service providers responsible for contents they display on the Net. In practice, the ISPs, ICPs and Internet café owners have set up their own monitors, known as "Big Mama", to censor the chatrooms and bulletin boards and to delete materials that are not in line with the laws which are broadly decreed. Nine categories of information are banned in creating, replicating, retrieving, and transmitting:[8]

1. Materials that oppose the basic principles established by the Constitution;

2. Materials that jeopardize national security, reveal state secrets, subvert state power, or undermine national unity;

3. Materials that harm the prosperity and interests of the state;

4. Materials that arouse ethic animosities, ethic discrimination, or undermine ethic solidarity;

5. Materials that undermine state religious policies, or promote cults and feudal superstitions;

6. Materials that spread rumors, disturb social order, or undermine social stability;

7. Materials that spread obscenities, pornography, gambling, violence, murder, terror, or instigate crime;

8. Materials that insult or slander others or violate the legal rights and interests of others;

9. Materials that have other contents prohibited by laws or administrative regulations.

Two more categories of prohibited content were added in Article 19 of the Provisions on the Administration of Internet News Information Services released on September 25, 2005, by the State Council Information Office and the Ministry of Information Industry. These two additional categories are firstly "inciting illegal assemblies, associations, marches, demonstrations, or gatherings that disturb social order, and secondly, conducting activities in the name of an illegal civil organization.[9] Meanwhile, in the second item relating to "state secrets," other laws are specified. It declares that "...state secrets are all issues relating to the security and interests of the nation, determined in accordance with legally defined procedures, the knowledge of which is restricted to a defined scope of personnel for a defined length of time." (Article 2) As a result, state secrets include (Article 8):

Secret issues in significant decisions in national affairs;

1. Secret issues in the activities of national defense building and the strength of the armed forces;
2. Secret issues in the activities of diplomacy and foreign affairs;
3. Secret issues in the economic and social development of citizens;
4. Secret issues in scientific technology;
5. Secret issues in activities of maintaining national security and the investigation of criminal activity;
6. Any other state secret issues which the national secrecy protection work agencies determine should be preserved.

Those illustrations of "state secrets" or a broader definition of "state security" matters could deter Netizens' online activities for fear of breaching the laws. They serve to lay down warnings about the comprehensive limits to online activities and to deter potential offenders. It *de facto* deprives a significant portion of legal defense for cyber actions, such as virtual political debates and consultations against governments or authorities, since such conduct can be deemed illegal.

Great Net Firewall, Censoring and Blocking Web Sites

It is usually suggested that official surveillance of Internet (discussions) is especially tight at times of particular significance to the regime, e.g. the congresses of the Chinese Communist Party (CCP), or sessions of the NPC. The CCP's 17th National Congress in October 2007 is a recent example. The crackdown and closing of tens of thousands of Web sites one month prior to the important political gathering is believed to prevent Netizens from rigorously discussing sociopolitical and economic problems, making ISPs to disable online chatrooms and discussion forums as well as other online interactive feature that may "provide a platform for viewpoints unacceptable to the authorities." (Ford 2007)

The Chinese government has also maintained its control over the Net by means of physical Internet infrastructures. The four national networks,[10] for instance, require all direct international networking traffic to use international incoming and outgoing channels provided by China Telecom (Article 6), which functionally serves as a sort of "Intranet" that connects all ISPs within the country. Legally speaking, the *Provisional Regulations on the Administration of International Interconnection of Computer Information Networks*, enacted in February 1996 and amended later in May 1997, have formalized the control of "computer information network" (*jisuanji xinxi wangluo*). These government-led Internet gateways to the world are based in the metropolitan cities of Beijing, Shanghai and Guangzhou. In other words, the requirement that all ISPs must be registered with one of the four major networks is to ensure their global Internet access services, as opposed to home Internet (Chinese) contents, pass through the packet-level filtering software installed on the interconnecting networks.[11] Without doubt, the Chinese government has set up an ostensibly solid "Great Firewall," also known as *Jindun Gongcheng* (Golden Shield Project),[12] aiming to exert a tight grip on information to constrain what it perceives to be adversely liberalizing Net effects.

Besides, the Chinese government has allegedly begun to exploit the system of automated packet filtering from October 2002 that results in the slowdown of international connections as an extra stopover for each transmission is required (South China Morning Post, 5 March 2003; The Associated Press, 5 March 2003). Specifically, the "packet filtering" system (also known as patchwork system of control or packet-sniffer software) is integrated into the government-controlled international Internet routers that can terminate and/or block TCP packet transmissions when politically sensitive or controversial keywords are detected (Olesen 2007). At times, the program can re-direct Internet users trying to access certain domains that are deemed "inappropriate" to other "safer" or "politically neutral" Web sites (Zittrain and Edelman 2003). This move is obviously to embed a more centralized and sophisticated filtering device/software into policing global online contents. An Internet consultant based in Beijing described the current end-user impact of the "closed" routers as being as if all of China's online population were "…breathing through the same tiny air hole." (Walton 2003)

Moreover, the Chinese authorities have trained and employed above 50,000 virtual police on cyber-patrol for the purpose of maintaining socio-political order in Chinese cyberspace, mostly monitoring the content and usage of the Web, in addition to tracing cyber dissidents (Mooney 2004). As one senior cyber policeman claims, "People should pay attention to their behavior when they are surfing on the Net." (Shanghai Daily, 5 January 2006) The Party's *People's Daily* acknowledges that the special cyber police force is intended "…to intensify real-time monitoring, to intercept and delete harmful information and to capture and check illegal server data." (People's Daily, 1 April 2003) The breach of individual virtual privacy, the use of surveillance and even the imposition of criminal penalties have taken place in the PRC for the sake of national and public security. Indeed, it has often been claimed that the government practices comprehensive Internet censorship and has already blocked many Web sites deemed subversive and undesirable. The first in-depth scholarly report to argue this in any systematic way was the 2004-2005 survey by the OpenNet Initiative. With the help of a computer program, the Study argued that "China operates the most extensive, technologically sophisticated, and broad-reaching system of Internet filtering in the world." (Zittrain and Edelman 2005).

Breach of Privacy and Criminal Penalty

National security, as well as concern of public security, has historically overridden efforts to protect the privacy of personal communications not merely in communist countries such as China.[13] As Jose Caral observes, there has been a steady increase in government regulation of the Internet in the US since 1996. "Civil libertarians are disturbed by the intrusive nature of emerging Internet regulation, particularly those granting security agencies wider powers of surveillance." (Caral 2004: 2) As information and communications technologies advance rapidly, Internet users' privacy—the collection, storage, dissemination, communication and the use of information—is to a varying degree violated when a state's security is at stake.[14] The breach of individual privacy, the use of surveillance and even the imposition of criminal penalties have taken place for the sake of national and public security. David Lyon vividly claimed that people are nowadays living in an "electronic panopticon," in which traces of the use of electronic communications can be recorded, compiled, and even compared as a personal record of people's online activities. (Lyon 1994)

In the Chinese case, there is no doubt that many officials believe that they can and will continue to control online activities partly by infringing Internet users' rights of privacy and freedom of expression through administrative measures. Control over the Internet and censorship of Web contents may be achieved from the government's view by applying regulatory measures as well as licensing procedures to the parties of ISPs and ICPs. ISPs are required to store all users' detailed personal information and keep a record of users' online activities, including Web sites visited, for at least sixty days and render them to public security officials when requested. In a similar vein, those ICPs are obliged to store contributions to any Internet chat rooms, discussion boards and disclosed to authorities on requested. Both ISPs and ICPs are required to report any of their patrons that violate relevant laws and regulations.

However, do ISPs, ICPs and the (local) police force rigorously enforce the laws to monitor and report all incidents of violations in online and offline activities? The fieldwork behind this project suggests that the authorities may charge or sentence a few offenders in the increasingly adept Internet population, thereby setting an example for other potential violators. In other words, it is called "killing chickens to frighten monkeys," (*sha ji jing hou*) as the Chinese proverb goes. In part this is because the government is unlikely to maneuver and mobilize all its physical resources to check the Net at all time. Selective prosecution of cyber-offenders, accordingly, seems a practical way to deter those who might intend to violate the laws. Arresting and detaining a few cyber-dissents has thus been one of the government's plausible measures to tackle the problem.[15]

The London-based Amnesty International has claimed that Chinese Internet users are at risk of arbitrary detention, torture and even execution by the authorities (Amnesty International 2002). Among those who have been arrested and detained is the Shanghai-based computer engineer, Lin Hai, who was allegedly the first victim of stringent Internet regulations in China. His breach of regulations in 1998 brought him two years in jail because he provided some 30,000 e-mail addresses to the pro-democracy "VIP Reference," (*Dacankao*), an underground electronic newsletter, run on a daily basis by Chinese dissidents and diasporas based in the United States (Amnesty International 2002). Another prominent figure, Huang Qi, a Chengdu-based Internet entrepreneur, also operated a pro-democracy Web site (www.6-4tianwang.com), which was provocatively established to defy the Chinese governments' atrocities in the 1989 Tiananmen Square massacre. He was consequently charged with subverting state power and sentenced to five years in prison in May 2003. In

the meantime, journalist Shi Tao was sentenced to ten years in prisons in April 2005 for his allegedly leaking "state secrets" overseas. He was accused of using his Yahoo! email account to post a summary of a government order instructing the Chinese media on how to carefully take in charge the 15th anniversary of the 1989 crackdown on pro-democracy activists (Human Rights in China, 2 January 2008). Reportedly, in 2007, China arrested six bloggers, and detained Internet users with a total of 51 cyber-dissidents, making China the "world biggest prison for Internet users" as the Reporters Without Borders proclaims. (Reporters Without Borders 2008).

Internet Café Regulations and Crackdown

The June 6, 2002 fire in Beijing's "*lanjisu*" cybercafé that claimed 25 lives with 12 other people injured severely alarmed the authorities about inadequate governance and supervision of Internet cafés. As such, the tragedy provoked a rapid order to crack down on illegal cybercafés and it made all cafés re-register. Take China's capital city for example. Beijing has 2,200 out of the total 2,400 Net bars operating illegally (People's Daily, 17&18 June 2002), where people usually associate with the strictest Internet governance of Net cafés.

In fact, official sources suggest that only 46,000 out of 200,000 Internet cafés are licensed, which means less than one quarter of them are legally registered (People's Daily, 29 June 2002; Beech 2002). The nationwide overhaul of Internet cafés simply resulted in the closure of some 3,300 illegal cafes over the six months following the fatal arson (People's Daily, 27 December 2002). Yet there are many other cafés still thriving across the country in all sorts of guises,[16] since the demand from young people, students, the badly-off and rural residents, for instance, remains high. Many of them cannot afford to buy the required computer facilitates or they are not better off by gaining Internet access at home. Still, they are keen on trying to get access to the Internet partly because the Internet may provide them with future opportunities to get prospered or they can also show that they can live like others in the information age.

In addition, the Chinese government at all levels has staged periodic raids on Internet cafés, not only because they are worried about online pornography or violent online games that pose a moral hazard to young people, but also largely because of the "reactionary" or "undesirable" materials readily available on the Internet that have long plagued the government. The former head of the Chinese Ministry of Information Industry, for instance, has warned that moral standards in China are being severely challenged by the rapid flow of information emerging from the Internet. He states:

Due to historical and technical reasons, 90 percent of the information available on Internet is in English and the overwhelming majority of it generated from developed countries, whereas developing countries are mostly information receivers. As information flows across borders and developing countries are absorbing advanced technological and cultural information, their cultural traditions, moral standards and values have been severely challenged (Rollnick 2002).

More importantly, Internet cafés in China are under the management of multiple government departments. Such a governing structure has often resulted in loose coordination between the different departments or bureaus. Specifically, to acquire the legal licenses to operate Internet cafés, four governmental organizations and three procedures are usually involved: firstly, special business and cultural permits issued by the Public Security Bureau and the Cultural Bureau; secondly, an Internet information service business permit issued by the Telecommunications Bureau (under the Ministry of Information Industry), and thirdly, a business license issued by the Administration of Industry and Commerce. Often an Internet

bar will have some but not all of these required licenses; there are frequently one or two missing. A report has revealed it usually takes one to two months to obtain one of the licenses, and it is thus rather difficult to attain all of the licenses for an Internet café in less than half a year (Deng and Wu 2002). Before granting the operational licenses, an owner first needs to have the rental contract approved, then get the cybercafé inspected and approved by the fire department, and also make sure that all the computer facilities have been purchased in advance. During the waiting period whilst the acquisition of licenses is pending, an owner will spend tens of thousands of *renminbi* in rent. Because of the poor inter-governmental coordination and excessively time-consuming and complicated approval procedures, the so-called "*hei wangba*" (literally "black bars" in Mandarin, illegal Internet cafés) has to seek for ways to survive. That precisely provides more channels for those who pursue rent-seeking behavior.

More recently, the Chinese authorities have since March 2005 furthered Internet control by rigorously enforcing the "Computer Information Network and International Internet Security Protection and Administration Regulations," (*Jisuanji Xinxi Wangluo Guoji Lianwang Anquan Baohu Guanli Banfa*)[17] which require that all Website operators register their sites with the local Public Security Bureau within thirty days of operation. The Ministry of Information Industry was aiming at those Web sites that are not business enterprises, including the free personal Web sites and the blogs.[18] As a result, thousands of Internet Web sites have been shut down for failing to register as required. Shortly in July 2005, newer regulations were issued to request instant message users as well as bloggers to use their "real names"[19] when engaging online for the purpose of national security and social stability. Take the instance message services of the QQ network for instance, there are reportedly more than 100 million active users and 8 million among them are QQ group founders and administrators. Requested by the Ministry of Education, all the university BBS's in Chinese higher education, such as Beijing University and Fudan University, adopted the "real name" system, and turned to campus intranet platforms only. A government official discloses that Web sites registration is like giving a Website a "*hukou*" (residence permit) under a *hukou* system. The Ministry of Information Industry requires real names from non-enterprises Web sites to continue to better improve their monitoring of the Internet (Li and Yu 2005).

In sum, because all ISPs, ICPs, and Internet café owners in China, are responsible for reporting any patron who violates the laws and regulations, the stringent but ambiguous regulations have a profound impact upon Internet entrepreneurs; they promote self-censorship and set up their own monitors, known as "Big Mama," to censor the chatrooms, bulletin boards and Internet cafés lest they may incur "severe penalties for content violations by third parties on their network, site, or server." (American Chamber of Commerce in the People' Republic of China 2002) In so doing, they may keep in line with the laws that are broadly decreed.[20] Given these constraints on Internet operators, they have publicly committed themselves on several occasions to adhere to (Internet) media controls put forward by the CCP on the one hand, and continue to utilize the leeway provided to Internet-based commercial portals unavailable to other media on the other hand. The Chinese government has tried hard to keep the average Internet users on a tight leash, regardless of self-censorship or penalties among Internet users. Some commentators contend that "the possibility of being shut down by the government has encouraged self-censorship [and discipline] by Internet companies—which in turn has dampened [democratic potentials of] online political communication." (Kalathil 2001; Kalathil and Boas 2003; Hughes and Wacker 2003).

The Politics of the Governing the Internet in China

It is often argued that free exchange of information poses fundamental challenges to authoritarian states that depend heavily upon social and political control to strengthen their legitimacy and maintain regime stability (Baehr and Richter 2004). It is also usually the premise that (new) technology can transform the mode of political communication and that this in turn alter the nature of political participation as well as the milieu in which political discussions are made (Barber 1984). Proponents of the Internet-derived democracy and democratization in authoritarian state like China have often held that the ICTs have empowered grass-roots citizens there to acquire, disseminate and exchange (alternative) information from outside (Hill and Sen 2005; Sheff 2002; McCaughey and Ayers 2003; Chase and Mulvenon 2002), which is usually unavailable from the official and mainstream media. The rise of online public opinion stands one spectrum of the Internet politics in those authoritarian regimes.

When information and communication technologies (ICTs), principally symbolized by the Internet, converge on the political environment in most authoritarian and developing countries, ICTs allow the possibilities of the public gaining more latitude in expressing opinions. China is a particularly significant country in this respect. In part it is because the (mass) media was traditionally used by the authorities to serve as tools of propaganda (*xuanchuan jiaoyu*) and for purposes of agenda-setting (*yulun daoxiang*) (Zhao 1998; Pei 1994; Wu 2001: 45-67). Entering the Internet age, the state is seen to assure its economic competitiveness in a globalized context where information largely drives global and domestic economy. The government there keenly bolsters the development of information and network technology, but at the same time, it has been persistently attempting to minimize the undesirable social and political effects that the Internet has brought about since it was introduced in the early 1990s. Because the political impact of the Internet has caused the Beijing government unease as it threatens its long-held monopoly over the flow of information, the regime has adopted a variety of strategies to harness it, limit the impact of the new technology to an acceptable degree, and hopefully to turn it to the government's benefit, particularly in the prospects of e-commerce and e-government (He 2004: 117-48; Damm and Thomas 2006).

In the area of e-commerce, the communications and information revolutions are key enablers of economic globalization. To grasp the opportunities provided by an increasingly globalized world, Chinese government has supported the development of the telecommunication infrastructure, such as telephone and cable lines, hoping to create the crucial means for the Internet diffusion (Zhou 2006). The fundamental reason is that progress in the range of wired and wireless technologies makes for faster and cheaper flows of (business) information across the globe. Besides, "IT exports"[21] have been in recent years an integral component of the economy in China.

Specifically, the government in China has initiated a so-called "twin-track strategy" that ambitiously integrates industrialization into the grand process of informatization (Dai 2002: 144). Informatization is treated as the key in promoting industrial advancement, industrialization and modernization in China. The key importance and role of the Information Technology will principally be to serve as the basic, pioneering, supporting and strategic industry of the national economy, and increasingly play a pivotal part in promoting the domestic economy, national safety, the welfare of citizens, and social development.[22]

For the development of electronic government, i.e., the initiatives of the governments that use information technology to deliver their informa-tion and services, the Chinese government has not merely rhetorically underlined the magnitude of the correct "agenda-setting," but has also vigorously shaped it in the "Government Online Project" since 1998. The project can, in one sense, be interpreted as the authorities' proactive effort to restore the propaganda machinery that has been weakened throughout the reformist period, and particularly the gradual reduction in the state's monopoly over the provision of information and communications (Dai 2000:151-2).

The Chinese government has nevertheless long been torn by the ambivalence brought about by the Internet. It regards the Internet as an engine to drive economic growth on the one hand, and as a subversive challenge to undermine the ruling Communist Party on the other hand. As soon as ICTs were introduced and Web sites mushroomed, the Party was so determined to harness the new medium to assure the Internet's economic and scientific benefits. As a consequence, controls other than stifling ICTs would be critical for the CCP's agenda to achieve the century-long modernization process and in the meantime, consolidate its power. As Wu Jichuan, ex-Minister of MII, recognizes, "Network and information safety can not be overemphasized, as it has a bearing on the sovereignty and economic security of a nation. Any improper handling of the relations will hamper the overall economic development (Rollnick 2002)."

In other words, while the government can exert certain control over the Web contents and messages/information posed online, the state control over the new medium is indeed diminishing, when particularly compared with traditional press and mass media. The Internet has incrementally created a shift in mass communication that allows the public to speak en masse (Hung 2003:1-38; Tai 2006). There are political implications as will become more apparent when the Internet penetration rate increases in the near future.

REFERENCES

1999: The Government Online Year. (1999, January 3). *People's Daily* (p. 4).

American Chamber of Commerce in the People's Republic of China (2002). *2002 White Paper: American Business in China.* Section on the Information Technology. Retrieved February 29, 2004, from http://www.amcham-china.org.cn/publications/white/2002/en-20.htm

Amnesty International. (2002, November 26). *China: Internet Users at Risk of Arbitrary Detention, Torture and Even Execution.* Retrieved January 5, 2003, from http://web.amnesty.org/library/index/ENGASA170562002

Amnesty International. (2002, November 26). *State Control of the Internet in China* (pp. 3-4). Retrieved June 5, 2008, from http://www.amnesty.org/en/library/asset/ASA17/007/2002/en/dom-ASA170072002en.pdf

Amnesty International. (2006, July). *Undermining Freedom of Expression in China: The Role of Yahoo!, Microsoft and Google* (p. 14). Retrieved January 7, 2008, from http://irrepressible.info/static/pdf/FOE-in-china-2006-lores.pdf

Baehr, P., & Richter, M. (Eds.). (2004). *Dictatorship in History and Theory: Bonapartism, Caesarism, and Totalitarianism.* Cambridge: Cambridge University Press.

Barber, B. R. (1984). *Strong Democracy: Participatory Politics for a New Age.* Berkeley: University of California Press.

Beech, H. (2002, July 22). Living It Up in the Illicit Internet Underground. *Time*, 160(4), 4.

Beijing Internet Cafes Ordered to Stop Operation for Rectification. (2002, June 17). *People's Daily.* Retrieved June 18, 2002, from http://english.peopledaily.com.cn/200206/17/eng20020617_97950.shtml

Caral, J. (2004). Lessons from ICANN: Is Self-Regulation of the Internet Fundamentally Flawed. *International Journal of Law and Information Technology*, 12(1), 2.

Castells, M. (2000). *The Rise of the Network Society.* (p. 1). Oxford: Blackwell Publisher.

Castells, M. (2001). *The Internet Galaxy.* (p. 1). Oxford: Oxford University Press.

Chase, M. S., & Mulvenon, J. C. (2002). *You've Got Dissent! Chinese Dissident Use of the Internet and Beijing's Counter-Strategies.* Santa Monica, CA: RAND.

China Closes over 3,300 Internet Cafes in Six Months (2002, December 27). *People's Daily.* Retrieved April 20, 2008, from http://english.peopledaily.com.cn/200212/27/eng20021227_109161.shtml

China Orders Unlicensed Internet Cafes Closed Nationwide. (2002, June 29). *People's Daily.* Retrieved April 30, 2008, from http://english.peopledaily.com.cn/200206/29/eng20020629_98774.shtml

China Seeks to Build Boundary on Internet (2003, April 1). *People's Daily.* Retrieved April 17, 2008, from http://english.peopledaily.com.cn/200304/01/eng20030401_114386.shtml

China's Eleventh Five Year Plan (2006-2010)—Information Industry, Ministry of Information Industry. (2007, March 1). Retrieved January 14, 2008, from http://www.mii.gov.cn/art/2007/03/01/art_3986_1936.html

China's Web Surveillance Slows Access even as Government Promotes Internet Use. (2003, March 5). *The Associated Press.* Retrieved April 2, 2008, from http://www.clearwisdom.net/emh/articles/2003/3/7/zip.html#34

Chung, Y. (2002). *Anatomy of the Decision-Making Process in China: (De)Concentration of the*

Internet Industry. (p. 55). Unpublished master thesis, Seoul National University, South Korea.

Coyle, M. (2001, October 1). September Attacks Prompt Sharp Debate on Scope of Surveillance Law. *Law.com.* Retrieved December 18, 2003, from http://www.law.com/cgi-bin/nwlink. cgi?ACG=ZZZ1NUCI6SC

Cullen, R., & Choy, P. D. W. (1999). The Internet in China. *Columbia Journal of Asian Law,* 13(1), 116.

Cyber Police to Guard All Shenzhen Web sites. (2006, January 5). *Shanghai Daily.* Retrieved November 2, 2007, from http://www1.china.org. cn/english/government/154200.htm

Dai, X. (2000). *The Digital Revolution and Governance.* (pp. 151-2). Aldershot: Ashgate.

Dai, X. (2002). Towards a Digital Economy with Chinese Characteristics? *New Media and Society,* 4(2), 144.

Damm, J., & Thomas, S. (Eds.). (2006). *Chinese Cyberspace: Technological Changes and Political Effects.* London and New York: Routledge.

Deng, K., & Wu, C. (2002, June 21). Wangba he Tade Shengcun zhi Dao (Internet Cafés and Their Ways of Existence). *Nanfang Zhoumo* (Southern Weekend). Retrieved June 4, 2008, from http:// www.people.com.cn/GB/it/48/297/20020621/75 8428.html

Diffie, W., & Landau, S. (1998). *Privacy on the Line: The Politics of Wiretapping and Encryption.* Cambridge, MA. & London: MIT Press.

Fire Prompts Tight Control on Internet Cafes in China. (2002, June 18). *People's Daily.* Retrieved April 18, 2008, from http://english.peopledaily. com.cn/200206/18/eng20020618_98048.shtml

Ford, P. Why China Shut Down 18,401 Web sites. (2007, September 25). *The Christian Science Monitor.* Retrieved January 9, 2008, from http://www.csmonitor.com/2007/0925/p01s06-woap.htm

Hachigian, N. (2002). Telecom Taxonomy: How Are the One Party States of East Asia Controlling the Political Impact of the Internet? In J. Zhang & M. Woesler (Eds.). *China's Digital Dream: The Impact of the Internet on Chinese Society* (pp. 35-67). Bochum: The University Press Bochum.

Hachigian, N. (2002). The Internet and Power in One-Party East Asian States. *The Washington Quarterly,* 25(3), 41-58.

Harwit, E. (1998). China's Telecommunications Industry: Development Patterns and Policies. *Pacific Affairs,* 71(2), 175-194.

He, Q. (2004). *Zhongguo zhengfu ruhe kongzhi meiti* (How the Chinese government controls the media). (pp. 117-48). Retrieved January 25, 2008, from http://www.ir2008.org/PDF/initiatives/Internet/Media-Control_Chinese.pdf

Hill, D. T., & and Sen, K. (2005). *The Internet in Indonesia's New Democracy.* London and New York: Routledge.

Hu, A. (2002). *Zhongguo Zhanlue Gouxiang* (Strategy of China). (p. 15). Hangzhou: Zhejiang Renmin Chubanshe.

Hughes, C. R., & Wacker, G. (Eds.). (2003). *China and the Internet: Politics and the Digital Leap Forward.* London and New York: RoutledgeCurzon.

Hulianwang Xinwen Xinxi Fuwu Guanli Guiding (Provisions on the Administration of Internet News Information Services). Retrieved March 12, 2008, from http://www.cnnic.cn/html/ Dir/2005/09/27/3184.htm

Hulianwang Xinxi Fuwu Guanli Banfa (2000, October 1). Retrieved February 12, 2004, from http://past.people.com.cn/GB/channel5/28/20001001/257566.html

Human Rights in China. (2008, January 2). Press Advisory: HRIC Launches 2008 Take Action Website and Calls on China To Release Shi Tao.

Retrieved January 7, 2008, from http://hrichina. org/public/contents/press?revision%5fid=46424 &item%5fid=46414

Hung, C. F. (2003). Public Discourse and "Virtual" Political Participation in the PRC: The Impact of the Internet. *Issues & Studies, 39*(4), 1-38.

Hung, C. F. (2005). The Interaction between Internet Entrepreneurs and the Chinese Authorities: Possible Implications for Civil Society. *Issues & Studies, 41*(3), 145-80.

Hung, C. F. (2006). The Politics of Cyber Participation in the PRC: The Implications of Contingency for the Awareness of Citizens' Rights. *Issues & Studies, 42*(4), 142-5.

Kalathil, S. (2001). The Internet and Asia: Broad Band or Broad Bans? *Foreign Service Journal, 78*(2). Retrieved April 18, 2001, from http:// www.ceip.org/files/Publications/internet_asia. asp?p=5&from=pubdate

Kalathil, S. (2003). Dot Com for Dictators. *Foreign Policy, 135*, 42-49.

Kalathil, S., & Boas, T. C. (2001). The Internet and State Control in Authoritarian Regimes: China, Cuba, and the Counterrevolution. *First Monday, 6*(8). Retrieved June 8, 2008, from http://www. firstmonday.org/issues/issue6_8/kalathil/

Kalathil, S., & Boas, T. C. (2003). *Open Networks, Closed Regimes: The Impact of the Internet on Authoritarian Rule.* Washington, DC: Carnegie Endowment for International Peace.

Lam, W. (2008). Stability Trumps Reform at China's Parliamentary Session. *China Brief, 8*(6), 2-4. http://www.jamestown.org/terrorism/news/ uploads/cb_008_006a.pdf

Li, L., & Yu, L. (2005, August 18). 14 Buwei Lianhe Jinghua Hulianwang. (Fourteen Departments United to 'Purify' the Internet) *Nanfang Weekend.* Retrieved June 6, 2008, from http://

www.nanfangdaily.com.cn/zm/20050818/xw/ szxw1/200508180019.asp

Lu, X. (Ed.). (2002). *Zhongguo Xinxihua* (China's Informatization). (p. 53). Beijing: Electronics Industry Publisher.

Lynch, D. C. (1999). *After the Propaganda State: Media, Politics, and "Thought Work" in Reformed China.* (p. 173). Stanford: Stanford University Press.

Lyon, D. (1994). *The Electronic Eye: The Rise of Surveillance Society.* Cambridge: Polity Press.

McCaughey, M., & Ayers, M. D. (Eds.). (2003). *Cyberactivism: Online Activism in Theory and Practice.* London and New York: Routledge.

Mooney, P. (2004, April). China's "Big Mamas" in a Quandary. *YaleGlobal.* Retrieved June 8, 2008, from http://yaleglobal.yale.edu/display. article?id=3676

Nathan, A. J. (2003). Authoritarian Resilience. *Journal of Democracy, 14*(1), 6.

Naughton, B. (1997). The Patterns and Logic of China's Economic Reform. In the Joint Economic Committee, Congress of the United States (Ed.), *China's Economic Future: Challenges to U.S. Policy* (p. 1). New York and London: M.E. Sharpe.

Negroponte, N. (1995). *Being Digital.* London: Coronet Book/Hodder & Stoughton.

Net Users Angry Over Slow Connections. (2003, March 5). *South China Morning Post.*

Olesen, A. (2007, October 20). China's Internet Controls Tightened for Politics' Sake. *The Associated Press.*

Pei, M. (1994). *From Reform to Revolution: The Demise of Communism in China and the Soviet Union.* Cambridge, Mass: Harvard University Press.

Reporters Without Borders, *The 2007 Annual Report—Dictatorships Get to Grips with Web 2.0*. Retrieved June 5, 2008, from http://www.rsf.org/rubrique.php3?id_rubrique=675

Reporters Without Borders. (2008, January 4). Retrieved January 12, 2008, from http://www.rsf.org/article.php3?id_article=24946

Rollnick, R. (2002, January 14). Telecommunication Summit: China Concerned at Electronic Threats to Moral Standards. *Earth Times News Service*. Retrieved November 23, 2002, from http://earthtimes.org/jan/telecommunicationchinajan14_02.htm

Sheff, D. (2002). *China Dawn: The Story of a Technology sand Business Revolution*. New York: Harper Business.

Shoesmith, B., & Hearn, K. (2004). Exploring the Roles of Elites in Managing the Chinese Internet. *Javnost: The Public*, 11(1), 101-14.

Sohmen, P. (2001). Taming the Dragon: China's Efforts to Regulate the Internet. *Stanford Journal of East Asian Affairs*, 1(1), 18.

Tan, Z. (1999). Regulating China's Internet: Convergence toward a Coherent Regulatory Regime. *Telecommunications Policy*, 23(3), 261-76.

The Internet Timeline of China, Part II. *China Internet Network Information Center*. Retrieved May 12, 2004, from http://www.cnnic.net.cn/html/Dir/2003/12/12/2001.htm

Triolo, P. S., & Lovelock, P. (1996). Up, Up, and Away—With Strings Attached. *The China Business Review*, 23(6), 28.

Walton, G. (2001). *China's Golden Shield: Corporations and the Development of Surveillance Technology in the People's Republic of China*. Retrieved June 2, 2008, from http://www.ichrdd.ca/english/commdoc/publications/globalization/goldenShieldEng.html

Walton, G. (2003, March 10). *Great Wall, Small World*. Congressional-Executive Commission on China (CECC). CECC Open Forum. Retrieved May 26, 2008, from http://www.cecc.gov/pages/roundtables/031003/walton.php

Wu, Guoguang. (2001). One Head, Many Mouths: Diversifying Press Structures in Reform China. In C. C. Lee (Ed.), *Power, Money, and Media: Communication Patters and Bureaucratic Control in Cultural China*. Evanston, Ill.: Northwestern University Press.

Xinhua News Agency. (2008, January 17). Retrieved January 20, 2008, from http://news.xinhuanet.com/newscenter/2008-01/17/content_7439151.htm

Yang, D. L. (2001). The Great Net of China. *Harvard International Review*, 22(4), 65.

Zhang, J. (2003). Network Convergence and Bureaucratic Turf Wars. In C. R. Hughes & G. Wacker (Eds.), *China and the Internet: Politics of the Digital Leap Forward* (p. 85). London and New York: RoutledgeCurzon.

Zhang, J., & Woesler, M. (Eds.). (2002). *China's Digital Dream: The Impact of the Internet on Chinese Society*. (pp. 35-67). Bochum: The University Press Bochum.

Zhao, Y. (1998). *Media, Market, and Democracy in China: Between the Party Line and the Bottom Line*. Urbana and Chicago: University of Illinois Press.

Zhengfu Shang-wang Gongcheng: Huigu yu Zhanwang (Government Online Project: Reviews and Prospects). Retrieved March 6, 2001, from http://www.gov.cn/govonlinereview/6future/01.htm

Zhou, R. (2002, December 4). Internet Cafes Percolating despite Legal Clampdown. *China Daily*. Retrieved January 16, 2003, from http://www3.chinadaily.com.cn/en/doc/2002-12-04/content_146503.htm

Zhou, Y. (2006). *Historicizing Online Politics: Telegraphy, the Internet, and Political Participation in China.* Stanford: Stanford University Press.

Zittrain, J., & Edelman, B. (2003). *Internet Filtering in China.* Harvard Law School Public Law. Research Paper, No. 62. (pp. 70-7). Retrieved January 24, 2008, from http://unpan1.un.org/intradoc/groups/public/documents/apcity/unpan011043.pdf

Zittrain, J., & Edelman, B. (2005, April 14). *Internet Filtering in China in 2004-2005: A Country Study.* Retrieved June 6, 2008, from http://opennet.net/sites/opennet.net/files/ONI_China_Country_Study.pdf

ENDNOTES

[1] The "Golden Projects" consist of several sub-projects, including primarily, Golden Bridge—a national public economic information communication network aiming to connect ministries and state-owned enterprises and to build the infrastructure backbone over which other information services will run; Golden Card—an electronic money project which aims at setting up a credit card verification scheme and an inter-bank, inter-region clearing system; Golden Customs—a national foreign economic trade information network project; Golden Marco—a national economic macro policy technology system; Golden Tax—a computerized tax return and invoice system project; Golden Gate—a foreign trade information network aimed at improving export-import trade management; Golden Enterprise—an industrial production and information distribution system; Golden Intelligence—the China Education and Research Network (CERnet); Golden Agriculture—an overall agricultural administration and information service system; Golden Info—a state statistical information project, and Golden Cellular—a mobile communications production and marketing project.

[2] Zhengfu Shang-wang Gongcheng: Huigu yu Zhanwang (Government Online Project: Reviews and Prospects). Retrieved March 6, 2001, from http://www.gov.cn/govonlinereview/6future/01.htm

[3] The Internet Timeline of China, Part II. China Internet Network Information Center. Retrieved May 12, 2004, from http://www.cnnic.net.cn/html/Dir/2003/12/12/2001.htm

[4] According to official Xinhua News Agency, by the end of 2007, the Web population in China was merely 5 million behind the United States. Nevertheless the Chinese online community was said to surpass the U.S. in 2008, becoming the world's largest cyber world. See Xinhua News Agency. (2008, January 17). Retrieved January 20, 2008, from http://news.xinhuanet.com/newscenter/2008-01/17 /content_7439151.htm

[5] Retrieved January 25, 2009, from http://www.cnnic.cn/uploadfiles/pdf/2009/1/13/92458.pdf (p. 11)

[6] For details on the conflict between the MPT and MEI, see, for example, Harwit, E. (1998). China's Telecommunications Industry: Development Patterns and Policies. Pacific Affairs, 71 (2), 175-194. For confrontation between the MPT and the State Administration of Radio, Film and Television (SARFT), see Zhang, J. (2003). Network Convergence and Bureaucratic Turf Wars. In C. R. Hughes & G. Wacker (Eds.), China and the Internet: Politics of the Digital Leap Forward (p. 85). London and New York: RoutledgeCurzon.

[7] Ministry of Information Industry. Retrieved September 10, 2003, from http://www.mii.gov.cn/mii/bmjs.htm

8   Hulianwang Xinxi Fuwu Guanli Banfa (Measures for Managing the Internet Information Services) was promulgated on October 1, 2000 by the State Council. Retrieved February 12, 2004, from http://past.people.com.cn/GB/channel5/28/20001001/257566.html

9   Hulianwang Xinwen Xinxi Fuwu Guanli Guiding (Provisions on the Administration of Internet News Information Services). Retrieved January 12, 2008, from http://www.cnnic.cn/html/Dir/2005/09/27/3184.htm

10 The four interconnecting networks are (1) CHINANET (Zhongguo Gongyong Jisuanji Hulianwang; China Network), (September 1994) formerly owned by the MPT and now by MII since 1998; (2) CSTNET (Zhongguo Keji Wang, China Science and Technology Network), (April 1995) owned by the Chinese Academy of Sciences. (3) CERNET (Zhongguo Jiaoyu he Keyan Jisuanji Wang, China Education and Research Network), (July 1995) owned by the State Education Commission and (4) CHINAGBN (Zhongguo Jinqiao Xinxi Wang, China Golden Bridge Network) (September 1996) formerly owned by the MEI, but now by MII. See Article 7 of the Provisional Regulations. It has been expanded into eight international Internet gateways, including additional four networks, CIETNET (Zhongguo Guoji Jingji Maoyi Hulianwang, China International Economy and Trade Network) (January 2000), owned by China International Economy and Electronic Commerce Center (Zhongguo Guoji Jingji Dianzi Shangwu Zhongxin)

11 Remarks made by Michael Iannini, general manager of Nicholas International Consulting Services Inc. in Beijing. Cited from Walton, G. Great Wall, Small World. (2003, March 10). CECC Open Forum. Retrieved May 26, 2008, from http://www.cecc.gov/pages/roundtables/031003/walton.php

12 The Golden Shield Project is itself a national online surveillance and censorship project that China's Ministry of Public Security operated as early as in 2003. Reportedly it is the project implementing to construct a communication network and computer information system for the Chinese police to improve their capability, efficiency, and control over cyberspace. The Ministry of Public Security later in November 2006 announced the completion of the essential tasks of constructing the first stage of its Golden Shield Project. Retrieved December 2, 2007, from http://www.mps.gov.cn/cenweb/brjlCenweb/jsp/common/article.jsp?infoid=ABC00000000000035645. For more reference on this Project, see Walton, G. (2001). China's Golden Shield: Corporations and the Development of Surveillance Technology in the People's Republic of China. Report from the Rights & Democracy. Retrieved June 2, 2008, from http://www.ichrdd.ca/english/commdoc/publications/globalization/goldenShieldEng.html

13 Marcia Coyle, for instance, argues that even a democratic country like the United State has few provisions currently in place protecting American citizens from unconstitutional forms of surveillance. Such situation may be further deteriorating in the wake of the September 11 (2001) terrorist attacks on the World Trade Center and the Pentagon. See Coyle, M. (2001). September Attacks Prompt Sharp Debate on Scope of Surveillance Law. Law.com Retrieved December 18, 2003, from http://www.law.com/cgi-bin/nwlink.cgi?ACG=ZZZ1NUCI6SC

14 For a more general discussions about the state's meddling with the privacy because of the needs for national interests, national security or law enforcement, see, for example, Diffie, W. & Landau. S. (1998). Privacy on the Line: The Politics of Wiretapping and Encryption. Cambridge, MA. London: MIT Press.

[15] For more information about Chinese individuals currently detained for online political and religious activity, see, for example, the full list compiled by the organization of Digital Freedom Network (DFN) at <http://www.dfn.org/focus/china/netattack.htm>, accessed July 22, 2003. According to DFN, there are currently 34 Chinese individuals detained and only three have been released.

[16] Some cafés operate under the guise of other businesses, such as "labor skill training center," which is regulated by the labor department, and some have a private school as their front, which falls under the jurisdiction of the education department. Other cafés even discreetly place cameras overlooking the pavement, and monitoring any movement toward the door. See Zhou, R. (2002). Internet Cafes Percolating despite Legal Clampdown. China Daily (December 4). Retrieved January 16, 2003, from http://www3.chinadaily.com.cn/en/doc/2002-12/04/content_146503.htm

[17] Article 12. Retrieved March 20, 2006, from http://news.xinhuanet.com/newmedia/2003-02/08/content_897720.htm

[18] As a matter of fact, the Internet regulations for the non-enterprise Website had already been initiated as early as in 2000. The 2000 "Internet Information Service Administration Regulations" (Hulianwang Xinxi Fuwu Guanli Banfa) specifies that while the government uses the permit system for information enterprises, registration system is employed for non-enterprise information services. See Article 4. For full text, see <http://www.china.org.cn/chinese/zhuanti/193663.htm>, accessed March 12 2006.

[19] The debate over the "real name" system firstly appeared in early 2003 when Li Xiguang, professor of media and communication of the Tsing Hua University in Beijing, proposed "the National People's Congress ought to legislate against people expressing things anonymously on the Internet." See <http://www.blogchina.com/new/display/8433.html>, accessed February 10, 2006.

[20] A Report by Amnesty International UK has also suggested that "…despite sophisticated technological filters, the effectiveness of censorship in China still rests on self-censorship, as [Internet] companies, institutions, and individuals seek to avoid punishments associated with crossing the [official] line. See Amnesty International. (July 2006). Undermining Freedom of Expression in China: The Role of Yahoo!, Microsoft and Google (p. 14). Retrieved January 7, 2008, from http://irrepressible.info/static/pdf/FOE-in-china-2006-lores.pdf. For more analysis about Internet entrepreneurs and the Chinese authorities, see, for example, Chin-fu Hung. (2005). The Interaction between Internet Entrepreneurs and the Chinese Authorities: Possible Implications for Civil Society. Issues & Studies, 41(3), 145-180.

[21] The IT exports or products usually include desktop PCs, notebook PCs, monitors, modems, PC peripherals, electrical parts, personal digital assistants (PDAs), and communications equipment.

[22] See China's Eleventh Five Year Plan (2006-2010)—Information Industry, Ministry of Information Industry. (2007, March 1). Retrieved January 14, 2008, from http://www.mii.gov.cn/art/2007/03/01/ art_3986_1936.html

Section III
Themes and Issues

Chapter X
Information Networks, Internet Governance and Innovation in World Politics

Claudia Padovani
University of Padova, Italy

Elena Pavan
University of Trento, Italy

ABSTRACT

Political processes are undergoing profound changes due to the challenges imposed by globalization processes to the legitimacy of policy actors and to the effectiveness of policy-making. Building on a socio-political approach to governance and focusing on global information policies and networks, this chapter aims at developing a better understanding of the possibility of change in world politics nowadays, by critically analysing two innovative elements: the reality and relevance of "multi-stakeholder" practices and the growing role of information technologies as a complementary support to actors' relations. Looking at Internet Governance debates in recent years, the authors reconstruct networks of interaction connecting actors in the virtual space, and look at actors' communication modes. Thus they analyze the extent to which technological, as well as processual and cognitive innovation, shapes actors' orientations and the structures within which they interact in the specific context of Internet Governance.

Political processes are undergoing profound changes due to the challenges imposed by globalization processes to the legitimacy of policy actors and to the effectiveness of policy-making.

These changes also affect the supra-national and trans-national conduct of politics, suggesting that emerging global governance arrangements may be conceived as answers to dynamics characterized by diversity and complexity in the post-Cold War era. Such developments are made even more critical by the role and relevance of communication, in its different forms.

Building on a socio-political approach to governance and focusing on global information policies and networks, this chapter aims at developing a better understanding of the possibility of change in world politics nowadays, by critically analysing two innovative elements: the reality and relevance of "multi-stakeholder" practices and the growing role of information technologies as a complementary support to actors' relations.

Looking at Internet Governance debates in recent years, we reconstruct networks of interaction connecting actors in the virtual space, and we look at actors' communication modes. Thus we analyze the extent to which technological, as well as processual and cognitive innovation, shapes actors' orientations and the structures within which they interact in the specific context of Internet Governance.

In the first two paragraphs we theoretically clarify the existing nexus between governance processes, information technology and information networks as emerging governance structures in world politics. The third paragraph sets the scene for the analysis, introducing contemporary debates on the governance of the Internet as a case study and explaining our methodological approach. In the fourth paragraph we articulate and critically assess the multi-stakeholder approach in relation to the notion of diversity in terms of actors involved, issues addressed, and knowledge produced. The fifth paragraph explores, through a comprehensive view of communication modalities, the complexity through which communication nurtures world politics and its dynamics: languages and frames, off-line and on-line interactions, innovation in processes through societal learning. In the Concluding remarks, we summarize the main findings from our investigation and introduce some open issues for further research.

CHALLENGES TO CONTEMPORARY WORLD POLITICS: DIVERSITY, DYNAMICS, COMPLEXITY

Addressing the complexities of political processes in the global landscape, we find a useful point of entry to our investigation in the socio-political approach to governance elaborated by Jan Kooiman, who suggests that governance in contemporary societies should be conceived as essentially interactive: "a mix of all kinds of efforts by all manners of socio-political actors, public as well as private" (2003, p. 3) through which actors with governing roles assume shared responsibilities.

Kooiman underlines how governing mechanisms are societal responses to demands that emerge in a context characterized by diversity, dynamics and complexity; a situation where more traditional arrangements, centred on state-actors, are no longer capable to respond effectively. On the one hand "no single actor, public or private, has the knowledge and information required to solve complex, dynamic and diversified societal challenges; no governing actor has an overview sufficient to make the necessary instrument effective; no single actor has sufficient action potential to dominate unilaterally" (Kooiman, 2003, p. 11). On the other hand, governing arrangements differ from local to global and from one policy domain to another, while the necessary technical and political knowledge is dispersed and governing objectives are difficult to define, and challenging to realize.

Diversity, in Kooiman's terms, refers to the plurality of actors involved in relevant processes: they are shaped in the interaction and, at the same time, they shape interactions by defining their boundaries, setting the political agenda, framing issues, problems and solutions. Actors' diversity can be gained by giving actors the opportunity to play out their identities in interaction. If we translate this to the global context, diversity refers to the shifting from a centrality of state actors to a plurality of entities - states, non-governmental actors, global networks, inter and trans-national organizations – engaged in (more or less formal) interactive exchanges, producing (more or less binding) political outputs and playing out their identities in a variety of situations. Global governance literature has evolved over time stressing the emergence of networks of relations between state and non-state actors, forming large Webs of interactions between governments, IGOs, NGOs, TNCs and other interested parties (Held, 2004; Rosenau, 1995). Nevertheless, empirical research is needed to assess to what extent open and participatory modalities are actually played out in the global context.

Dynamics can be conceived as a reframing of what Deutsch (1963) named "the nerves of government": it refers to the dynamic quality of governing processes and it characterizes the choice between change and conservation in policy discourses and decision-making arrangements. Again, moving to the supra-national level of political processes, we can conceive dynamics as a shifting from a nexus between processual and structural political interactions framed within a state-centred logic (grounded on traditional diplomatic codes of secrecy and exclusivity within an understanding of political relations based on "hard power") to an approach that is characterized not only by actors' diversity, but also by societal requests for transparency, public scrutiny and legitimacy of political action; all of which bring the "soft side of power" into the picture (Nye & Owens, 1996). Moreover, the very possibility of actors' ideas, interests and perceptions being transformed through interaction[1] connects dynamics to the possibility of change in world politics. Again, it is only through the investigation of specific governance arrangements that change in supra-national power relations can be assessed.

Finally, complexity points to the multitude of interactions that take place in different forms, with different intensity, at different levels, with different outputs. This feature recalls the need, expressed by several authors (Held & McGrew, 2003; Held et al., 1999), to reconsider the conceptual distinction between domestic politics and international affairs, between internal and external mechanisms of decision-finding and decision-making. Furthermore, the complexity dimension calls for an "inclusive look" through which we - as observers as well as political actors - should be able to handle the multidimensionality of global processes. To ideate and implement empirical research in such a context, articulated and innovative methodologies are needed.

GOVERNANCE THROUGH INFORMATION NETWORKS

"Governance is achieved by the creation of interactive, socio-political structures and processes stimulating communication between actors involved" (Kooiman, 2003, p. 3). This centrality of communication in governance processes justifies our interest for the who, how and why of interaction in a specific policy domain – that of global communication - where the ideation, production, exchange and diffusion of information is, at the same time, a means, a process enabler and the very substance of political negotiations. We therefore integrate Kooiman's perspective with new components, inspired by recent scholarly attempts to look at the interplay between information technology and world politics (Braman, 2006; Kamarc & Nye, 2002; Rosenau & Singh, 2002).

Information and communication technologies are transforming political processes in practical as well as cognitive terms. Rosenau stresses the fact that it is people and collectivities that infuse values into information technologies as enabling or constraining elements to innovative political processes. According to him, among the ways in which communication and ICTs contribute to restructure world affairs, we find an altering in individual skills where people, being exposed to a plurality of information flows, become more and more aware and capable to analyse, imagine and react to global trends. We also find changes in the logic according to which individuals and groups relate to each other, with a growing importance of horizontal modes of exchange that supplement, not necessarily replace, traditional hierarchies. Moreover we find an evolution of political structures, through the emergence and consolidation of "networks" of actors (Keck & Sikkink, 1998; Reinecke & Deng, 2000). Finally, we see the emergence of global networks operating around communication and information as a policy domain: information and technologies are no longer just the instruments through which global networks are created ad sustained. They constitute the very substance of specific networks' political activity.

Rosenau talks about "the traditional world of anarchical states (...) supplemented by a second world of world politics comprised of a variety of non-governmental, trans-national and sub-national actors" (2002, p. 284); a "second world" that shows a degree of internal consistency that allows its co-existences with the state-centred world, which is not shrinking but shifting roles and functions. Thus, thanks also to information technologies, trans-national information network structures become new protagonists on the global scene.

Yet this decentralization of authority produces a "sum of crazy-quilt patterns among unalike, dispersed, overlapping and contradictory collectivities seeking to advance their goals" (Rosenau, 2002, p. 285). Differently expressed, "we are still in a period of creative disorder concerning governance" (Kooiman, 2003, p. 5): it is not clear at all if and in what manner actors' power relations are actually being transformed.

J.R. Singh's reflection (2002) on the changing scope of power and governance in information networks adds another useful piece to our puzzle. In conceiving global politics as essentially relational, Singh identifies three types of power and discusses the implication of technological developments for each. At the level of traditional

instrumental power—the ability to influence outcomes—technologies enhance the capacity of traditional actors, but they also empower new actors. At the level of structural power—the ability to affect rules and institutions—technologies, as the Internet governance case suggests, contributes to shaping structures (and institutions) while being also shaped by them. But there is another level where power relations should be analysed, one that Singh calls "meta-power", referring to how ICTs enable formerly underprivileged groups to play a role in global politics. In this view, interactions in global networks appear as constitutive elements of actors and issues; while the very nature of power changes, due to the fact that "the collective meaning that actors hold about themselves, or meanings imposed upon them, are shaped by networks and in turn influence networks" (*ibidem*, p. 15).

Networks are a very powerful image for portraying growing complexity in contemporary societies where policy outcomes can be seen as the result of interactions among a plurality of agents "generated within multiple-actor-sets in which individual actors are interrelated in a more or less systematic way" (Kenis & Schneider, 1991, p. 32). As the networking logic "modifies operations and outcomes in processes of production, power and culture" (Castells, 1996, p. 469), the landscape of politics, including that of world affairs, is being transformed.

Distributed intelligence and competence represent complicate patterns of enmeshment between state and society (Laumann & Knoke, 1987; Knoke et al., 1996), and we cannot avoid considering that this mixing up has consequences over power distribution in political processes: in possessing and exchanging information and resources, like public support, and in acting within and through networks, global actors may actually reshape global power structures. Yet, analyses are seldom capable of empirically investigating the functioning of such networks and to assess if diffused and participatory arrangements that are taking place on the global scene show any capacity to transform global power relations. Thus, there is a strong call for empirical investigation of network structures beyond the prevailing metaphorical use of the term (Katz & Anheier, 2006).

INTERNET GOVERNANCE: CASE STUDY AND METHODOLOGICAL APPROACH

According to Singh "Information networks are facilitating a new social *epistéme* that not only changes the definition of issues in question … but also allows for new actors to start playing key roles in global politics" (2002, p. 15): information technology has the potential to bring about new intellectual configurations that are also new sources of power. Differently stated "International governance of IT (through global information networks) may epitomize the new forms of governance arising in global politics" (Singh, 2002, p. 18). We believe the Internet Governance case offers an interesting opportunity to observe if and how the abovementioned trends are actually shaping global information politics.

The issue of Internet Governance (IG) is as recent as the development of the Internet itself and has just recently gained importance on a global scale[2]. After being for a decade a matter for technicians and Internet developers with no appeal to the broad public, Internet Governance came to be recognized as a strategic and potentially controversial issue at the end of the 1990s, at the time of the institution of the Internet Corporation for Assigned Name and Numbers (ICANN), the organization which manages key resources, such as Top Level Domains. More recently, on the occasion of the UN promoted World Summit on the Information Society (WSIS), IG was addressed within an ad hoc Caucus[3], to quickly become a hot topic in the overall process. During the first phase of the WSIS (Geneva 2002-2003), when it appeared clear that no "common vision of the information society" could be elaborated

without taking into consideration the challenges posed by the management of Internet resources, actors could not come to an agreement on what governance of the Net could possibly mean. A Working Group on Internet Governance (WGIG) was then officially established[4] bringing together actors of different nature – governments, private entities and civil society organizations - to provide a working definition of IG. The WGIG final Report defined IG as "the development and application by governments, the private sector and civil society, in their respective roles, of shared principles, norms, rules, decision-making procedures, and programmes that shape the evolution and use of the Internet" (WGIG, 2005). Such definition served as a starting point for the official negotiations during the concluding phase of the WSIS (Tunis 2005) when the decision was adopted, by world governments, of setting up a specific process focused exclusively on IG. Greece offered to host, in 2006, the first meeting of the Internet Governance Forum[5] (IGF), an open space where individuals and organizations concerned with IG-related issues could bring their contributions and share ideas. The IGF received an official mandate from the WSIS[6] and is to be understood not as a decision-making structure, but as "a new forum for a multi-stakeholder dialogue on Internet Governance" (UN Department of Public Information, 2006).

During the first IGF meeting in Athens (November 2006) and on the way to the second meeting in Rio (November 2007) several *Dynamic Coalitions* (DCs) were launched. Such coalitions are informal groups that reflect the multi-stakeholder approach[7] initiated within the WSIS process[8], formalized through the WGIG experience and the IGF mandate and then articulated throughout the Forum structure. Dynamic Coalitions gather actors from governments, private sector and civil society organizations and are aimed at shaping common discourses on specific issues within the overall IG framework[9]. DCs have no explicitly defined membership criteria nor procedural guidelines: the only requirement being their multi-stakeholder composition. Besides this, DCs bring together individuals and organizations from all over the world showing how, a part from the physical IGF meetings, most of the interactions taking place among globally dispersed actors is inevitably carried on through Internet devices, especially e-mail and electronic platforms.

Dynamic Coalitions, despite their informal organizational patterns, have gained official recognition by the UN Secretariat for Internet Governance as shown by the formally devoted time slots during the IGF meeting in Rio in 2007. Moreover they are effectively contributing to a more articulated understanding of IG-related issues as well as to the enlargement of the IGF agenda. They can certainly be conceived as "trans-national information networks", in Singh's terms, since they are publicly recognized multi-actor structures operating trans-nationally and focusing on a variety of issues that pertain to the development, diffusion and usage of communication technologies. Furthermore, they provide evidence of the fundamental role played by ICTs in fostering sustainable trans-national interactions.

In identifying DCs as concrete networks, we acknowledge the interplay between off-line and on-line interactions; but we also recognize that the off-line IG debate takes place mainly in a context – that of official IGF meetings - that is hardly accessible to those who do not have the time, knowledge or financial resources to travel and attend meetings in Geneva, Tunis, Athens or Rio de Janeiro. At the same time, because of the very nature of Internet resources, we assume virtuality as an inner component of Internet functioning as well as governance. Furthermore, it is often suggested that constraints to participation of different actors, operating in different geographical and cultural contexts, may be less dramatic in the online world: once appropriate platforms are set up, technical requirements are met and basic skills are provided, potentially everyone is able to contribute to a plural online conversation.

Finally online interactions also contribute to the creation of a discursive space where issues are framed and actors's interests shaped (Padovani & Pavan, 2007).

For all these reasons in this chapter we concentrate our attention on two aspects directly related to on-line interaction: *we look at the thematic networks deployed by actors in the on-line space to assess diversity, dynamics and complexity in IG debates and we look at the modus communicandi adopted by those actors through their use of the Web in order to clarify if and how the virtual space is actually perceived and inhabited as a dynamic relational space that offers true potential for change.*

As far as the first aspect is concerned, we make use of digital harvesting software, and adopt the language and conceptual tools of the network approach, in order to trace and visualize thematic networks that develop, through linkages and virtual ties in the Web-sphere, among actors engaged in the IGF process[10]. What issue networks on the Web show is a thematic space of discourse that takes shape in the Web-sphere and parallels off-line debates where the management of Internet strategic resources is being discussed and defined. The nature of the nodes in the on-line networks we observe (Figure 1) might vary: they can be Websites of organizations, online documents, databases, single Web pages such as news or wikis or blogs. Ties among nodes represent links between actors in the online space, in other words signs of recognition amongst actors within a discursive space. It is the construction, destruction or the removal of links in the Web-sphere that renders the on-line discursive interaction around the IG issue.

As far as the *modus communicandi* is concerned (paragraph V), we look at how organizational actors involved in IG debates position themselves in the Web-sphere through their adoption and usage of technical functionalities, if and how they invest in interactive modalities, if and how they do this consistently with their innovative view of technology and the guiding principles they publicly promote. Summing up, we investigate if they translate their awareness of the dynamic potential offered by ICTs into an intentional strengthening of networking relations aimed at fostering new configurations of power.

INTERNET GOVERNANCE: INNOVATING WORLD POLITICS THROUGH A MULTI-STAKEHOLDER APPROACH?

Multi-stakeholderism can be defined as "Processes which aim to bring together all major stakeholders in a new form of communication and decision-finding (and possibly decision-making) structure on a particular issue; are based on the recognition of the importance of achieving equity and accountability in communication between stakeholders; involve equitable representation of three or more stakeholder groups and their views; are based on democratic principles of transparency and participation; aim to develop partnerships and strengthen networks between and among stakeholders" (Hemmati, 2002, p. 19).

The multi-stakeholder concept has recently entered international debates and has been formalised in policy documents; it has almost become a *passé-partout*, widely adopted in political discourses, often with the implicit assumption that a consensus exists on how participatory political processes should be organized and managed. In fact, as we have outlined elsewhere, the multi-stakeholder concept is a highly contested notion: "Different actors hold very different perspectives as to how stakeholders should be conceived, who is to be included and who is excluded and how their interaction should lead to information exchange, deliberation or decision" (Cammaerts & Padovani, 2006). It is growingly evident that stakeholders' participation risks becoming a rhetoric exercise aimed at neutralising criticism through the adoption of an unproblematic consensual understand-

ing of political life. Moreover it is crucial to take into consideration the objective constraints and necessary preconditions to full and effective participation, such as financial and knowledge resources, or the available power base on which actors define their positions in governance processes (Padovani, 2005a). We believe that a fruitful way to better articulate the multi-stakeholder notion, is to relate multi-stakeholderism to the very concept of diversity, to be conceived as a matter of actors involved, issues addressed, knowledge produced and, in the end, power relations.

The analysis presented in the following paragraphs are grounded in our interpretation of the thematic networks elaborated, one of which is visualized in Figure 1.

Actors Diversity in Internet Governance Debates

First we look at what kind of actors are involved in the Web-based "conversation" about IG. Is there a meaningful diversity among them, so that we can actually speak of a multi-stakeholder conversation? Which actors occupy central positions in IG thematic networks and what kind of power relations can be inferred? An initial answer to these questions can be given by looking at the typology of nodes in the networks, identified through their domain extension. Our maps show that ".org" nodes are the prevailing type of actors animating the conversation. This is, according to our investigation, a feature of the IG debate that has not changed over time: there are other kinds of actors as well (identified as .edu, .int, .info, .com or local domains) but in a very small proportion if compared to the .orgs. Nevertheless, this does not imply homogeneity in the field, since the .org extension can refer to a variety of different subjects.

If we take a closer look at these organizations, we find at least three different types of .org actors engaged in the conversation. On the one side, composing a well connected cluster on

Figure 1. Internet Governance thematic network elaborated through Issue Crawler on March3, 2007 (starting points: Dynamic Coalitions launched around the first IGF meeting; Iteration 2; Crawl depth 2; Analysis Mode: by page).

the left side of our figure, we find organizations that have traditionally dealt with (and de facto managed) the governance of the Internet in the past decades: ICANN, IANA, IETF and the likes. They represent the "traditional" nongovernmental-mainly-technical-approach to IG: one that stemming from the structural evolution of the Net and related infrastructures has developed "naturally" overtime, with its own logic, receiving little attention from the side of other actors, including governments (a part from the obvious involvement of the United States into the ICANN). Then we have, though not really articulated into visible clusters of interactions, organizations such as the ITU, WIPO, the UN and UNESCO: international organizations that are supposed to orientate the management of global resources, and here represent the traditional logic of intergovernmental decision-making. Finally in the .org category we find organizations such as IP Justice or Computer Professionals for Social Responsibility (CPSR), that are expressions of civic engagement in the IG discussion: the so-called "civil society".

Alongside the above mentioned technical management cluster, we find an identifiable cluster in the centre-to-right side of the figure composed of nodes that relate to the Internet Governance Forum (IGF) conceived as a process[11]. Interestingly the IGF has acted as a catalyst for many organizations, favouring an increase in actors' diversity: our analyses have shown that the "pre-Athens virtual space" (2006) was mainly inhabited by technical and institutional actors[12], while the "post-Athens" environment, and even more so the 2007 Rio edition of the Forum, were much more diversified. In particular, organizations from the so-called civil society have grown in visibility and connectedness, while a number of nodes that relate to the "dynamic coalitions" – such as a2k-igf.org or Internet-bill-of-rights.org – appear as the thematic "homes" of specific issues that have come to compose IG as a policy field. New nodes

and new links in recent network visualizations give a sense of the dynamism in the on-going process. Indeed, maps realized after the Rio edition of the IGF confirm that the relevance of old nodes playing a central role in the discussion over Internet Governance, such as the ICANN, is challenged by new protagonists: realities such as the Electronic Frontier Foundation, Privacy International or the Association of Progressive Communication not only have entered the discussion but stand at its very core[13].

In terms of relevance in the network, and therefore potential influence in the governance of the Internet, we suggest that interconnected clusters, and central nodes within them, probably play a more relevant role in the debate than single non-strongly-connected nodes and actors. Connected clusters entail a coherent language, a shared logic, history and vision, and therefore a likely stronger capacity to impact the trans-national conversation; at the same time they do not seem to have a broad understanding of the overall network and tend to be self-referential.

This potentially more powerful position is partly counterbalanced by the presence of other actors involved in the debate, though peripheral in our maps, such as Internetpolicy.net. The higher the distance from core actors in the network, the less likely these nodes' relevance in the conversation. Nevertheless, some of these actors can be considered representative of alternative interests, different values and, possibly, emerging issues. Moreover, some of them play bridging roles in the network, fostering connections among otherwise dis-connected clusters and nodes and therefore contributing to the consolidation of overall thematic networks.

New questions emerge from our investigation into diversity of actors: what is the role of academia in this virtual conversation[14]? How should we explain the absence of the private sector from a debate that clearly touches strategic interests of business oriented entities[15]?

Issue Diversity in Internet Governance

Moving to the content of debates, we looked at prevailing issues in the IG debate. Can thematic clusters be traced in the Web? Who promotes which issues as priorities? What kind of power relations can be inferred from the articulation of substantial elements? We believe on-line conversations contribute to the definition and framing of issues, and to the structuring of relevant language. Thus, though thematic networks on the Web do not immediately reveal hierarchies in the status of issues, some observations can be made on the basis of a general understanding of the IGF process.

The fact that traditional actors (technical and institutional) and newer actors co-exist in the virtual space, especially after the first IGF (Athens 2006), and (at least partly) recognize each other as legitimate parties in the debate, suggests that issue diversity has in fact grown over time. The presence in the thematic networks of the Dynamic Coalitions and of civic organizations such as Consumer Project on Technology (cptech.org) or Reporters sans Frontieres (rsf.org), alongside those organizations who have been historically engaged with the governance of the Internet, parallels the widening range of themes included in the umbrella concept of IG. From a prevailing focus on technical matters the discussion has gradually opened up to issues concerning human rights promotion and defence (foeonline.worldpress.org), universal access to knowledge and resources (a2k-igf.org), free software and knowledge sharing, multi-stakeholderism as a basic principle for cooperation (igf2006.info).

Finally, if we assume that central positions in the network reflect, to some extent, a more powerful status in the field, the relevance of traditional actors indicates the prevalence of issues traditionally connected with an infrastructural view of IG, such as the management of critical resources, security problems and technical standards. At the same time, the necessity of articulating positions on controversial matters, such as the promotion of freedom of expression or the defence of privacy and security rights, has led newer actors in the field to privilege networking activities among themselves in order to collectively develop shared positions and promote them more effectively in the debate.

Overall what emerges from IG issue networks on the Web is a dynamically growing plurality of themes and positions. Issue priority and actors' capacity to foster specific views, as well as transformations in issue framing due to actors' interaction, should be assessed through in-depth analysis of off-line interactions.

How Global is Internet Governance?

Given the strategic relevance of Internet resources for societal developments worldwide and, even more important, given the attention posed in the IG discourse on ICTs' potential for reducing economic and knowledge divides globally, a final key question concerns actors' capacity to (re)present and express their differences in the debate, from a geographical, linguistic and cultural point of view. If we are to fully appreciate the opportunities and constrains that are linked to multi-stakeholder practices, diversity in the conversation should be assessed not only looking at trans-national and supra-national actors, but also at subjects coming from different cultural contexts as well as local constituencies, who should be given the opportunity to express specific views and needs. From this point of view, can we actually talk about a globally rich and diverse conversation?

Our analysis shows that rarely local or national domains enter Web-based issue networks[16]. Sometimes a local initiative is included: in the map shown above, for example, we find the national Belgian ISOC chapter or a specific initiative organized in Australia. But in general we see very little national, not to speak about local, interventions in the online debate. Slightly more visible is the regional level, which is brought into the conversation through the presence of regional registrars, connected to the technical cluster of traditional managers of Internet resources. When more politically oriented actors from different regions seem to contribute to the debate, this happens through a very institutional channel: it is through UNECA or UN Habitat that African needs find their way into the discussion.

We finally underline as problematic the fact that almost the totality of actors included in our thematic networks as contributing to the definition of how Internet resources should be managed in the future come from North-Western areas of the world. Furthermore, beside the overly unbalanced geographical representation, the dominance of English as the language through which issues are framed, definitions are given and relevant knowledge is produced is another problematic fact: it is quite evident that this global unbalance does not contribute to a rich and articulated understanding of worldwide realities, needs and expectations.

INTERNET GOVERNANCE: INNOVATING WORLD POLITICS THROUGH COMMUNICATION?

International politics has traditionally been characterized by secrecy of information and limited accessibility to communicative structures such as intergovernmental organizations; as well as by exclusive diplomatic rituals, behind-closed-doors decision-making processes, hierarchies among actors and centralization of power resources (Hockings, 2006). New technologies, among other factors, are challenging this situation through a number of inherently built potentialities: easier and more affordable access to information leads to higher expectations in terms of transparency, for instance through electronic forums and consultations. This, in turns, translates into broader opportunities for participation, and raises demands

for more open and democratic decision-making processes.

These transformations suggest that a more explicit scholarly attention to processes of communication in the supra-national/trans-national space is timely: as technologies are transforming political communication within state boundaries (Blumler & Coleman, 2001), we believe it is no longer possible to keep under-estimating the reality and relevance of political communication beyond "the national".

In looking at contemporary political processes, characterized by diversity, dynamics and complexity, it is necessary to adopt a comprehensive approach to communication modalities, one which is able to combine and properly weight the different modes through which communication creates and nurtures world politics: languages and frames, off-line and on-line interactions, innovation in processes through societal learning.

(International) Political Communication Revised

In order to set the bases for a conceptual framework that focuses on (global) political communication through the nexus between technology and politics, an interesting contribution is offered by Bijker (2006). He underlines how technology and politics matter one another especially when information technology is at stake. Technology depends on politics in terms of resources allocation and its promotion as part of political strategies; information technology has also become a highly politicized domain among decision-makers[17], though it remains marginal in public discourses. At the same time, politics depend on technology by "shaping the means of political debate: the arena, the communication links, the agenda" (Bijker, 2006, p. 696). Contemporary information technology can be seen as a communication channel, a political message and a powerful symbol, all at once: it provides an infrastructure for political discussion that offers new means to address complexity through diverse and dynamic interactions. Thus we understand technology as a multi-layered concept that combines together *artefacts* (physical objects such as computers and infrastructures), *human activities* (designing, making and handling machines and networks) and *knowledge* (what people know about and do with technologies) (*ibidem*, p. 682). The relationship between technology and politics becomes an articulated one: to the extent that technology is not just about artefacts but also about their societal potential and deployment, it matters politically, it reframes political spaces and it serves political interests.

This articulation resonates with Singh's tripartite idea of power relations—where instrumental, structural and "meta" power is at play—and can fruitfully be applied towards an innovative conceptualization of political communication in world affairs. Technology as artefacts is a resource for instrumental power in global communicative settings since, as Rosenau recalls, it "sets the range within which ends and means are framed, alternatives pondered and choices made" (2002, p. 275): technological devices, considered as tools, can address challenges of legitimacy and effectiveness in policy-making and foster innovation through transparency and participation in supra-national political processes.

Technology as human activities defines the rules and creates the conditions for shaping institutional structural power to takes shape and for global information governance networks to develop: the deployment and actual use of such devices show how far the innovative potential is translated into practices of less hierarchical and more horizontal interaction.

What people know and do, i.e. the informational logic that runs through technology and evolves into actors-in-interaction, becomes a cognitive resource—or "meta-power"—for innovation in political processes. Knowledge can be conceived as the combination of the competence and understanding needed to translate the innova-

tive potential into reality, as well as the result of a self-reflective activity from the side of actors.

Transparency in information sharing that generates greater public awareness, horizontal interaction that helps overcoming traditional hierarchies and the *nexus between off-line and on-line relations that produce new knowledge* can therefore be seen as three spheres of potential innovation in the conduct of world politics as well as three dimensions through which actors' communication can be analyzed.

Political Communication Revised in the IGF Context

Labelling such a comprehensive view of communication as actors' *modus communicandi*, in analysing the on-line dimension of Internet Governance debates we had the following questions in mind: how should we investigate and assess organizational actors' communication modes? Do these actors fully exploit the potentialities brought by information technologies in terms of access to information, interaction and collective production of content? To what extent actors' typology contributes to the explanation of existing differences in technology use? Do actors involved in IG debates actually plan their communicative spaces with the aim of strengthening the networks they are engaged within? Finally, what features characterize the interplay between on-line and off-line interactions?

For each dimension of actors' communication - (the way of conceiving) transparency the possibility of horizontal exchanges and the interplay between off-line and on-line interactions – our Website-focused analysis of actors engaged in IG allows to describe emerging trends and identify critical aspects.

The first dimension we explored is transparency and, more precisely, the different ways in which actors understand this concept and push it into political processes. Indeed, if it is true that information technologies favour greater transparency in political mechanisms, it should not be taken for granted that all actors understand and value transparency in the same way, nor that it constitutes a priority for all of them. Our analysis, in fact, reveals that in most cases actors belonging to the so-called "civil society" make a plea for enhanced transparency in political processes, differently from governmental actors who tend to stress the idea of control and never include transparency explicitly in their discourses.

This distinction between governmental and non-governmental actors is a general one, since the actual situation being more diversified. As far as governmental attitudes towards the transparency issue, we should not forget how important access to information and transparency in policy processes have become in recent years, especially within strategies and programmes for e-government and e-democracy (Padovani et al., 2007): institutional actors, at different levels, have transformed their public discourses to include transparency as a central element in the effort to redefine and improve their relations to the citizen, in search for consensus and legitimacy. In spite of these developments, the commitment to more open and accessible modes of political conduct do not seem to have expanded to supra-national spheres of information governance.

Even among non-governmental actors different emphasis in fostering transparency can be found: private sector entities do not show any interest for the issue and appear quite similar to governments in their understanding of actors' interaction; but even among public interest groups, transparency does not evenly appear as a priority. This confirms what we have underlined above (and, more extensively, in Padovani & Tuzzi, 2006): the galaxy of the so-called civil society is complex and diversified and should not be conceived as a single interlocutor.

Secondly, we analyzed horizontality in communication exchanges characterizing the Dynamic Coalitions on IG. These structures were created with the aim of fostering dialogue among

actors of different nature—governmental and non-governmental—in specific subfields of the IG domain. Given the trans-national space in which DCs operate, the role played by ICTs has been determinant in favouring information exchange as well as in creating new ties among previously disconnected actors. This can be inferred by the expansion in actors' mutual recognition as well as in the progressive consolidation of DCs in the Web-sphere. Nevertheless, our analyses also show that only in few cases organizations that are members to one or more IGF Dynamic Coalitions link to and make explicit mention of such horizontal structures (and even their involvement in the IGF process) on their Websites: there seems to be a gap between the active involvement of organizations in the off-line Internet Governance debate and the relevance those same actors publicly attribute to this involvement in their online everyday life; as if the innovative horizontal mode of interaction that has developed off-line has not yet found an equivalent in the Web-space.

This finding rises the question of whether and to what extent DCs can effectively be understood as horizontal networks of exchange and cooperation or if they rather constitute ad-hoc spaces to catalyze efforts under specific circumstances, namely during official IGF meetings. This ambiguity between the potential and the actual realization of horizontal communication patterns questions the rhetoric according to which the multi-stakeholder loose arrangements experienced in the IGF context could be transferred to other policy fields. After all, it is actors' sincere conviction (and investment) in multi-stakeholder processes that makes a difference in actualizing innovation.

The potential offered by technology to foster horizontality could be played out more explicitly in support to off-line dynamics; yet it appears clearly that it is not only diversity in the nature of actors that hampers the full realization of horizontal exchanges, but also the fact that actors are characterized by different velocities in innovating existing practices. Actors engaged in

IG information networks clearly proceed at different speeds for what concerns the creation and use of new languages and new technologies as tools to support their engagement in world politics. This reflects different levels of competence in approaching technologies, but it also outlines different motivations for innovating through technologies between those actors who seek to maintain traditional quotas of power and those new actors who see technological innovation as a source of "new power". To the extent that different actors walk at different speeds, horizontality of exchanges, which could be enhanced and strengthened through the use of "artefacts" and sustained by ad hoc "human activity", remains more a vision than a reality. The necessary knowledge to innovate is slowly been produced, but much reflection is needed to turn experiments into "protocols".

The last element we have analysed, linked to the previous one, is the coherence that actors show in coordinating online and offline initiatives. More particularly, we have examined to what extent actors validate and strengthen their online action through the offline and vice versa. Results show a low level of coordination and consistency among the two worlds, which do not appear as complementary but rather as disconnected: in very few cases initiatives started in the off-line world are complemented or supported by on-line action. Generally speaking, initiatives are seldom supported by remote participation mechanisms, nor online petitions are publicized offline.

This gap appears even more problematic for the fact that traditional mass means of communication, such as television or radio, diffused among the broad public, are never taken into consideration as yet another "artefact" to be used to broaden the debate. One problematic consequence of this lack of integration in media usage, is that the vast majority of people, who cannot access official meetings nor the Internet or are e-illiterate, will hardly become aware of the stakes connected to (IG) global discussions. It is therefore very unlikely that they become part of any global conversation and, as a consequence, the kind of soft power that informed general publics may potentially exert, amplifying the horizontality of exchange beyond few existing avant-gardes, is not likely to be produced.

CONCLUDING REMARKS. INNOVATING WORLD POLITICS THROUGH KNOWLEDGE?

We started our journey by indicating diversity, dynamics and complexity not just as characterizing dimensions of contemporary globality, but also as defining features of governing arrangement that aim at addressing and handling such globality. We have proposed an understanding of related political challenges as brought about and enhanced by the interplay between information technology and the conduct of world affairs. In conceiving the very possibility of innovation in world politics as the capacity to adapt and respond to such challenges, we focused on the case of contemporary Internet Governance debates to investigate how the three dimensions of diversity, dynamics and complexity are played out in reality and how processual as well as technological innovation may affect actors' orientations, as well as the structures within which they operate and interact. We have adopted an approach that combines a network vision of societal transformations with the growing relevance of on-line interactions among actors, focusing our analysis on Web-based issue networks around Internet Governance and the on-line communication modes adopted by the actors involved. We can now conclude by outlining some evidence from the analysis and issues that are left open.

As far as innovating world politics through the multi-stakeholder approach, our analysis of on-line thematic networks, and their transformation over time, shows that we can in fact talk about a shift from a centrality of state actors in supra-national settings to growing actors' diversity. Different agents are entering the space of global debates, bringing different visions and interests and different ways of conceiving such space. Institutional settings, such as the IGF, position themselves as opportunities for convergence and facilitators of a plural debate more than as decision-making spaces. What remains to be assessed is the actual impact of different actors: if and how new actors, occupying growingly central and brokerage positions, will in fact be able to influence the discursive process over time and what kind of power do they actually exert.

Internet Governance debates show changes in actors' centrality in on-line networks over time, particularly favouring non-institutional entities; a result which can be interpreted as an incremental recognition of actors' plurality, and their legitimacy to intervene, which can partly be explained through the "necessary knowledge" they bring into the process. It should be outlined that conversations taking place in the Web-sphere do not directly affect the actual decision-making; nevertheless they may contribute to a better articulation and understandings of the issues involved, thus influencing future political decisions on the basis of that understanding. This is visible in the constitution of the IGF Dynamic Coalitions, which can be considered trans-national information networks. They play a relevant role in framing issues, potentially orientating the global agenda.

The most controversial aspect remains one of inclusion and exclusion, directly related to complexity. The Global South, and in particular its localities, with their languages and cultural ways of expressing different concerns and needs, have not yet found adequate space in the on-going conversations in the Web-sphere. Moreover, the multi-level characterizing feature of contemporary governance arrangements does not appear clearly in IG debates, nor we find a clear indication of interconnection between IG-related issues and other global issues, such as sustainability in

development or the broader landscape of media-related policies. Overall, diversity and dynamics are evident features of contemporary global information networks; the same cannot be said for complexity, which does not seem to be fully addressed, certainly not in the governance of the Internet as it has been traced in cyberspace.

As for innovation in world politics through communication, we suggested that a comprehensive understanding of political communication could be grounded on a tri-dimensional vision of technology, complemented by an explicit focus on power relations. Our starting point was the idea that technology as artefact can promote transparency in policy processes, thus overcoming traditional diplomatic secrecy and enlarging the basis for the exercise of instrumental power; technology as human activity can favour horizontal exchanges thus challenging traditional hierarchies and fostering change in structural power; technology as knowledge would translate this potential into reality by constituting cognitive spaces inclined towards diffused forms of authority.

Our analysis of how actors involved in Internet Governance conceive and make use of technologies does not allow a very optimistic conclusion in terms of world politics innovation through communication, at least not for the time being.

In spite of the expectation that actors engaged in Internet Governance would constitute a creative avant-guard, due to their better understanding of the potential offered by technologies for transparent and participatory processes, evidence from actors' on-line communication modes show that transparency is an issue that is only taken up and promoted by those actors who perceive the challenge of instrumental power as vital: those civil society organizations who have few other resources than knowledge to play out in the global context but have a thorough understanding of the challenges related to transparency and access to information, if they are to become meaningful interlocutors in policy contexts. Instrumental power, which can certainly be supported by technological developments, needs intentionality and attitude towards innovation to become a fully appreciated feature of "new global politics".

Also the shift from hierarchic modes of interaction to more horizontal practices, though put in practice in the off-line context of the IGF through the Dynamic Coalition arrangement, does not seem to have become a model, nor a strategic objective, for most actors in their on-line operations. Only few of them show an intentionality in their use of on-line resources to actually give visibility and strengthen the horizontal networks developed within such Coalitions. This appears problematic as far as transformations in structural power, if we consider that actors' capacity to influence rules and institutions - including making innovative practices perceived as viable precedents to transform world politics - is closely linked to their ability to communicate such experiences outside the restricted networks. This implies making effective use of technology inwards (towards the specific process they are engaged in) as well as outwards, towards broad publics.

This leads to our final point: the whole issue of strategic usages of communication, by taking advantage of different platforms and languages, channels and symbolic forms, shows that the translation of the communicative potential into realities of innovation is yet to be grasped. Recalling Singh (2002), network interactions can be constitutive elements of actors and issues, while the very nature of power can change, due to the fact that "the collective meanings actors hold about themselves are shaped by networks". This constructivist understanding of actors-in-networks would require a consistency between off-line and on-line interactions that is not a given: the direct experience of participation in policy contexts by underprivileged groups, and the effort of developing assessment mechanisms, can be consolidated through sharing such experiences in public. Open strategies of self-reflection, supported by a strategic use of the Web, could generate new understanding and knowledge concerning participation in world politics, thus leading to new intellectual configurations and new sources of power. These challenges do not seem to be fully addressed yet and therefore remain as open issues, for practitioners and scholars.

REFERENCES

Appadurai, A. (1990). Disjuncture and difference in the global cultural economy. *Theory, Culture and Society*, (7), 295-310.

Bijker, W. E. (2006). Why and how technology matters. In R. E. Goodin & C. Tilly (Eds.), *The Oxford handbook of contextual political analysis* (pp. 681-706). Oxford: Oxford University Press.

Blumler, J. G., & Coleman, S. (2001). Realising democracy online: A civic commons in cyberspace. *IPPR Citizens Online Research Publication*, (2), March 2001. Retrieved from http://www.citizensonline.org.uk/site/media/documents/925_Realising%20Democracy%20Online.pdf. Last access 31st May 2008.

Braman, S. (2006). *Change of state: Information, policy, and power*. Cambridge: MIT Press.

Cammaerts, B. (2008). Civil society participation in multi-stakeholder processes: In between realism and utopia. In D. Kidd, C. Ropdriguez, & L. Stein (Eds.), *Making our media: Mapping global initiatives toward a democratic public sphere*. Cresskill: Hampton Press.

Cammaerts, B., & Padovani, C. (2006, July). *Theoretical reflections on multi-stakeholderism in global policy processes: the WSIS as a learning space*. Paper presented at IAMCR conference, Cairo, Egypt.

Castells, M. (1996). *The rise of network society*. Oxford: Blackwell Publishers.

Deutsch, K. W. (1963). *The nerves of government*. New York: Free Press.

Held, D. (2004). *Global covenant: The social alternative to the Washington consensus.* Cambridge: Polity Press.

Held, D., & McGrew, A. (2003). *The global transformations reader.* Cambridge, UK: Polity Press.

Held, D., McGrew, A., Goldblatt, D., & Perraton, J. (1999). *Global transformations: Politics, economics, and culture.* Cambridge, UK: Polity Press.

Hemmati, M. (2002). *Multi-stakeholder processes for governance and sustainablility: Beyond deadlock and conflict.* London: Earthscan

Hockings, B. (2006). Multistakeholder diplomacy: Forms, functions and frustrations. In J. Kurbaljia & E. Katrandjiev (Eds.), *Multistakeholder diplomacy: Challenges and opportunities* (pp. 13-32). Diplo Foundation.

Hoffmann, J. (2006). *Internet governance: A regulative idea in flux.* Retrieved from http://duplox.wz-berlin.de/people/jeanette/texte/Internet%20Governance%20english%20version.pdf. Last access March 1, 2008

Kamarc, E. C., & Nye, J. S. (Eds.) (2002). *Governance.com: Democracy in the information age.* Washington DC: Brookings Institution Press.

Katz, H., & Anheier, H. (2006). Global connectedness: The structure of transnational NGO networks. In M. Glasius, M. Kaldor, & H. Anheier (Eds.), *Global civil society 2005/6* (pp. 240-65). London: Sage.

Keck, M. E., &. Sikkink, K. (1998). *Activists beyond borders: advocacy networks in international politics.* New York: Cornell University Press.

Kenis, P., & Schneider, V. (1991), Policy network and policy analysis:Scrutinizing a new analytical toolbox . In B. Marin & R. Mayntz (Eds.), *Policy networks: Empirical evidence and theoretical considerations* (pp. 25-62). Boulder, CO: Westview Press

Kleinwächter, W. (2004). Beyond ICANN vs. ITU? How WSIS tries to enter the new territory of Internet governance. *Gazette, 66* (3-4), 233-251

Kleinwächter, W. (2007). The history of Internet governance. In C. Möeller & A. Amoroux (Eds.), *Governing the Internet* (pp.41-66). OSCE Representative on Freedom of Media.

Knoke, D., Pappi, F. U., Broadbent, J., & Tsujinaka, Y. (1996). *Comparing policy networks.* UK: Cambridge University Press.

Kooiman, J. (2003). *Governing as governance.* London: Sage Publications.

Kratochwil, F. (1989). *Rules, Norms and Decisions: on the Conditions of Practical and legal Reasoning in International Relations and Domestic Affaires.* Cambridge, UK: University Press.

Laumann, E. O., & Knoke, D. (1987). *The organizational state.* Madison: University Winsconsin Press.

Mueller, M., Brenden, K., & Pagè, C. (2004). *Reinventing media activism: Public interest activism in the making of U.S. communication-information policy 1960-2002.* Retrieved from http://dcc.syr.edu/ford/rma/reinventing.pdf. Last access March 1, 2008.

Nye, J. S., & Owens, W. A. (1996). America's information edge. In *Foreign Affairs, 75*(2), 20-36.

Padovani, C. (2005a). WSIS and multi-stakeholdersim. In D. Stauffacher & W. Kleinwachter (Eds.), *The World Summit on the Information Society: Moving fro the Past into the Future* (pp. 147-155). UN ICT Task Force.

Padovani, C. (2005b). Civil society organizations beyond WSIS: Roles and potential of a "young" stakeholder. In O. Drossou & H. Jensen (Eds.), *Visions in process II: The World Summit on the Information Society. Geneva2003-Tunis 2005* (pp. 37-45). Berlin: Henrich Boell Foundation.

Padovani, C., & Pavan, E. (2007). Diversity reconsidered in a global multi-stakeholder environment: insights from the online world. In W. Kleinwächer (Ed.), *The power of ideas: Internet governance in a global multi-stakeholder environment* (pp. 99-109). Berlin: Germany Land of Ideas.

Padovani, C., & Tuzzi, A. (2004). WSIS as a world of words: Building a common vision of the information society?. In *Continuum. Journal of Media and Society, 18*(3), 360-379.

Padovani, C., & Tuzzi, A. (2006). Communication governance and the role civil society. Reflections on participation and the changing scope of political action. In J. Servaes & N. Carpentier (Eds.), *Towards a sustainable information society beyond WSIS* (pp. 51-79). Bristol & Portland: Intellect.

Padovani, C., Tuzzi, A., & Nesti, G. (2007). Communication and (e)democracy: Assessing European e-democracy discourses. In B. Cammaerts & N. Carpentier (Eds.), *Reclaiming the media: Communication rights and democratic media roles* (pp. 9-30). Bristol & Portland, UK: Intellect.

Onuf, N. (1989). *World of our making: Rules and rule in social theory and international relations.* Columbia: University of South Carolina Press.

Raboy, M., & Landry, N. (2004). *La communication au coeur de la gouvernance globale.* Enjeux et perspectives de la société civile au Sommet mondial sur la société de l'information. Montréal: Département de Communication, Université de Montréal. Retrieved from http://www.lrpc.umontreal.ca/smsirapport.pdf. Last access May 31st, 2008.

Reinecke, W. H., & Deng, F. (2000). *Critical Choices. The United Nations, networks and the future of global governance.* Ottawa: International Development Research Centre.

Rosenau, J. N. (1995). Governance and democracy in a globalizing wold. In D. Archbugi, D. Held, & M. Mohler, (Eds.), *Re-imagining political community: studies in cosmopolitan democracy* (pp. 28-57). Cambridge, UK: Polity Press.

Rosenau, J. N. (2002). Information technologies and the skills, networks, and structures that sustain world affairs. In J. N. Rosenau & J. P. Singh (Eds.), *Information Technologies and Global Politics* (pp. 275-88). Albany: State University of New York Press.

Singh, J. P. (2002). Introduction: information technologies and the changing scope of global power and governance. In J. N. Rosenau & J. P. Singh (Eds.), *Information technologies and global politics* (pp. 1-38). Albany: State University of New York Press.

Smith, S. (2001). Reflectivist and constructivist approaches to international relations. In J. Baylis and S. Smith (eds.), *The globalization of world politics. An introduction for international relations* (pp. 224-51). UK: Cambridge University Press.

United Nation Department of Public Information, News and Media Division (2006). *SG/A/1006. PI1717.* Retrieved from http://www.un.org/News/Press/docs/2006/sga1006.doc.htm. Last access March 1st, 2008.

United Nations General Assembly. *Resolution 56/183.* Retrieved from http://www.un.org/News/Press/docs/2006/sga1006.doc.htm. Last access March 1, 2008.

Wendt, A. (1992) Anarchy is what states make of it: The social construction of power politics. *International Organizations* 46: 391-425

WGIG (2005). *WGIG final report.* Retrieved from http://www.wgig.org/WGIG-Report.html. Last access March 1, 2008.

ENDNOTES

[1] As suggested by a constructivist understanding of processes in world politics, a perspective that can be traced back to the work of Kratochwil (1989), Onuf (1989) and Wendt (1992), and can be situated at the 'intersection between rationalist and reflectivist approaches' to the study of International Relations (Smith, 2001, p. 242).

[2] On the development of Internet Governance as a controversial issue see Kleinwächter, 2004 & 2007; Hoffman, 2006.

[3] Caucus was the term used to indicate thematic working groups in which the Civil Society Sector was organized within the WSIS process.

[4] WSIS *Declaration of Principles* par. 50 and *Plan of Action* par. 13.

[5] In 2007 the second IGF was hosted in Rio de Janeiro (Brazil) followed by Hyderabad (India), in 2008.

[6] WSIS *Tunis Commitment* par. 72.

[7] On multi-stakeholderism within the WSIS process see Cammaerts, 2008; Kleinwachter, 2004; Padovani, 2005a & b; Raboy e Landry, 2004.

[8] Where "governments, international organizations, private entities and civil society organizations are invited to effectively participate in the processes" (UNGA Res 56/183, December 2001).

[9] At the time of writing, there are 13 coalitions focused on: spam, privacy, open standard, access to knowledge, Internet bill of rights, linguistic diversity, access for remote and dispersed communities, online collaboration, freedom of expression online, gender, framework of principles for the Internet, child online safety, accessibility and disability.

[10] We adopted the Issue crawler software developed by Govcom.org and accompanying tools. The issue networks constituted online, and traceable through the crawling software, are the result of linking strategies among URLs, and offer a visual image of the degree of reciprocal recognition among actors operating around specific issues, in the Web sphere. The software operates through co-link analysis, creating maps where visualized nodes/sites are those linked by at least two of the starting points identified for the analysis. The content of the ties between Websites might vary inside the same network, as motivations that justify the linking among Web resources might vary. It should furthermore be noticed that issue networks are not neutral just because are "placed" in a virtual space: they rather integrate our knowledge of debates that take place offline, thus offering a better understanding of how the framing of issues takes place in "spaces of place" as well as in "spaces of flow" (Castells, 1996). For further information see: www.govcom.org and www.issuecrawler. net. Other visualizations can be retrieved from the Issue Crawler Archive.

[11] The official Website – www.intgovforum. org- is central in the map both in terms of its positioning in relation to other nodes and for the number of links it receives from the network.

[12] It should also be recalled that these actors were represented in the map as belonging to separated clusters, the IGF playing a connecting role among them. The map is available in the Issue Crawler archive as IntGovForum2006_1.

[13] These results are drawn by the map GDC_ ON_gennaio08 retrievable at the URL http://issuecrawler.net/index.php?requestact ion=changepage&changepagename=Netwo rkDetailsSummary&network_id=310282 .

[14] Only the Cyberlaw Centres at Stanford and Harvard Universities and the Internet Governance Project from Syracuse University appear in the maps.

15   Search conducted in the Web-sphere at large shows that the private sector has indeed great interests in IG; yet very few actors from the sector seem to be actually engaged in IGF Web-related structures.

16   The most remarkable exception being the Greek Web site - igfgreece2006.gr - clearly due to the fact that the first IGF was hosted in Athens.

17   "Information and communication have emerged as one of the newest and most internationalized areas of public policy and institutional change" (Mueller et al., 2004).

Chapter XI
Who Governs Cyberspace?
Internet Governance and Power Structures in Digital Networks

Mauro Santaniello
University of Salerno, Italy

ABSTRACT

The Internet Governance debate has, for a long time, been influenced by a well-defined characterization of information networks. The depiction of a decentralized network, governed on a consensual basis by distributed forms of authority, has led to focus little attention on the configuration strategies that are implemented in the network architecture by a particular set of parties. Interests of these actors are rarely explicated in the Internet governance debate or in institutional plans and policies inspired by it. It follows that some important structures of network government are not publicly recognized as constitutive places where processes of economic, political and social shaping on technology application occur. On the contrary this chapter will be dedicated to the analysis on those geo-strategic issues relating to international flows of data and to remote control activities deployed by a small group of software houses and hardware manufacturers.

INTRODUCTION

The Internet Governance (IG) debate has, for a long time, been influenced by a well-defined characterization of information networks. An ideal-typical representation of a network controlled by a decentralized system of equivalent nodes[1], produced by a transnational multi-stakeholder partnership[2] and administered by a rich panorama of technical authorities with shared responsibilities[3], has fulfilled a dual function.

On the one hand, this has helped to define the scope of the issues selected for the institutional discourse about network control, focusing the debate on the regulation of network use and marginalizing those issues related to the regulatory trim of the network architecture. From this point of view, the protection of property rights, enforcement of national laws, taxation of online transactions, conflicts of jurisdiction, arbitration of international disputes, privacy and freedom of expression have been institutionalized as central themes in the debate on IG[4]. Alongside these issues, on the initiative of the so-called emerging countries, the discourse has been extended to such infrastructural issues as control over IP addresses and domain names.[5]

By framing the IG debate on a neutral image of networks, geo-strategic issues relating to international flows of data and remotely controlled activities deployed by a small group of software houses and hardware manufacturers have been excluded. Thanks to the depiction of the network as a shared and widespread set of resources organized in a non-hierarchical manner, the de-politicization of the discourse found in the terminological shift from "government" to "governance" took place, determining the delimitation of the policy field around uncontroversial themes from a geopolitical point of view.

The interests of national governments in regulating on-line interactions that involve questions of sovereignty have led the discourse on network control to focus on less conflictual aspects such as the government of how cyberspace is used, rather than configured.

The marginalization of issues related to network resources that are not only scarce but also tend to be monopolized by global players like multinational corporations, explains the "impressive degree of consensus on most issues" addressed in the IG debate, as it has been found, for example, by the European Commission member responsible for Information Society and Media.[6]

The second purpose of this neutral image of network control, produced and promoted by those who Vincent Mosco calls the powerful "mythmakers" of the digital age[7]—including software houses and hardware manufacturers as well as journalists, politicians and academics—has been

to support the legitimacy of policies adopted by governments and international organizations with an articulated rhetorical repertoire that, in its most advanced recommendations, even reaches an explicit eschatological perspective of redemption. The myth of the digital revolution, conjuring up scenarios of a global "information society" governed by an "electronic democracy" and aimed at development and welfare by paradigms of a "new economy", supported the adoption of networks by political, economic and social institutions almost everywhere in the world. Plans for institutional re-engineering modeled on the "virtual state" theorized by Jane E. Fountain, with administrative and political processes increasingly "dependent on the Internet and Web"[8], have been justified and promoted in the perspective of economic development and social empowerment[9], political participation[10], institutional accountability and responsiveness[11].

These plans, articulated on a scale ranging from local municipalities to international organizations, have helped to set in motion a process of conversion to the Digital. Alongside the enthusiasm for new technologies, this has generated an uncritical and unproblematic approach to the deep structures of control and regulation in cyberspace.

The depiction of a decentralized network, governed on a consensual basis by distributed forms of authority, has therefore led to focus little attention on network architecture that is implemented in a conflictual scenario by a set of parties whose interests are rarely explicated in the IG debate or in institutional plans and policies inspired by it. It follows that some important structures of network government are not publicly recognized as constitutive places where processes of economic, political and social shaping on technology application occur.

Without a proper reconnaissance of such structures, the IG agenda looks unable to understand and systematize the entire landscape of configurations and processes that help to establish a system of network government, nor its interconnections with the broader cyclical and structural dynamics of the contemporary world-system.

CYBERSPACE: AN INDEPENDENT OR A GEOPOLITICAL SPACE?

In the Declaration of Independence of Cyberspace, following the Telecommunications Act of 1996, John Perry Barlow wrote to the governments of the world:

You have no moral right to rule us nor do you possess any methods of enforcement we have true reason to fear. [...] Cyberspace does not lie within your borders. [...] Your legal concepts of property, expression, identity, movement, and context do not apply to us. They are based on matter, There is no matter here.[12]

Despite the considerable charm exercised by Barlow's declaration during this decade, the prophecies contained therein were never fulfilled. Since the second half of the 1990s, in fact, it is possible to observe a proliferation of constraint mechanisms implemented in order to govern and control interactions in cyberspace.

Enforcement mechanisms work in a digital network with two main modalities. A first form of regulation is that exercised by a territorial authority, which sanctions the behavior in cyberspace of a person subject to its laws. This type of regulation is expressed through the recording, tracking and analysis of data. These operations, that make it possible to locate and identify the perpetrator of the illegal behavior, are implemented by those who usually deal with the handling of data on a national scale. With respect to these procedures of *ex post* control, the political order and international relations of the different countries seem to be insignificant, as no territorial authority interested in the process of network interconnection has renounced control over the digital data produced or used by its own citizens.

Such control activities have been concentrated in certain areas of cyberspace, and particularly in the net access zones: Internet Service Providers and national telecommunication networks. In almost all countries, courts or even governments have access to the data handled by these subjects, even when these no longer constitute monopolies controlled by the state. They in fact physically operate – with offices, servers, transmitting plants, computing centers and commercial activities – on the national territory and are therefore directly subjected to state jurisdiction. The blurring of illegal sites and Weblogs, the detection of crimes perpetrated through networks and the identification of those responsible, the analysis on data for purposes of national security or tax justice, the filtering of data flagged as relating to peer-to-peer connections on which unauthorized exchanges of copyrighted material occur, are examples of regulation and control now abundant in the chronicles of both democratic countries and states ruled by authoritarian regimes.

A first element of fallacy in the libertarian vision of Barlow is therefore traceable to the proposition regarding the immateriality of cyberspace, which has too quickly led to underestimating, or even excluding, the ability of territorial authorities to enforce what happens in cyber places.

A second form of regulation is implemented by those that the American constitutionalist Lawrence Lessig calls "coding authorities", software houses and hardware manufacturers that have the possibility to elaborate information code that excludes specific behaviors or that activates preventive procedures of control[13]. This type of *ex ante* regulation is performed via algorithms that instruct systems about operations to be carried out on data. Algorithms, above and beyond those operations strictly functional to the elaboration process, could include instructions about filtration, duplication, forwarding, manipulation and even destruction of data.

The information code, unlike the legal code, works directly in cyberspace to determine its features and structures, representing the main tool of regulation and control on remote interactions. The instructions introduced within the source code of programs or in the firmware of hardware components regulate data by modalities that are inaccessible to the common net user. If at the dawn of digital networks a lot of the instructions were codified directly by subjects who were designers and administrators of those same networks, the "migration" of common users towards the boundaries of cyberspace has involved the routinization of interaction processes, which is increasingly entrusted to software agents like programs and user interfaces.

Facilitating access to information superhighways for millions of individuals has resulted, therefore, in the deployment of a "black box"[14] architecture, a "world taken for granted"[15] with respect to which the user perceives only the functional surface of spaces. The user sends e-mails, surfs Websites, chats, downloads multimedia products, pays taxes, participates in discussions, receives services, do business, play on-line games. The universe of algorithms that regulates and monitors these interactions is largely precluded to the user, and the information that he or she produces or enjoys are automatically managed according to instructions codified by third parties.

POWER STRUCTURES IN DIGITAL NETWORKS

Thanks to software agents, the user has the perception of being in direct contact with the area that he perceives, the interface. This process of routinization, aimed at the simplification of human-network interaction, generates a large gap that separates human commands from their implementation, users' intentions from the logics that produce, record, organize and transmit data.

To whom is the control of these processes entrusted? Who instructs software agents? Who are the coding authorities and what are their relations with the territorial authorities in different countries? When drawing up a response it will be necessary to take into account the complexity of today's control and communication networks.

The regulatory algorithms that are encoded by software houses operate in functionally different areas of processing centers and within the same machine at hierarchically structured levels. With regard to the functional areas of cyberspace, we can distinguish at least three main groups of computers:

i. Machines with structural functions, utilized by the mass user in order to gain access to the Internet, or to remotely produce and exchange data with other subjects;

ii. Machines with infrastructural functions, administrated by connectivity providers in order to manage data envelopment;

iii. Machines with supra-structural functions, used by digital content producers in order to host data and make them accessible.

On these machines, that have different functions in the overall network, regulatory algorithms operate on a well-defined functional hierarchy of processing levels, the most important of which are the following:

i. The firmware level, and particularly the CPU firmware;

ii. The operating system level;

iii. The application level.

As instructions always respond to a functional project, regulatory algorithms are generally organized in coherent sets, produced by well-defined software houses. Individualization of those software installed on the main normative levels of computers in the three cyberspace areas could thus allow us to identify the most prominent command and control centers operating in the whole system.

"Personal" Computers and Internet Appliances

The structural machine most frequently used nowadays in order to gain access to networks is the so-called personal computer. Despite the adjective "personal", very few things in a common PC are under the direct control of its owner, as a large part of the operations that the machine executes respond to instructions encoded in its hardware and software.

The encoding of CPU firmware in the more than one billion personal computers deployed in the world is entrusted to a global duopoly, consisting of two Californian transnational organizations: Intel, based in Santa Clara and AMD in Sunnyvale.[16]

Norms encoded by these two subjects present the highest level of juridical efficacy, as they are free of intermediaries between their own algorithms and the machine hardware. These algorithms also apply to the wider digital "case in question", being a gateway for all data processed by personal computers. It is then possible to affirm that, even if the great number of personal computers connected to networks allows a delocalization of processing capacity, it is not possible to find a decentralization of regulation and control activities, as norms operating in distant geographic locations are encoded by only two normative centers located in a unique region.

Control of algorithms on the operating system level belong to a monopoly that is *de facto* owned by Microsoft, a software house based in Redmond, in the state of Washington. More than 90% of personal computers connected to Internet use the Windows operating system, while the main alternative OSs have very small shares: MacOS with 3.33%, MacIntel with 2.82%, Linux with 0.77%.[17] At the application level, by contrast, there is a more articulated field of coding authorities. This is, in fact, a very fragmented area, in which a great number of coding authorities operate. On the other hand, among the millions of applications circulating in Internet there is a core of software that is adopted uniformly by most global users. Within this core, composed almost exclusively of proprietary source code software, some programs have become a standard for certain functions. It is the case, for example, of Microsoft Word for text editing, Microsoft Power Point for creating presentations or Adobe Photoshop for image manipulation. Many of these programs are produced by Microsoft itself, and they form part of a single regulatory project belonging to the Redmond software house, constituting an extension at the application level of the Microsoft operating system. Other mass software is produced by large software houses, mostly based in U.S., particularly in California, including Adobe Systems in San Jose, Blizzard Enterteinment in Irvine, Electronic Arts in Redwood City, Autodesk in San Rafael, Symantec in Sunnyvale and McAfee in Santa Clara.

Ultimately, personal computers are governed by a hierarchical system, with a small group of actors at the summit (Microsoft, Intel, AMD), located on a geographical axis of just 1.400 kilometers, holding the power to regulate almost all data produced on every PC in the world. The base of the pyramid is formed instead by a wide range of coding authorities, which in addition to Microsoft hold special positions of power related to certain application functions.

The structure of the Internet is not shaped by personal computers alone. A growing number of other devices have features that allow them to connect to networks. Among these, the most popular are the so-called smart phones, portable phones that use the connectivity functions of the IP protocol in order to obtain mobile access to the network. These devices, as well as game consoles like Xbox (Microsoft) and Playstation (Sony), while presenting a functional articulation in levels, similar to that of the PC, are characterized by the presence of a single regulatory project that from the processor encompasses the applications. Manufacturers of this kind of device have control over the encoding of all regulatory algorithms, disabling the possibility of executing even simple programs written by third parties. In other terms, the device executes only those applications expected and approved by its manufacturer and does not permit unplanned functions.

Unlike personal computers, characterized by high functional complexity and by the generality and indeterminacy of their use, the so-called Internet Appliances are relatively simple to use, and this makes them very attractive to the mass market. Finding all the necessary software already installed, users effectively delegate the configuration of the appliance to its producer; the impossibility of running third party programs reduces the risks of malfunction and tampering that are typical of personal computers.

The spread of entirely proprietary devices is therefore gradually transforming the structure of the Internet, leading to a configuration of terminals that is even more closed than in the case of personal computers.[18] This process of simplifying the landscape of coding authorities by using code configurations that exclude the implementation of rules codified by third parties is currently one of the most important political challenges to the legal and regulatory structure of the Internet.[19]

The Decline of the "Stupid Network"

Infrastructural machines have, for a long time, been ignored in the debate over network government. The idea of a neutral infrastructure which merely carries bits from point A to point B, without any intervention on the data, has helped to blur the dynamics of control and conflict that are commonly found in this area.

Currently, the dispute about the so-called "network neutrality" shows that the "end-to-end argument"[20] and the paradigm of the "stupid network"[21] are no longer able to sustain a vision that is consistent with the reality of telecommunications systems. In fact, they presuppose an infrastructural model in which "intelligent terminals" are able to process, from the periphery of the infrastructure, all required information to connect and communicate with other computers, without the need for action on the part of an infrastructural computer at the centre of the system. But with a simple operation tracking the path of data, it is possible to observe that between any two points on the network there are numerous machines that act as nodes of mediation between the sender and the recipient of the message, and many of these perform this function in a continuous fashion. The regulatory algorithms installed on these computers placed in strategic positions are able to apply any regulation process to information in transit.

Amongst those computers with infrastructural functions, those on which the so-called routing tables are stored have a prominent role. These tables convert the domain names typed by users in verbal form (e.g. www.unisa.it) into the typical numeric IP addresses (eg 193.205.160.14).

Since September 18th 1998, the authority which is responsible for managing the addressing system of the Internet is the Internet Corporation for Assigned Names and Numbers (ICANN), which is formally an international nonprofit entity based in Marina Del Rey (California), but which in fact works in continuity with previous Internet Assigned Numbers Authority (IANA) delegated by the Department of Commerce of the United States of America.

Computers that host routing tables are organized on the basis of a three-level hierarchy. On the highest level, there are the DNS root servers, located in about 130 locations on the planet, and managed by nine organizations over ICANN: Veri-Sign, Information Sciences Institute in Marina Del Rey, Cogent Communications, the University of Maryland, NASA, the Internet Systems Consortium, the Network Information Center of the U.S. Department of Defense, the Research Lab of the Army of the United States, Autonomica / NORDUnet, RIPE (Réseaux IP Européens), and WIDE Project.[22]

Between DNS root servers and the mass of computers that require constant updates of the addresses is a brokerage area composed of the so-called authoritative servers, servers that are directly associated both with the root servers, and with a number of structural machines that use its database to obtain the address of each given URL.

DNS root servers, and thousands of interconnected authoritative servers, are not the only computers inserted into Internet infrastructure. Other machines with infrastructural functions are used by Internet Service Providers (ISP) in order to grant their subscribers access to the network. These servers, as well as the DNS servers, carry out their infrastructural tasks through an architecture that is conceptually similar to that of structural computers, but with a configuration that presents a substantial normative consistency from the firmware of machinery to the applications used.

The servers, in fact, being dedicated to a limited cluster of functions, and being run by experts, involve a set of coding authorities that is much more limited than that of a personal computer on the application level, and a wider dissemination of open source code that the administrator of the machine can see and adapt to his or her needs. In this case, however, the ability to read and amend the rules used by a machine is not extended to the whole set of users whose interactions are subject to control, but remains a possibility reserved only to the owners and administrators of a machine.

Another similarity between the DNS servers and ISP servers is the hierarchical organization of access. There are first-level ISPs, communicating by means of redundant connections at very high speeds, that form the backbone of the Internet. This network, called Tier 1, includes just nine organizations, including 8 U.S. and one Japanese: AOL Transit Data Network (ATDN), AT & T, Global Crossing (GX), Level 3, Verizon Business, NTT Communications, Qwest, Savvis, SprintLink.

Network T1 ISPs, often known as carriers, exchange data by peering connections, without imposing charges to traffic and with mechanisms for mutual assistance if one node becomes overloaded. ISPs connected by a peering connection with just some T1 nodes, and buying connectivity from others, belong to the category T2. The third-level ISPs (T3) connect to these second-level ISPs, usually operating on a local scale, and buying connectivity from T2 ISPs in order to distribute it to end users. This hierarchical organization of interconnections and the bandwidth available between T1 nodes have the effect of pushing long distance data traffic, such as intercontinental communications, towards infrastructural computers located on the territory of the United States of America.

Moreover, with regard to the internal configuration of these machines, in many cases we find the same coding authorities found on the deeper regulatory levels of structural computers, and typically Intel, AMD and Microsoft. Adding to these subjects, in this area we can also find other producers of high-end hardware and software, such as IBM (based in the New York State) and California-based companies such as HP (Palo Alto), Sun Microsystems (Santa Clara), Oracle (Redwood City) and Cisco Systems (San Jose).

In conclusion, in the infrastructural area of cyberspace, many devices and computers help to make the network's infrastructure more intelligent, increasing the degree of governability and establishing a clear hierarchy of regulatory centers in its control activities. On the lowest levels of the hierarchy, such as network T2 and T3 and DNS authoritative servers, we noted that action by the territorial authorities of all interconnected countries is significant and aimed at punishing illegal behavior by their citizens in cyberspace. At the highest levels, by contrast, only the U.S. authorities can implement such a control strategy, as top-level infrastructural computers are almost all located—as well as produced and codified—on its territory.

The Data Suprastructure: Centralizing Trends

A quite similar situation is found in the suprastructural area of cyberspace, in relation to computers intended to host data. The so-called hosting servers present an internal configuration of coding authorities often indistinguishable from that of servers used by ISPs, with software and hardware produced by the same companies. There are instead substantial variations in the ownership and administration of servers. Data hosting is a rapidly-growing sector. But it does not involve as many competitors as is commonly supposed. If there are about one hundred and fifty million Websites, with a monthly growth rate of 5%, about half of all new registered sites are hosted on servers owned by just three companies: MySpace (owned by Fox Interactive Media), live.com (Microsoft) and Blogger (Google). [23]

Myspace, Blogger and Live are the most popular blog platforms that offer free server space for hosting data and for the registration of domain names under their general domain. These platforms, although free, derive substantial economic benefits by hosting users' data.

First of all, these platforms host blogs managed and promoted by millions of users, with a very high commercial value: even tough every blog receives, few visitors on average, their total amount allows the owners of the platform to vehicle advertisement to a wide public. [24]

The second main benefit that accrues to the owners of blogging and social software platforms by hosting the algorithms and information supplied by users stems from the possibility of subjecting large amounts of data produced by these users to the regulatory algorithms that reside on its own machines. The management of hosting servers, in fact, is organized on two levels. The blog administrator operates only that server partition that corresponds to the pages that he or she manages, and he or she can only intervene in relation to these data and algorithms, which are nevertheless conveyed to the server after being processed by a graphic user interface. The platform administrators can introduce regulatory algorithms at a deeper level of server architecture, operating on OS and software that generate the interface used by the blog's administrator.

The platforms that host content produced by users currently represent one of the most important phenomena of the Internet superstructure, both for the aforementioned methods of data production – which have earned the name of Web 2.0 – and for the traffic they generate. A quantity of visitors and hits that is only matched by the traffic generated by the most popular search engines, such as Yahoo!, Google, Microsoft's MSN and Baidu.

The popularity of the services offered by search engines has grown over time, in proportion to the increasing amount of information in the superstructure of the Internet, and is therefore closely related to the explosion of content that social software has helped to generate.

Search engines face the need to organize access to information when the latter goes beyond the limits of data that are manageable by a structural processor. In fact, hosting servers owned by search engines contain periodically updated versions of all Web pages that their requests (typically managed by software called crawlers or spiders) encounter when browsing the Web link by link.

The information collected by crawlers are then sent to huge databases on which DataBase Management Systems (DBMS) work with their regulatory algorithms. These algorithms make sure to categorize the collected content and to return it in an organized manner in order to meet end users' demands. But their functions are not limited to this. With the stated purpose of increasing the relevance of the resources provided in response to users' queries and creating a general assessment system to determine index rankings, the biggest search engines trace paths (clickstreams) that users follow starting at Search Engine Results Pages (SERPs). [25]

The main search engines, that John Battelle calls "database of intentions" because they record the queries and paths of millions of users, now manage such enormous quantities of data that their control policy assumes significance in relation to the whole architecture of the network. Although their configuration is conceptually similar to that of common PCs, these machines represent key means for controlling the Internet and their political role is difficult to subvert in a bottom-up way. The success of the major search engines in fact does not depend on the quality of their indexing algorithms, but it resides on the barriers to entry represented by expensive hardware whose resources in terms of bandwidth, calculation and storage are crucial for the competitiveness of the service that is offered. In other words, it is not enough to devise a search algorithm that is functionally superior to PageRank and its derivatives, in order to scrape Google's power in this area. Only an equivalent investment of capital would give the new competitor the means of production necessary for this kind of mission.

The notion of networks that lack central nodes, in which all processing activities take place at the periphery of the system, guided by end users themselves, is just a representation, and a powerful one in guiding public debate on information technology. It is nevertheless deceitful when compared with the empirical reality of networks such as Internet. Here, data generated on a processing structure that is largely controlled by a few authorities are conveyed through an intelligent infrastructure, where many machines apply their own regulatory algorithms to data in transit. The latter tends to accumulate and to concentrate itself on hosting servers owned by an identifiable group of companies, whose investments in computing power, bandwidth and storage capacity have reached a dimension that promotes oligopoly also in the supra-structural area of the network.

CODING AUTHORITIES AND NATIONAL GOVERNMENTS

Since 1996, the legal code produced by territorial authorities and the information code produced by coding authorities have increasingly worked together to make cyberspace a regulated and reliable environment. The WIPO treaties of that year – the Copyright Treaty (WCT) and the Performances and Phonograms Treaty (WPPT) – constitute the starting point for a long alliance between the powers of the interstate system and those of the global communication system, united by the urgency to subject digital networks to control. The most relevant aspect of the international legal framework determined by these two treaties, ratified recently also by China, is the introduction of legal protection accorded to mechanisms of enforcement encoded by the coding authorities directly within digital communication devices and media in order to manage their intellectual property rights. But if consensus on legal protection of intellectual property in the international arena has been painstakingly reached, the geopolitical location of the territorial authorities has resulted in different approaches to the issue of network government and to the form of regulation exercised by coding authorities. Most of those able to encode instructions in the focal points of the network are, as we noted, U.S. organizations. This means that the legal code of the United States enjoys much more effective protection by the information code and a more concrete possibility of enforcement in network spaces, when compared to the laws of other countries. Thus relationships between single governments and the U.S. government become fundamental in determining the cybernetic strategy of a country.

In this respect, we can discern two main positions. The so-called emerging countries may be distinguished from the European Union and other allies of the U.S. in relation to the problem of government of the information code. An open dispute has been recorded, for example, during the World Summits on the Information Society (WSIS) held in 2003 and 2005. The proposal put forward in Geneva by countries such as China, Brazil, India and Syria regarding an internationalization of control on IP protocols and DNS, with the passage of competences from ICANN to a UN agency, was effectively rejected two years later in Tunis, with arguments focusing on the inherent neutrality of technical administration of networks.[26]

But if the debate on higher levels of Internet infrastructure, partly supported also by the EU, has registered one of the most evidently antagonistic position on the part of some countries, far more significant clues indicate the articulation of complex strategies oriented to the long term.

Brazil, for example, by refusing Microsoft's offer of half a million operating systems at a facilitated price, but with functional limitations implemented in the source code of the OS, gave impetus to a fast conversion of public systems towards open source software.

The Câmara de Implementação do Software Livre, focusing public spending on publicly accessible code, has not only fostered the birth of an autochthonous software economy, reducing dependence on foreign countries, but has also managed to achieve economies of scale that have facilitated an expansion in Brazilian policies against the digital divide.[27]

For a long time India has also facilitated the construction of a domestic software economy by developing coordinated policies of intervention in key areas such as training and research, and promoting the adoption of open source code at the level of the operating system. But the most complex and comprehensive strategy is that implemented by the Government of the PRC.

In addition to the development of its own code for applications and operating systems, the Chinese authorities have for over a decade engaged in research and production of hardware systems on which they could encode their firmware. By producing their own models of CPU, which as we

have observed represent the deepest and the most efficient level of a machine's processing activities, the Beijing authorities are able to use computers whose code responds exclusively to their control. Given the successes of China's space program, it is possible to assume a relative Chinese independence also from T1 backbone network in order to handle intercontinental data traffic.

Restoring China's technical supremacy and thereby reviving the Middle Kingdom has been a constant goal for Chinese leaders since Sun Yat-sen, who gave up his presidency of the Republic of China to become the Minister of Railways out of a belief in the power of technology.[28]

The activism shown by such countries as Brazil, India and China in code control activities counterbalances the position of EU countries. In this case, in fact, most initiatives focus on Web sites and portal production, which are located in the supra-structural area of the network, while control of structural computers' code – the code installed on personal computers belonging to citizens and to public and private organizations – is quite nil. In these countries, public spending has been uncritical with respect to the control of the information code and a short-sighted fragmented strategy for technological development has led to the spread of proprietary, and foreign, source code that is difficult to reverse.

The fines imposed on Microsoft for its abuse of a dominant position, with economic arguments aimed at protection of a market where Europe is almost entirely absent, are accompanied by a total lack of political control by the EU on the code inserted in millions of computers scattered on its territory, computers to which strategic functions are assigned in such sectors as public administration, large industry, finance, trade, energy, education and defense.

The EU, embracing the prospects of democratic deficit reduction that are offered by e-government and by e-democracy, appears to concentrate its regulative activities in the most peripheral areas of the network, increasingly depending on the U.S. for enforcement of its laws in the deepest areas of cyberspace.

CONCLUSION

The deployment of digital data processing and transport systems, and their convergence towards the communication standards of Internet Protocol Suite, have as the more evident and significant effect the establishment of a global space of interaction that now represents one of the key geopolitical and economic elements of the contemporary capitalism.

The strategy of U.S. authorities to interconnect, within an information space, organizations and individuals that are geographically distant from one another, began in the early 1970s as part of a wider and more general systemic reorganization process. As David Harvey argues, this process consists in the combined acceleration of the two basic strategies for acquiring capital gain described by Marx. An extensive strategy based on absolute value is deployed in space with the movement of corporate capital from regions with high wages to low-wage regions and the systematic search for opportunities in new markets. An intensive strategy, based on relative value, is pursued via massive investments in technology that have the effect of isolating and protecting companies with high rates of fixed assets. These companies can thus obtain competitive advantages of a monopolistic nature in areas with a faster rotation of investment, over the mass of companies with lower capitalization levels. The latter must fight in an arena already full of competitors facing decreasing profit margins with a single weapon: the reduction of labor costs.[29] These two kinds of expansive processes lead the capitalist world system to incorporate new areas and to encourage technological innovations.[30]

With regard to these two strategies, whose cyclical acceleration indicates, according to Harvey, the historic transition from the Fordist system of accumulation to a flexible accumulation system, the space of interconnected networks is a fundamental and indispensable instrument for implementation on a global scale. Many indeed are the areas of synergy between the processes of digitalization of human experience and transitions taking place in the capitalist world-system.

It is useful to consider the effects of e-banking and e-trading on the financialization processes of the economy, of market enlargement resulting from the standardization of e-commerce systems, of the relocation of production processes allowed by remote work and by corporate networks that enable new forms of transnational organizations to continuous and instantly monitoring subcontract chains issued worldwide, of e-government and e-democracy policies aimed at reinvigorating the fundamental institutions of the interstate system which constitute the political superstructure of capitalism, of the increased potential for control on the workforce and for the public order arising from the deployment of heterogeneous surveillance systems, and of the use of robotics in both domestic and foreign conflicts.

In such a scenario, old and new powers of the global economy are confronted with the aim of obtaining strategic controlling positions in that network which is configuring itself as a huge system of interconnection between the means of production, consumption and trade.

If some countries seem to be more ready than others to build their own areas of control, that might limit and contain the American one, the process of network deployment and interconnection has tended to take on the traits of inevitability, for governments and public opinion, that have in the past characterized other systemic reorganization processes, such as industrialization.[31]

Like the industrial revolution, the digital one seems to operate with the canonical mode of what Immanuel Wallerstein calls an "organizing myth", a "meta-history" able to dictate categories, units of analysis, concepts and reference contexts for the organized knowledge, regardless of its institutional forms.[32] The mythological nature of revolutionary characterization of these processes would conceal, according to Wallerstein, the true character of a planned reform, of a controlled and gradual process, aimed at an institutional reorganization rather than at radical social upheaval.

From this point of view, the digital revolution, rather than subverting the world-system's political and economic balance (as the term revolution would suggest), appears to be a formidable process of reorganization. On the one hand, this is clearly attributable to U.S. hegemony, whilst on the other, it indicates the hegemonic transition that is underway. Above all, Chinese strategies in cybernetics as well as in other geo-strategic key sectors, appear to constitute the most solid and far-sighted alternative to unilateral initiatives by the North American authorities.[33]

The future of the government of cyberspace will thus be largely determined by configurations derived from the global system of power relationships in which ultimately cyberspace was created. Far from being independent, its architecture will continue to evolve, following the political and economic dynamics of the world system to which it is deeply, physically, interconnected.

The absence of fundamental geo-strategic issues from official debates on Internet Governance is the signal that conflict between the subjects who are emerging as cybernetic powers is oriented to take place through information code and hardware investments, rather than through ICANN's advisory bodies or the public meetings of WSIS.

REFERENCES

Andreson, C. (2006). *The long tail: Why the future of business is selling less of more.* New York: Hyperion.

Arndt, H. W. (1987). *Economic development. The history of an idea.* IL: The University of Chicago Press.

Arrighi, G., & Silver, B. J. (1999). *Chaos and governance in the modern world system.* Minneapolis: University of. Minnesota Press.

Barlow, J. P. (9th February 1996). *A cyberspace independence declaration.*

Battelle, J. (2005). *The search. How Google and its rivals rewrote the rules of business and transformed our culture.* New York: Portfolio.

Coleman, S. (2005). *Direct representation. Towards a conversational democracy.* Institute for Public Policy Research.

Coleman, S., & Gøtze, J. (2001). *Bowling together Online public engagement in policy deliberation.* London: Hansard Society.

EU Commission (2005). *I2010 – A European information society for growth and employment.* COM 229 final {SEC(2005) 717}.

Fountain, J. E. (2001). *Building the virtual state. Information technology and institutional change.* Brookings Institution Press.

Frick, M. M. (2005). *Parliaments in the digital age.* Geneve: E-Democracy Centre.

Gillett, S. E., Lehr, W., Wroclawski, J. T., Clark, D. D. (2001, October). *A taxonomy of Internet appliances.* Sloan Working Paper 4186-01 eBusiness@MIT Working Paper 112.

Gillet, S. E., & Kapor, M. (1996). The self-governing Internet: Coordination by design. In B. Kahin & J. Keller (Eds.), *Coordination of Internet.* MIT Press.

Harvey, D. (1990). *The condition of postmodernity.* Basil Blackwell.

Isenberg, D. (1997, August). *Rise of the stupid network.* Computer Telephony.

Jedlowski, P. (1994). *Il Sapere dell'Esperienza.* Milano: il Saggiatore.

Keinwachter, W. (2007). *The power of ideas: Internet governance in a global multi-stakeholder environment.* Berlin: Marketing fur Deutschland GmbH.

Kenis, P., & Schneider, V. (1991). Policy network and policy analysis: Scrutinizing a new analytical toolbox, (pp. 25-62). In B. Marin & R. Mayntz (Eds.), *Policy networks. Empirical evidence and theoretical considerations.* Boulder, CO: Westview Press.

Mosco, V. (2004). *The digital sublime. Myth, power, and cyberspace.* Cambridge, MA: MIT Press.

Latour, B. (1987). *Science in action. How to follow scientists and engineers through society.* Cambridge, MA: Harvard University Press.

Lessig, L. (1999). *Code and other laws of cyberspace.* New York: Basic Books.

NetDailogue (2005). *Clearing house on international Internet governance.* Harvard Law School's Berkman Center for Internet and Society (Berkman Center) and Stanford Law School's Center for Internet and Society (CIS).

Qiu, J. L. (2003). *The Internet in China: Data and issues.* Annenberg Research Seminar on International Communication.

OECD (2001). *Citizens as partners OECD handbook on information, consultation and public participation in policy-making.*

Reding, V. (2005). *Opportunities and challenges of the ubiquitous world and some words on Internet governance 2005.* European Commission.

Saltzer, J. H., Reed, D. P., & Clark, D. D. (1981). *End-to-end arguments in system design.* M.I.T. Laboratory for Computer Science.

Trechsel, A. H., Kies, R., Mendez, F. & Schmitter, P. C. (2003). *Evaluation of the use of new technologies in order to facilitate democracy in Europe: E-democratizing the parliaments and parties of Europe (STOA research report).* Strasbourg, European Parliament.

UN General Assembly (2003). *Resolution 57/295.*

Wallerstein, I. (1991). *Unthinking social science: The limits of nineteenth century paradigms.* Cambridge: Polity.

WSIS (2005). *Tunis agenda for the information society.* WSIS-05/TUNIS/DOC/6(Rev.1)-E

Zittrain, J. (May 2006). The generative Internet. *Harvard Law Review, 119,* 1974-2040.

ENDNOTES

[1] Gillet, Sharon Eisner and Kapor, Mitchell (1996) The Self-governing Internet: Coordination by Design in Kahin, Brian and Keller, James (eds.) Coordination of Internet. MIT Press.

[2] Desai, Nitin (2007) Foreward to Keinwachter, Wolfgang (ed.) The Power of Ideas: Internet Governance in a Global Multi-Stakeholder Environment. Berlin: Marketing fur Deutschland GmbH. Nitin Desai is Special Adviser to the UN Secretary-General on Internet Governance.

[3] Kenis, Patrick and Schneider, Volker (1991), Policy network and policy analysis: scrutinizing a new analytical toolbox, pp. 25-62 in B. Marin and R. Mayntz (eds.), Policy networks. Empirical evidence and Theoretical considerations. Boulder: Westview Press.

[4] NetDailogue (2005). Clearing house on international Internet governance. Harvard Law School's Berkman Center for Internet

and Society (Berkman Center) and Stanford Law School's Center for Internet and Society (CIS) W: http://www.netdialogue.org/

[5] WSIS (2005). Tunis Agenda for the Information Society. WSIS-05/TUNIS/DOC/6(Rev.1)-E. W: http://www.itu.int/wsis/docs2/tunis/off/6rev1.pdf

[6] Reding, Viviane (2005). Opportunities and challenges of the Ubiquitous World and some words on Internet Governance 2005. European Commission. Available at: http://ec.europa.eu/commission_barroso/reding/docs/speeches/ubiquitous_world_20051017.pdf

[7] Mosco, Vincent (2004). The Digital Sublime. Myth, Power, and Cyberspace. Cambridge, MA: MIT Press.

[8] Fountain, Jane E. (2001). Building the Virtual State. Information Technology and Institutional change. Brookings Institution Press, p. 4.

[9] See for example: OECD (2001). Citizens as Partners OECD Handbook On Information, Consultation And Public Participation In Policy-Making; UN General Assembly (2003) Resolution 57/295; EU Commission (2005). I2010 – A European Information Society for growth and employment. COM 229 final {SEC(2005) 717}.

[10] See for example: Coleman, Stephen (2005). Direct Representation. Towards a conversational democracy. Institute for Public Policy Research; Coleman, Stephen and Gøtze, John (2001). Bowling Together Online Public Engagement in Policy Deliberation. London: Hansard Society.

[11] See for example: Trechsel, A. H., Kies, R., Mendez, F. e Schmitter, P. C. (2003) Evaluation of the use of new technologies in order to facilitate democracy in Europe: e-democratizing the parliaments and parties of Europe (STOA research report), Strasbourg, European Parliament; Frick, M.M. (2005). Parliaments in the Digital Age. Geneve: E-Democracy Centre.

[12] Barlow, John Perry (9 Febbraio 1996). A Cyberspace Independence Declaration. W: http://w2.eff.org/Censorship/Internet_censorship_bills/barlow_0296.declaration

[13] Lessig, Lawrence (1999), Code and Other Laws of Cyberspace. New York: Basic Books.

[14] Latour, Bruno (1987). Science in Action. How to Follow Scientists and Engineers through Society. Cambridge, MA: Harvard University Press.

[15] Jedlowski, Paolo (1994). Il Sapere dell'Esperienza. Milano: il Saggiatore.

[16] PcPitstop Research. W: http://www.pcpitstop.com/research/

[17] Marketshare. W: http://marketshare.hitslink.com/report.aspx?qprid=2 . See also XiTi Monitor. W: http://www.xitimonitor.com/en-us/Internet-users-equipment/operating-systems-august-2007/index-1-2-7-107.html.

[18] Gillett, Sharon; Lehr, William; Wroclawski, John T.; Clark, David D. A Taxonomy of Internet Appliances. Sloan Working Paper 4186-01 eBusiness@MIT Working Paper 112. October 2001: 2.

[19] Zittrain, Jonathan (May 2006). The Generative Internet. Harvard Law Review, Vol. 119, pp. 1974-2040.

[20] Saltzer, J.H., Reed, D.P. and Clark, D.D. (1981) End-to-end arguments in system design. M.I.T. Laboratory for Computer Science.

[21] Isenberg, David (1997) Rise of the Stupid Network. Computer Telephony, August 1997, pp. 16-26.

[22] W: http://www.root-servers.org

[23] Netcraft (October 2007) Web Server Survey. W: http://news.netcraft.com/archives/web_server_survey.html.

[24] Andreson, Chris (2006) The long tail: why the future of business is selling less of more. New York: Hyperion.

25 Battelle, John (2005). The Search. How Google and Its Rivals Rewrote the Rules of Business and Transformed Our Culture. New York: Portfolio.

26 Supra, nota 5, Paragraph 69.

27 Software Livre no governo do Brasil. W: http://www.softwarelivre.gov.br/

28 Qiu, Jack Linchuan (2003). The Internet in China: data and Issues. Annenberg Research Seminar on International Communication, October 1.

29 Harvey, David (1990) The Condition of Postmodernity. Basil Blackwell.

30 Wallerstein, Immanuel (1991). Unthinking Social Science: The Limits of Nineteenth Century Paradigms. Cambridge: Polity.

31 Arndt, H.W. (1987) Economic Development. The History of an Idea. The University of Chicago Press.

32 Supra, nota 30.

33 Arrighi, G., Silver, B.J. (1999). Chaos and Governance in the Modern World System. Minneapolis: University of. Minnesota Press.

Chapter XII
Measuring ICT:
Political and Methodological Aspects

Diego Giannone
University of Salerno, Italy

ABSTRACT

Technology embeds the ideology, politics and culture of the society in which it was created. Working on this assumption the highlights the historical and political link between the rise of the neoliberal paradigm, which has occurred since the seventies of the twentieth century in Western industrialized capitalist countries, and the dissemination of ICT. More specifically, it analyses the issue of the measurement of ICT, which emerged functionally to the need to identify new tools to legitimize the hierarchy of development, attributing some countries with the tag of "most advanced" and the others of "developing" or "underdeveloped". Indeed, the measurement, acting as a scientific justification for the Western superiority, is a part of those structures of knowledge which constitute an essential element in the functioning and legitimacy of the political, economic and social structures of the existing world-system. This contribution analyses the methods of knowledge developed at international level, by the institutions and organizations who have taken a leading role in defining and measureming of ICT: the OECD, the ITU, and the World Bank.

MEASURING ICT AND THE NEOLIBERAL PROJECT

The rapid diffusion of new information and communication technologies (ICT) is crucially linked to the early years (1970s) of their development. The paradigmatic changes occurring in that period in the industrialized countries of the capitalist West played a key role both in the dissemination of ICT and for their legitimacy as a criterion for measuring the progress of society. During the 70s, Western societies experienced a period of political, economic and social crisis becaus the Keynesian and welfare models, according to which they had been governed, were losing legitimacy because of creating ungovernable democracies. Therefore these models were deemed by then social systems unable to cope with the growing issues of the social body (Crozier, Huntington & Watanuki, 1975).

In the battle for a new hegemony, neoliberalism emerged as the best model for responding to the attempts to restore class power by the economic and political world elite (Harvey, 2005). Within a few years, the neoliberal doctrine, which had achieved a fundamental reversal of the existing balance in the relationship between the individual and society, economics and politics, public and private, was able to impose worldwide, even as regards the man in the street, a new vision of the world (Harvey, 2005).

There are many reasons why neoliberalism has been a suitable political-ideological terrain for the spreading of ICT and a justification and ideological legitimacy of their development. One possible explanation lies in the fact that, unlike the preceding welfare model, this doctrine maintains that:

the social good will be maximized by maximizing the reach and frequency of market transactions, and it seeks to bring all human action into the domain of the market. This requires technologies of information creation and capacities to accumulate, store, transfer, analyze, and use massive databases to guide decisions in the global marketplace. (Harvey, 2005, p. 3)

The trajectories of neoliberalism have therefore intersected with the incentive to produce new technology infrastructure (software and hardware),because it has been: (a) a new area of prolific development of the capitalist economy, (b) an effective solution to the decrease in produc-

tion costs and the acceleration both of economic transactions and financialization of the economy, (c) an appropriate solution to the imperative of a more streamlined and less expensive statehood, (d) an ideological tool to reaffirm on a global scale the superiority of some countries over others. This last point deserves particular attention. Indeed, it is necessary to remember that:

the new information technologies were developed in, by and for highly advanced capitalist economies – that of the USA in particular. It is to be expected, therefore, that these technologies are now being employed single-mindedly to serve market objectives. Control of the labour force, higher productivity, capture of the world markets, and continued capital accumulation are the propelling influences under which the new information technologies are developed. (Schiller, 1985, p. 37)

Their legitimacy as a key element of the development of society took place in the historical period that marks the transition in terms of international relations from the framework of the "development project" to that of the "globalization project" (McMichael, 2004). This new frame of development dissemination, which the neoliberal doctrine put in place especially through the central role of international institutions (World Bank, International Monetary Fund),

succeed[ed] the development project, partly because the latter failed and partly because the former became a new exercise of (market) power across the world (as transnational firms and banks grew and as neoliberal ideology took hold, restructuring states and societies everywhere). (McMichael, 2004, p. XXXIX)

Within this new historical-political framework different rhetorical descriptive discourses have been structured, among which the most persistent being the information society. This label which also had a stronghold in the seventies, emerged together with "post-Fordism" and "post-modernism" as one of three potential versions of post-industrial society (Kumar, 1995), prevailing at that time surreptitiously in international organizations (Mattelart, 2003) and becoming the reference code of their development policies in the field of ICT. ICT have become both the instrument of development of these (alleged new) societies and the modern measurement criterion of their development.

The problem of measurement emerged when it became necessary to identify the tools to legitimize the hierarchy of development, attributing some countries with the tag of "most advanced" and the others of "developing" or "underdeveloped". The need to obtain data, information, sound knowledge on the state of ICT was therefore certainly a strong motive for the development of methods of measurement, but it is clear that the framework within which it was included basically transformed it into a political problem and project. Measurement, in fact, as a scientific justification for Western superiority, is a part of those structures of knowledge that constitute an essential element in the functioning and legitimacy of the political, economic and social structures of existing world-system (Wallerstein, 2006). Therefore, the development of specific measuring instruments is part of the battle for "the creation of a hegemonic apparatus, [because this], as it creates a new ideological terrain, determines a reform of consciences and *methods of knowledge*, it is a matter of knowledge, a philosophical question" (Gramsci, 1997, p. 285, the translation and italics are mine).[1]

A reform of the methods of knowledge was needed because the old tools and old indicators for the most part seemed inadequate to legitimize the spread of ICT as a development criterion. At international level, institutions such as the Organization for Economic Co-operation and Development (OECD), the International Telecommunication Union (ITU) and the World Bank, (seemingly non-partisan institutions, but

in reality controlled by the so-called "developed countries"[2]) have made a series of initiatives and developed a set of tools to address this problem. The convergence of approaches, methodologies harmonisation and standardisation of indicators have been the main objectives for the action of the international institutions. But the geo-political context within which they have tried to achieve these objectives is far from complying with the principles of a "universal universalism" (Wallerstein, 2006), being configured rather as a kind of "Western universalism". This, through the establishment of a single development model (the globalization project), the imposition of a specific political-economic doctrine (neoliberalism) and the articulation of a rhetorical descriptive discourse (information society) based on specifically Western economic, political, technological and social concepts (Hyder, 2005), defines a set of practices and guidelines only seemingly universal, but in actual fact complying with the interests of developed countries.

Before giving a summary account of these initiatives, aimed at defining the "who" and "how" of the measurement of ICT, I shall begin by focusing on the "what" and then highlight some still problematic nodes related to the "why" of measurement, in conformity with the theory outlined so far, which tends to regard as closely interrelated the methodological and political-ideological aspects of measurement. Subsequently, where I maintain a distinction between these two components it will be only for the purpose of presentation.

ICT: THE RHETORIC OF FREEDOM AND CONTROL POLICIES

The need to have data on ICT comparable at international level emerged when, during the second half of the '90s, the central role of ICT as a factor of economic growth was highlighted. In an effort to bridge the knowledge gap, several players moved: ITU[3] began to collect data on the Internet market, defining in a specific manual, *Indicators Handbook*, a number of indicators measuring the IS. In 1996, during a summit between ministers of the OECD the need was recognized to obtain new data to analyze the development and use of "information highways", in particular by recommending the establishment of a working party[4] with the task of developing "new indicators which identify, assess and monitor the emergence of the GIS [Global Information Society]" (OECD, 1996, p. 10) and the identification of "a common framework for indicators and standard definitions [...] to be developed, tested and shared among OECD countries" (OECD, 1996, p. 10).

In fact, while in specialized places work to find internationally shared ways of defining and measuring the Web was in progress, on a common sense plane, loaded discourse, fomented by the rhetoric of globalisation and the policies of deregulation and liberalization, was going in a diametrically opposite direction. The Internet was painted as an area of almost complete freedom for the end user, a space of unlimited access to knowledge, devoid of control and borders. The message appeared truly mystifying, because the discourse on the absence of boundaries on the Web, certainly linked to the redefinition of space-time coordinates that the virtual experience made possible, was deceptively linked to two other discourses. The first, more general, related to the border crisis experienced by nation states as a result of several phenomena: migration increasingly uncontrollable, the processes of political supranational integration (EU) and the globalization of markets. The border crisis was a crisis of territorial sovereignty of the State, highlighted in the announced end of its traditional role as collective mediator (Negroponte, 1995). The concept disseminated, in this case, was that virtual space was indifferent to national borders and national sovereignty (Goldsmith & Wu, 2006). It was namely, and this is the second mystifying discourse, an open space, "boundless", "huge",

difficult to associate with the concept of control. Although the implications of such an assumption were evident and it was stressed that cyberspace could easily be transformed from an area of freedom into an area of control simply by acting on its "code", i.e. the technological architecture[5] network (Lessig, 2006), the common user still finds it difficult to think of the Internet as a "finite space". While for consumer users the rhetoric of the first generation Internet continues to apply, making freedom and the lack of control two factors inherent to the World Wide Web, both companies and governments realised that the Web had to somehow be controlled and regulated. Private sector groups, in particular, soon realized the benefits related to the control of the network especially to achieve two objectives:(1) to ensure a high level of security in e-commerce; (2) to learn more about the consumer behaviour, habits and lifestyles of their customers.

Clearly, with regard to the private sector, to act on regulatory processes, to define standardised procedures for the creation of financial transactions, to ensure the existence of safe virtual spaces in order to create systems of identification of their customers and to draft profiles of the different consumers are processes that require a knowledge of the network, its controllability and measurability. The same is true for the public sector: governments must also know the network to be able to regulate, to define cases of new crimes and to prosecute illegal behaviour, to provide services to citizens (e-government) and to pave the way for new forms of participation in public life (e-democracy). This is, as Menou and Taylor maintain, "a grand challenge" (Menou & Taylor, 2006), especially when one considers how the World Wide Web changes rapidly, as a result of the creation of new pages and new links. However, the challenge has to be addressed because, as a maxim fairly widespread among experts on these issues explains, "you can only manage what you can measure": i.e. collecting relevant and reliable data and information on the complex world of the IS is imperative.

The Issues on the Agenda

The main issues involved in the measurement both of the IS and of its main component, ICT, can be circumscribed to the definitions of the following issues:

1. The universe to be measured (e.g. the IS);
2. Objects and phenomena to include in the universe (e.g. ICT, digital divide, e-commerce, e-democracy, etc..);
3. Background theories;
4. Units of measurements;
5. Data sources and collection;
6. Methods of analysis and construction of indicators;
7. Target audiences for measurements;
8. The purpose and utilization of measurements (Menou & Taylor, 2006).

These are questions involving both methodological aspects, relating to the "how" and "what" of measurement, and socio-political aspects, relating to the "who" and "why" of measurement.

Each issue listed requires detailed analyzes of the social, political and economic context and choices consistent both with the ultimate objective of measuring and the resources and data available. In addition, methodologically speaking, the process of measurement must develop through several stages: from the conceptualization of the units of analysis to the formulation of operational definitions and then to the choice of the dimensions to be measured and, consistently with them, the most appropriate indicators and variables arising therefrom.

The issue of the measurement of ICT appears crucial for the organization of scientific debate, for industrial development and for the implementation of public policies (Ricci, 2000; Servaes & Heinderyckx, 2002). The method chosen determines how data are collected and processed, which in turn influences the vision of the object of analysis. The choice of a particular tool also influences the scientific debate on ICT (the number and what benefits they bring?), provides specific market data to businesses and directs the agenda of the political debate (Ricci, 2000). Without the measurement of consumer participation and preferences and the nature of the business activities, no process of e-commerce could be implemented; in the same way, the digital divide could not be explained without measuring "what" divides "who" and "where" from what; and governmental strategies aimed at the development and economic growth of the ICT sector could not be designed without appropriate indicators.

Governing a country without knowing its vital statistics is inconceivable, like starting a business without being aware of the physical and human resources available. Even a simple calculation based on criteria of efficiency and effectiveness leads us to recognize that: "informed policies have a greater chance of success, as they can be better designed and targeted. Business decisions based on sound information concerning current and potential demand are more likely to be successful and produce desirable outcomes" (UNCTAD, 2003, p. 4).

Reliable and internationally comparable statistics are needed in order to inform government e-policies and business strategies, and to assist steering through the complex reality, as well as to shed light on the ultimate impacts of ICT on growth and wealth creation. Measurements are equally needed for benchmarking and for the assessment of comparative performances so that we can learn what works best by establishing cause and effect. They will also help to continue raising awareness about the real opportunities and challenges of the information society. (UNCTAD, 2003, p. 5)

Therefore, the production of comparable statistics on access and usage of ICT is crucial for formulating policies related to economic growth in the sector, cohesion and social inclusion (identifying the "haves" and "have nots " and how they are defined?), and to monitor and assess the impact of ICT on social and economic development. However, internationally comparable statistics are quite limited, particularly with regard to developing countries (Schulz & Olaya, 2005). To deal with this information gap, in 2004 the Partnership on Measuring ICT for Development,[6] was set up; its task was mainly that of defining standards, harmonizing statistics on ICT globally and developing a list of key indicators of ICT (Partnership on Measuring ICT for Development, 2005).

The need to measure ICT also derives from the consideration of the transformation their development led to economy, politics and society. On the one hand there is concern to understand and monitor the developments of this process, and on the other hand the aim is to build useful tools for addressing public and private action, so as to reduce the risks associated with social changes resulting from the spread of ICT. In this vein, the traditional indicators of industrial society have to be in order to verify their reliability.

From the Information Society to Information Without Society

Despite the fact that the study of IS involves numerous issues, ranging from ICT availability to access mechanisms, from the type of hardware and software to the applications used, from the quantity and dexterity of use to results of medium and long term impact, IS basically has to do with people, with their lives, i.e. with their voluntary or otherwise routine decision making. The study and measurement of the IS implies, or in theory should imply, simply the intermediate steps needed to classify, understand and predict human behaviour. However, one of the fundamental problems of existing measuring tools is their technology-centrism (Pruulmann-Vengerfeldt, 2006). "However, as many theorists [...] have pointed out, society is not exclusively—if at all—driven by technology; thus, measuring

computers, cables, and connections tells us very little about the actual state of society" (Pruulmann-Vengerfeldt, 2006, p. 303). The attention to the informational aspects leads to completely ignoring the social ones, while to understand the IS we should go beyond measuring the diffusion of pieces of hardware and analyze the social context within which these developments take place (Pruulmann-Vengerfeldt, 2006). In fact, "these technologies do not create the transformations in society by themselves; they are designed and implemented by people in their social, economic and technological contexts" (Mansell & When, 1998, p. 12). The application of ICT is subject to "social shaping" (Kubicek, Dutton & Williams, 1997), in that it is linked to factors such as previous technology, culture and the legal framework which cannot be neglected. Instead, amongst the existing instruments and approaches "the great absentee, as usual, is the ordinary citizen who is supposed to benefit from all these innovations and experience a more brighter life thanks to them" (Menou & Taylor, 2006, p. 265).

But despite being largely excluded from the measurement methodologies of IS, with the diversity and complexity of everyone's behaviour, the individual, at least theoretically, remains central to the coexistence of different perspectives of analysis of IS: social, technological, economic, spatial, occupational (Webster, 2002), and although they are all interlinked with one another (e-Business W@tch, 2005), there is a prevalence of technological and economic dimensions (mainly because this sector was the first to see the potential of the network), as can be evinced[7] by the presence of tools and indicators focused on technology and/or addressed to analyzing investment in ICT and productivity of e-commerce, or to *translate* the impact of ICT in a change in the market share of companies (Albright, 2005; Menou & Taylor, 2006).

The choice of a particular perspective therefore determines a distinctive definition of the object of analysis, underlining particular aspects at the expense of others and the use of specific indicators. From a theoretical and methodological point of view, each perspective reflects the use of one of the classical approaches in the social sciences, the quantitative and the qualitative, or is a dialectical synthesis.

Qualitative, philosophical, conceptual approaches to the analysis of the effects of the new technologies on our societies focus the definition of the «IS» on the analysis of the quantity, ubiquity and speed of information flows, on how these new patterns of communication are capable to «compress time and space» and on the impacts of modern technologies on the perception of reality/identity, on culture/social, alienation/integration, on the balance of powers within the political and social system and on the formation and manipulation of public opinion [...]. Quantitative, substantive, behavioral approaches either refer to the actual amount of information channeled to the masses, or to the emerging social scenarios (i.e. the effects on the economy, on employment, on political stability, on public order, on fundamental rights), or focus on the technological developments (such as the Internet and the hyperlinked databases of the World Wide Web) as the key agents of change, or on end users and their behaviour. (Ricci, 2000, pp. 142-144)

Whatever the approach, the construction of a system of indicators of IS needs to overcome a series of obstacles and restrictions. One of the first problems—which has already been mentioned—stems from the new and extensive nature of the phenomenon. This makes knowledge about IS rather rudimentary and thin, partly because the background lacks a strong theoretical framework of reference (Bianco, Lugones & Peirano, 2003; Menou & Taylor, 2006). A second difficulty is that we are facing a global but not homogeneous process: this means that it would take comparable indicators at the international level, but at the same time ones capable of bringing out the

diversity of local situations. In other words, the indicators should be able to preserve the right balance between the *intension* and *extension* (Sartori, 1970) of the concept of IS. Another critical factor lies in methods of data collection: the increase in demand for statistical data on the IS and the nature of the phenomenon itself result in a wide dispersion of data sources. In addition to the national statistical offices, in fact, often the data collected from institutions, organizations and companies are not always completely reliable and disinterested as concerns the production of this information. While it is unthinkable at the present time to propose a unique source of information, which for logistical and political reasons would be unable to cover all the issue of IS nor obtain consensus, the alternative to relying on the present disparate sources leaves important questions open on issues such as validity, comparability and reliability of data. The solution sought at an international level to ensure the consistency and quality of the information is the creation of a network strategy (of which the Partnership on Measuring ICT for Development is one of the most important representations) based on the cooperation of different sources, which need to reach an agreement on what should be measured and how (Bianco, Lugones & Peirano, 2003). The road ahead is still lengthy and problems are not lacking, especially with regard to the peripheral role in the process reserved for developing countries[8] (Albright, 2005). However, the difficulties which at present seem insurmountable are linked to the different reasons based on which the various stakeholders wishing to compare the performance of different countries in the world, choose the most suitable instruments and indicators. It is once again the inextricable mix of motivations of the subject client and the measurement tools available to determine the universe of reference of IS within which to make political, economic and methodological decisions. Despite these obvious obstacles and the lack of consensus on what to measure and the difficulty even in harmonizing the different methodologies, at an international level actors have been identified who, although not yet defined as the pattern-setters in the construction of a shared framework of measurement of IS, have taken a leading role in the debate on the issue. It is therefore to them and to some of the most reliable measurement methodologies on an international scale that our attention is now addressed.

ACTORS AND TOOLS

If one of the priority objectives is to achieve standardization of procedures for the measuring and harmonisation of indicators, it is clear that the most appropriate level of discussion is the international one. It is mainly through the work of international agencies and institutions that the on the measurement of ICT debate has been fueled. In particular the work by OECD is highly relevant; its members are among the main promoters of ICT development. According to art. 1 of its Convention, the OECD must pursue the objective of promoting policies designed to achieve "the highest sustainable economic growth and employment and a rising standard of living in Member countries" (OECD, 2002, p. 2). To carry out this policy, focus, mainly from the 90s onwards, was centred on the inadequacy of traditional economic indicators that were unable to take into account ICT as the greatest source of economic growth in recent years. OECD's work in this area has focused mainly on the recognition of the importance of developing appropriate statistical indicators to understand the changes taking place and to provide adequate information as a basis for public policy. This work was carried out through the Committee for Information, Computers Communication Policy (ICCP) Statistical Panel, originally set up with the objective of establishing "a set of definitions and methodologies to facilitate the compilation of internationally comparable data for measuring various aspects of the information society, the information economy and electronic commerce" (OECD, 1999, p. 2). The name of the Panel was changed in 1998 to the Working Party on Indicators for the Information Society (WPIIS), responsible for the production of key methodological documents by OECD on ICT indicators.

Included among the major results achieved by WPIIS is the development of a working definition of e-commerce, now widely used by all member countries (OECD, 2003) and the production of numerous guidelines and questionnaires to measure the use of ICT in the household and in business. Of particular importance, finally, was the publication of the book *Monitoring the Information Economy*, containing scores of quantitative indicators for mapping the IS in different countries (OECD, 2002), and especially the *Guide to Measuring the Information Society* (OECD, 2005).

At the same time as the OECD started its work on IS, the European Commission developed an action plan to meet the challenges posed by the introduction of ICT (European Commission, 1994). In 1999, a communication from the European Commission entitled "eEurope: an Information Society for All" (European Commission, 1999), submitted to the European Council in Lisbon in March 2000, was the basis for the subsequent adoption of the Action Plan "eEurope 2002: an information society for all" (Council of the European Union & Commission of the European Communities, 2000), a tool considered essential to achieve three objectives:

1. To create a cheaper, faster and more secure Internet;
2. To invest in human resources and training;
3. To promote the use of the Internet.

At the same time the Commission prepared a list of 23 indicators to monitor the progress of the action plan and the development of IS in member countries (Conseil De L'Union Européenne, 2000).

These initiatives were followed by the recent "*i*2010 – European Information Society 2010", the aim of which is to exploit opportunities for economic growth and jobs in Europe by promoting an open and competitive digital economy. It proposes three priorities for Europe's information society policies:

1. The completion of a Single European Information Space;
2. The strengthening innovation and investment in ICT research to promote growth and jobs through wider adoption of ICT;
3. The achieving of an inclusive European information society that prioritises better public services and quality of life (i2010 High Level Group, 2006).

The monitoring of progress is achieved through the use of flexible and timely indicators and benchmarking strategies.

Starting from the indicators used to benchmark eEurope, the i2010 list of indicators focuses on more complex issues of impact and use of technologies in the wider economy. The indicators are grouped under nine themes: (1) Developments in broadband,(2) Advanced services, (3) Security, (4) Impact in relation to the overall Lisbon objectives of growth and employment, (5) Investment in ICT research, (6) Adoption of ICT by businesses, (7) Impact of adoption of ICT by Business, (8) Inclusion, (9) Public services.

The European strategy for the promotion and dissemination of ICT is closely linked to the Lisbon strategy for economic growth: ICT are considered essential to the modernisation of the economy, since they drive productivity growth, create an open and competitive digital economy and stimulate innovation to tackle globalisation and demographic change.

Another major stakeholder is the above mentioned ITU. The agency, in addition to producing annual reports on the development of ICT (ITU, 2006; 2007c) and defining its own list of indicators

(ITU, 2007a; 2007b), also organizes international meetings concerning the debate on issues related to the identification and the definition, collection, processing and dissemination of indicators related to ICT and telecommunications. Of these the most important is certainly the World Summit on the Information Society (WSIS), a forum for discussing the many aspects of IS, including the measurement of ICT. As stated in the *Plan of Action* of the summit:

A realistic international performance evaluation and benchmarking (both qualitative and quantitative) through comparable statistical indicators and research results, should be developed to follow up the implementation of the objectives, goals and targets in the Plan of Action, taking into account different national circumstances. (WSIS, 2003, p. 13)

The document also calls on all countries to develop consistent and comparable indicator tools and systems at an international level. Finally, it suggests a series of indicative targets relating to the use of ICT in different areas (access, education, health, government, culture) to be achieved by 2015. In the Tunis summit of 2005 a common policy was established for the regular assessment of progress achieved, based on a shared methodology (WSIS, 2005).

In response to the results of the statistical workshop of the WSIS in Geneva in 2003, UNCTAD (United Nations Conference on Trade and Development) coordinates the creation of the Partnership on Measuring ICT for Development illustrated above, having three objectives:

1. To define a set of key indicators of ICT shared internationally as a basis for the construction of a database of statistics on ICT;
2. To improve the capacity of the national statistical offices of developing countries, in particular by strengthening knowledge on statistics related to ICT;

3. To develop a comprehensive database on indicators of ICT by making it available on the Internet.

Finally, mention should be made of the initiative *infodev*, an international partnership promoted by the World Bank to support the global sharing of information on ICT for Development (ICT4D). *Infodev*'s mandate is to maximise the impact of ICT for achieving the Millennium Development Goals especially through support and assistance in policy and in the design and implementation of programmes.[9]

How Should ICT be Measured?

As well as for actors, in recent years different models have emerged also for methods of measurement of ICT. Clearly, it is not possible here to outline a comprehensive framework of the various initiatives, mainly because the object of analysis is so broad as to nullify any attempt at synthesis. According to the objectives of measurement, depending on the knowledge question, the subject matter and the methodologies of analysis vary: for example, if the objective is to know which online services are provided by the government, the subject of measurement will be e-government and the leading approach should look to the supply and/or demand perspective, measure the quantity and/or quality of services offered as well as the level of satisfaction and/or use of them by users. In this case, within the abstract, theoretical concept of IS (background concept), e-government acts as a "systematized concept", i.e. it functions as the point of departure for assessing measurement validity, the specific conceptual referent against which to assess the adequacy of a given measure (Adcock & Collier, 2001). Within this concept it is possible to identify different sub-dimensions (accessibility, transparency, efficiency, etc..), which in turn are linked to more specific indicators that, for measurement purposes, are operationalized in variable terms .

Bearing in mind the specificities of each measurement, key trends and processes can be identified in the general measurement of ICT.

First, the various indicators of ICT can be grouped under different areas of measurement (infrastructure, access, use) and/or macro-target groups (households, individuals, businesses). The Partnership on Measuring ICT for Development, for example, dividing the various indicators by merging these two criteria, identifies (2006):

*   Core indicators on ICT infrastructure and access (e.g., Internet subscribers per 100 inhabitants, Internet access tariffs (20 hours per month), in US$, and as a percentage of *per capita* income);
*   Core indicators on access to, and use of, ICT by households and individuals (e.g., number of households with Internet access at home; number of individuals who have used the Internet (from any location) during the last 12 months);
*   Core indicators on use of ICT by businesses (e.g., number of businesses using the Internet; number of businesses with a Web presence);
*   Core indicators on the ICT sector and trade in ICT goods (for example, number of total business sector workforce involved in the ICT sector; ICT goods imports as a percentage of total imports).

Of course, if the cognitive interest concerns a specific area, the indicators can be grouped differently. In the case of economics we could have:

*   Indicators on the ICT sector, to measure the contribution of this sector to the overall economy;
*   Indicators on ICT investment, to measure the aggregate investments by firms in ICT;[10]
*   Indicators on ICT use, which focus on the adoption and use of ICT in firms and households;

- Indicators on ICT services, to measure the availability, price and quality of ICT services, particularly telecommunication services.

Another criterion of measurement takes into consideration the different stages of development of ICT in the various countries. This "ICT' index of maturity" is obtained through the identification of three general indicators, readiness, intensity and impact, which can be applied to the measurement of different subjects: from e-commerce to e-government or ICT. Readiness evaluates the potential use and access by measuring both the technological infrastructure available and the existing skills, the expected benefits and costs for the user; intensity analyzes the level of use of ICT, in particular their frequency and nature and their more specific application; impact looks at the perceived benefits and results in terms of efficiency and effectiveness on the general economic system of the country concerned (e-Business W@ tch, 2005; UNCTAD, 2001; 2003).

Another interesting proposal, which seeks to keep most levels together, provides a modular division of the IS, represented in a matrix diagram (Bianco, Lugones & Peirano, 2003). The Information Society Indicators Matrix (or Knowledge Society Indicator Matrix)[11] identifies four IS key areas: Education, Science and Technology, IT and high value-added Services and Telecommunications. These four areas define the operational framework within which the "Use and Diffusion of Information and Knowledge Sub-Matrix", which is the second area of the general matrix, acts. The general matrix consists of four main topics, or thematic strands (infrastructure, skills, investments & efforts, applications) intersecting with the four most important socio-economic stakeholders (businesses, households, government and other institutions).

The matrix, albeit not very well known because processed outside the circuit of the leading international players involved in the issue, holds together structural factors and factors related to the agent. The former, although not directly involved in the measurement of the IS, constitute a prerequisite for understanding and analyzing the (in)successes, since they define the cultural, political, regulatory, technological and economic background of the IS. Consequently, scarce developments of these areas could have an effect on the difficulties and obstacles encountered in the development of IS. As regards method, in this case the Authors prefer to use a quantitative approach based on the selection of the major sector indicators available at an international level, and to reinterpret traditional ones.

The Problems and Challenges Ahead

To really understand present and future scenarios in the measurement of ICT, its universe of reference should be considered, i.e. the IS itself, therefore going beyond the veil of rhetorical speeches that celebrate its therapeutic virtues and depict it as the unsurpassable horizon of human evolution (Mattelart, 2003).[12] The IS, through ICT and computer networks of communications, has also set up new models of power and hegemony and the applications that depend on it will emerge from the activities of human actors, bound by their power relations (Hague & Loader, 1999, p. 4). Consequently, the Internet, one of the most popular technologies of the IS, "with its different levels of applications, creates a space of human interaction that is not neutral. The Internet is non-neutral, not only in its content and logical layers, but also in its foundations and structure" (Barzilai-Nahon, 2006, p. 270). In the same way, the logic of working of the most used search engine of the Web, the algorithm that enables Google to propose the 10 most relevant answers to a specific request, meets certain criteria which actually tend to penalize sites that are isolated to the benefit of the most cited ones, namely those covered by the largest number of links on other sites. By applying a purely quantitative logic, any search by Google refers not to the main reference on the subject, but to the most cited. It is clear that this mechanism tends to penalize sites created for the short term, which can only gain visibility and legitimacy by attracting the attention of sites that are already established. This, in addition to questioning the democratic nature of the Web - an élite democracy in which the most influential stakeholders have greater power than the newcomers - leads to a real mobilization for the ownership of the Web, with the aim of creating links with other sites and obtaining recognition by everybody who is anybody (Lazuly, 2004).

In addition to the logical conceptual point of view, the non-neutrality of the Web is evident also by its structural context. Here, as we have seen, different structures of cyberspace, as determined by specific architectures (Lessig, 2006), affect the levels of freedom (some sites are inaccessible, others are payment related, others require authentication, etc..), regulate space access and create different balances between public and private space.

The battle for dominance of a specific structure of the Web is ongoing. It is concerned not only with the logic of working and regulatory structures, but also with the issues of ownership and control of the network. Indeed this is a political and social space compared to which the functioning of the mechanisms of monopolistic capitalism implement processes of private appropriation of social knowledge (Morris-Suzuki, 1986). A specific configuration of the ownership and control of the network could, for example, have the effect of determining default structures of speech, with consequent little chance, as demonstrated by the Google case, to give space to alternative points of view. "The price to pay for inclusion in the system is to adapt to its logic, to its language, to its points of entry, to its encoding and decoding" (Castells, 1997, p. 374).

With reference to the specific problem of measurement of ICT, several different issues should be emphasised.

First, from a conceptual point of view,

it is frequently argued that the reality of developing countries is different, and that therefore adaptations must be introduced to existing conceptual frameworks and bodies of knowledge prior to implementation. Just one issue, for instance, concerns the appropriateness of the household as a unit of observation. In the context of developing countries, the notion of a household may not be the same, considering the housing situation, the more communal attitudes of people and the generally larger family size. (UNCTAD, 2003, p. 17)

The export of technological models and use of Western statistical tools, therefore, require for the careful assessment of contextual socio-cultural variables, a real "cultural screening" (Smythe, 1994), because it can not be based solely on a technology-centric dissemination model. The definition and the choice of indicators to measure the spread and use of ICT must take into account the dual need to obtain data that are internationally comparable, require a certain level of abstraction, but at the same time valid, i.e. based on indicators that are sufficiently responsive to the general concepts they wish in some way to represent.

Secondly, the risk of relying on statistical data of private organizations who produce this type of information to make a profit, must be considered. The main objective of these actors is to sell the information produced: on the one hand, this leads to production concentrated mainly on data that have market demand (mostly data industry), thereby excluding many measurement issues, on the other to a often simplified presentation of the results of measurement, in order to make the product more attractive to potential buyers. Linked to this problem there is the question of reconsideration of the role of the public sector. Despite the fact that WSIS itself states that "governments have a *leading* role in developing and implementing comprehensive, forward-looking and sustainable national e-strategies [and] the private sector and civil society, in dialogue with governments, have an important *consultative* role to play in devising national e-strategies" (ITU, 2005: 27, my italics), no region of the world has identified the most effective way to the e-regulation of IS in the role of the public sector (Venturelli, 2002). By leaving this task in the hands of private organizations, there is the risk of neglecting the analysis and solution of problems, above all that of the digital divide. It would not make sense, in fact, to ask the private sector firms, that operate on the basis of cost-benefit analysis, to implement policies for reducing the gap; they, in fact, generally tend to invest where there is a potential demand already supported by the presence of infrastructure, whose development is incumbent upon the public institutions; the task of ensuring effective e-inclusion can only lie with the public sector, both because it is linked to its citizens by a bond of political responsibility, and because part of the functions of a democratic state is the implementation of policies to ensure universal access to ICT, as a fundamental individual right (Rodotà, 2004).

CONCLUSION

As I have tried to argue, from a political-ideological point of view the most important challenge in the measurement of ICT will be, for the foreseeable future, in the deconstruction of the rhetorical discourse that celebrates the virtues of IS by hiding its political essence behind the apparent neutrality of technology and the apparent impartiality of its measurement. As Smythe (1994) argues, any technology, whether it is manufacturing or informational, embeds the ideology, politics and culture of the society in which it was created. Therefore, any design and decision related to the measurement of ICT should be treated as political and social problems. In fact,

technical fixes are less about fixing a problem, rather they are imposing a particular definition of

what the problem is (and to which the technology represents a happy solution). [...] The technical fix is, in fact, a 'political' solution, in the sense that it seeks to propagate a particular view of the world and of the methods appropriate to ordering it. (Street, 1997, p. 34)

The measurement of ICT is therefore to be considered and examined not only as a technical and methodological question, but above all as a political problem, falling within the broader context of the framework of power that rules the scenario of international relations. Regardless of what direction they take, whatever prevails, i.e. an American imperial project (McMichael, 2004), a multipolar scenario characterized by macro-regional powers or an inexorable decline of American power (Harvey, 2003; Wallerstein, 2005), the analysis of the issues of power, control and interest remains central. Furthermore "the central question concerning the character of, and prospects for, the new information technology are our familiar criteria: *for whose benefit and under whose control it will be implemented*" (Schiller, 1973, p. 175).

Currently, what is possible to note is that the adoption of specific rankings for the dissemination and use of ICT in different countries, based on criteria for measuring mainly based on technology (ITU) and economic aspects (World Bank), as compared to the underestimation of other approaches (occupational, spatial, cultural) which was grossly neglected (Albright, 2005), binds to a Western hegemony. Throughout history, in fact, technology and wealth were the main measure used by Western civilization to demonstrate its superiority. But

the criteria used by international organizations to measure information's effect on information societies are inadequate for two reasons: firstly, measurements are not translatable between the nations of the world due, in large part, to cultural differences and, secondly, the development model assumed upon which are based measurements is precisely the one that contributes to and perpetuates the economic, political, and technological domination and exploitation of developing nations by the most powerful. (Hyder, 2005, p. 25)

"Intentionally or not, these criteria may indeed reflect an inherent bias that promotes the goals of developed nations, sometimes negatively impacting those in developing countries (Hyder, 2005, pp. 26-27).

The risk is that international institutions, which have assumed a leading role in the process of institutional reform of developing countries, as well as in developing tools for the measurement of ICT, make them the scientific tool to legitimize new forms of dependency. But "no scientific instrument gives us direct, unmediated access to the phenomenon to be measured. Methods of analysis reflect values, cultural propensities, and ideological positions that must be illustrated" (Amoretti, 2006).

The aim of this chapter is to contribute to the debate on the political and methodological problems connected to the measurement of ICT. Having demonstrated that the spread of ICT does not happen in an ideological vacuum, but responds to political and economic logics, specific to certain countries and certain sectors, there is a shift in the need to distinguish between the rhetoric that focuses on the freedom of the network and policies which aim at its monitoring and regulation. A brief overview of the international actors having a leading role in the measurement of ICT has been attempted, showing the main trends. In addition to the methodological issues, although relevant, it is argued that the measurement of ICT responds to a specific logic and is functional to the consolidation of the existing hegemonic relations on the international scene. Inserted in its historical and political context, measurement must be addressed by taking into account (geo)-political and ideological aspects on which is based the at-

tempt of the richest and advanced countries in the West to impose as universal and superior values what is functional to their economic, political and cultural interests (Wallerstein, 2006).

REFERENCES

Adcock, R., & Collier, D. (2001). Measurement Validity: A Shared Standard for Qualitative and Quantitative Research. *The American Political Science Review, 95*(3), 529-546.

Albright, K. S. (2005). Global Measures of Development and the Information Society. *New Library World, 106*(7/8), 320-331.

Amoretti, F. (2006). Benchmarking Electronic Democracy. In A. V. Anttiroiko, & M. Malkia (Eds.), *Encyclopedia of Digital Government* (3 Voll.), Hershey, PA: Information Science Publishing.

Barzilai-Nahon, K. (2006). Gaps and Bits: Conceptualizing Measurements for Digital Divide/s. *The Information Society, 22*, 269-278.

Bianco, C., Lugones, G., & Peirano, F. (2003, February). *A methodological proposal for measuring the transition to Knowledge Society in Latin American countries*. Paper presented at the Second Workshop on Indicators for the Information Society, Lisbon, Portugal.

Bhuiyan, A. J. M. S. A. (2008). Peripheral View: Conceptualizing the Information Society as a Postcolonial Subject. *The International Communication Gazette, 70*(2), 99-116.

Castells, M. (1997). *The Rise of the Network Society Volume 2: The Power of Identity*. Oxford: Blackwell.

Conseil De L'Union Européenne (2000). *List of eEurope benchmarking indicators*. 13493/00, Limite, ECO 338, Annex, Bruxelles, le 20 novembre 2000.

Council of the European Union, & Commission of the European Communities (2000, June 19-20). *eEurope 2002: an Information Society for All*. Action Plan prepared by the Council and the European Commission for the Feira European Council.

Crozier, M., Huntington, S. P., & Watanuki, J. (1975). *The Crisis of Democracy. Report on the Governability of Democracies to the Trilateral Commission*. New York: New York University Press.

Daly, J. A. (2004, September). *What is wrong with ICT expenditures as a measure of ICT penetration*. Paper presented at the ASIS&T/EC and AoIR workshop on Measuring the Information Society, University of Sussex, Brighton, UK.

e-Business W@tch (2005). *A Guide to ICT Usage Indicators. Definitions, sources, data collection*. Special Report, Bonn-Brussels: European Commission, & Enterprise & Industry Directorate General.

European Commission (1999). *eEurope: an Information Society for All*. Communication on a Commission Initiative for the Special European Council of Lisbon, 23 and 24 March 2000.

European Commission (1994). *Europe's Way to the Information Society – An Action Plan*. Retrieved February 10, 2008, from http://europa.eu.int/ISPO/docs/htmlgenerated/i_COM(94)347final.html

Goldsmith, J., & Wu, T. (2006). *Who Controls the Internet? Illusions of a Borderless World*. New York: Oxford University Press.

Gramsci, A. (1997). *Le opere. La prima antologia di tutti gli scritti*. Ed. Antonio A. Santucci. Roma: Editori Riuniti.

Hague, B. N., & Loader, B. D. (Eds.). (1999). *Digital Democracy. Discourse and Decision Making in the Information Age*. London & New York: Routledge.

Harvey, D. (2005). *A Brief History of Neoliberalism*. Oxford: Oxford University Press.

Harvey, D. (2003). *The New Imperialism*. Oxford: Oxford University Press.

Hyder, S. (2005). The information society: Measurements biased by capitalism and its intent to control-dependent societies - a critical perspective. *The International Information & Library Review*, 37, pp. 25-27.

i2010 High Level Group (2006). *i2010 Benchmarking Framework*. Retrieved May 4, 2008, from http://ec.europa.eu/information_society/eeurope/i2010/docs/i2010_high_level_group/i2010_benchmarking_framework.doc

ITU (International Telecommunication Union). (2007a). *Telecommunication Indicators Handbook*. Retrieved April 14, 2008, from http://www.itu.int/ITU-D/ICT/handbook.html

ITU (2007b). *World Telecommunication Indicators database*, Retrieved January 20, 2008, from www.itu.int/ITUD/ICT/publications/world/world.html

ITU (2007c). *World Information Society Report 2007. Beyond WSIS*. Geneva: ITU.

ITU (2006). *World Telecommunication/ICT Development Report 2006: Measuring ICT for social and economic development*. Retrieved January 13, 2008, from http://www.itu.int/ITUD/ICT/publications/wtdr_06/index.html

ITU (2005). *World Summit on the Information Society. Outcome Documents. Geneva 2003-Tunis 2005*. Retrieved January 21, 2008, from http://www.itu.int/wsis/outcome/booklet.pdf

Kubicek, H., Dutton, W. H., & Williams, R. (Eds.). (1997). *The Social Shaping of Information Superhighways*. Frankfurt: Campus Verlag.

Kumar, K. (1995). *From post-industrial to postmodern society. New Theories of the Contemporary World*. Oxford: Blackwell Publishing.

Lazuly, P. (2004). Il mondo secondo Google. In AA. VV., *Il pensiero unico al tempo della rete* (pp. 44-46). Numero speciale fuori serie di *Le monde diplomatique/il manifesto*.

Lessig, L. (2006). *Code. Version 2.0*. New York: Basic Books.

Mansell, R., & When, U. (Eds.). (1998). *Knowledge societies: Information technology for sustainable development. Report for the United Nations Commission on science and technology for development*. New York: Oxford University Press.

Mattelart, A. (2003). *The information society: An introduction*. London, Thousand Oaks, CA, & New Delhi: Sage.

McMichael, P. (2004). *Development and social change. A global perspective*. Thousand Oaks, CA, London & New Delhi: Pne Forge Press.

Menou, M. J., & Taylor, R. D. (2006). A «Grand Challenge»: Measuring information societies. *The Information Society*, *22*, 261-267.

Morris-Suzuki, T. (1986). Capitalism in the computer age. *New Left Review*, *160*, 81-91.

Negroponte, N. (1995). *Being digital*. New York: Vintage.

Nulens, G., & Van Audenhove, L. (1999). An information society in Africa? An analysis of the information society policy of the World Bank, ITU and ECA. *International Communication Gazette*, *61*, 451-471.

OECD (Organization for Economic Co-operation and Development). (2005). *Guide to measuring the information society*. Paris: Oecd.

OECD (2003). *A framework document for information society measurement and analysis*. Paris: Oecd.

OECD (2002). *Measuring the information economy*. Retrieved March 20, 2008, from http://www.oecd.org/dataoecd/16/14/1835738.pdf

OECD (1999). *Report on the activities of the working party on indicators for the information society*. Retrieved March 18, 2008, from http://www1.oecd.org/std/nameet99/Docs/na99_48e.pdf

OECD (1996). *Global information infrastructure and global information society, (GII-GIS). Statement of policy recommendations made by the ICCP Committee*. Retrieved April 10, 2008, from http://www.oecd.org/dataoecd/3/58/1896739.pdf

Partnership on Measuring ICT for Development (2005). *Core ICT indicators*. Retrieved February 15, 2008, from www.itu.int/ITU-D/ICT/partnership/material/CoreICTIndicators.pdf

Ricci, A. (2000). Measuring information society. Dynamics of European data on usage of information and communication technologies in Europe since 1995. *Telematics and Informatics*, *17*, pp. 141-167.

Rodotà, S. (2004). *Tecnopolitica. La democrazia e le nuove tecnologie della comunicazione*. Bari-Roma: Laterza.

Sartori, G. (1970). Concept misformation in comparative politics. *The American Political Science Review*, *64*(4), 1033-1053.

Schiller, H. I. (1985). *Strengths and weaknesses of the new international information empire*. In P. Lee (Ed.), *Communication for All* (pp. 3-23), New York: Orbis.

Schiller, H. I. (1973). *The mind managers*. Boston: Beacon.

Schulz, C., & Olaya, D. (2005). *Toward an information society measurement instrument for Latin America and the Caribbean: Getting started with census, household and business surveys*. Santiago, Chile: United Nations.

Servaes, J., & Heinderyckx, F. (2002). The 'new' ICTs environment in Europe: Closing or widening the gaps? *Telematics and Informatics*, *19*, 91-115.

Smythe, D. (1994). *Counterclockwise: Perspective on communication*. Ed. Thomas Guback. Boulder, CO: Westview Press.

Street, J. (1997). Remote control? Politics, technology and "electronic democracy". *European Journal of Communication*, *12*(1), 27-42.

UN (United Nations). (2003). *Expanding public space for the development of the knowledge society*. Report of the Ad Hoc Expert Group Meeting on Knowledge Systems for Development. 4-5 September 2003. New York: United Nations.

UNCTAD (United Nations Conference on Trade and Development). (2003, July 3). *Information society measurements: The case of e-business*. Background Paper by the Unctad Secretariat TD/B/COM.3/EM.19/2.

UNCTAD (2001). *Electronic commerce and development report 2001*. New York & Geneva: United Nations.

Venturelli, S. (2002). Inventing e-regulation in the US, EU and East Asia: Conflicting social visions of the Information Society. *Telematics and Informatics*, *19*, 69-90.

Wade, R. H. (2005). Failing states and cumulative causation in the world system. *International Political Science Review*, *26*(1), 17-36.

Wallerstein, I. (2006). *European universalism: The rhetoric of power*. New York: The New Press.

Wallerstein, I. (2005). After developmentalism and globalization, what? *Social Forces*, *83*(3), 1263-1278.

Webster, T. (2002). *Theories of the information society 2nd Edition*. London: Routledge.

WSIS (World Summit on the Information Society). (2005). *Tunis agenda for the information society*. Retrieved April 18, 2008, from http://www.itu.int/wsis/docs2/tunis/off/6rev1.html

WSIS (2003). *Plan of action*. Retrieved April 16, 2008, from http://www.itu.int/dms_pub/itus/md/03/wsis/doc/S03-WSIS-DOC-0005!!MSW-E.doc

ENDNOTES

[1] "La realizzazione di un apparato egemonico, [dal momento che questa,] in quanto crea un nuovo terreno ideologico, determina una riforma delle coscienze e dei metodi di conoscenza, è un fatto di conoscenza, un fatto filosofico". The quote is taken from Quaderni del carcere [Prison Notebooks], notebook 10, Part II, paragraph 12.

[2] For example, the OECD is composed of 30 member states, which correspond to the 30 more advanced market democracies. Only two countries, the United States and Japan, account for about 48% of its funding; the World Bank, through the mechanism of shares, under which each member of the Bank operates a share of votes equal to that of its equity subscriptions, is basically controlled by the more developed countries. Only countries of the G8 (the United States, Japan, Germany, France, United Kingdom, Canada, Italy and Russia) hold approximately 46% of the total votes. Finally, with regard to ITU, "because in the 1990s the willingness of the member states to fund the ITU declined, the organization was obliged to open up towards the private sector […]. This trend is criticized by the developing countries, arguing that these companies are already represented by their respective home countries" (Nulens & Van Audenhove, 1999, p. 461).

[3] The ITU is the leading United Nations agency for ICT. It is globally one of the main points of reference for governments and the private sector, particularly in the areas of

radio communication, standardization and development.

4   This group, the Working Party on Indicators for an Information Society (WPIIS), became operational in 1997.

5   Lessig wrote that: "The differences in the regulations effected through code distinguish different parts of the Internet and cyberspace. In some places, life is fairly free; in other places, it is more controlled. And the difference between these spaces is simply a difference in the architectures of control – that is, a difference in code. If we combine the first two themes, then, we come to a central argument of the book: The regulatory issues described in the first theme depends on the code described in the second. Some architectures of cyberspace are more able to be regulated than others; some architectures enable better control than others" (Lessig, 2006, p. 24).

6   The Partnership is currently composed of the following members: Eurostat, ITU, OECD, UNCTAD, UNESCO (United Nations Educational, Scientific and Cultural Organization), UIS (Institute for Statistics), UN ICT Task Force, World Bank and four regional commissions of the United Nations (UNECA, UN Economic Commission for Africa, UNECLAC, UN Regional Commission for Latin America and the Caribbean, UNESCAP, UN Economic and Social Commission for Asia and the Pacific e UNESCWA, UN Economic and Social Commission for Western Asia. See http://measuring-ICT.unctad.org.

7   See also par. "How should ICT be measured?"

8   One of the greatest risks is that new technologies become the tool for creating new forms of dependence ("e-dependency") on the part of developing countries in favour of advanced countries. For example, by adopting the software and hardware tools, as well as standards, developed and disseminated by the most important Western multinationals, developing countries undermine their national sovereignty; as well as using a technology based on a software protected at source, namely that cannot be developed and/or amended by themselves, these countries restrict their capacity for growth, delegating the task to external actors, mostly Western ones (Wade, 2005)

9   Further details on this initiative and its activities are available at the website www.infodev.org/en/Publication.166.html

10   As noted John Daly (2004), using this indicator, based on monetary value, could be misleading for comparison at international level especially because it does not take into consideration the use of free open source software, as well as the effects of dissemination of software piracy.

11   To use, as the Authors do, the terms "information society" and "knowledge society" as synonymous is not without repercussions. It means in fact to take that information and knowledge are the same thing (which is not) and that it is sufficient to disseminate ICT to build the knowledge society (See UN, 2003).

12   Among the most critical radical positions of the ideological use of the term "information society" there arealso those who, like Bhuiyan (2008), maintain a close connection between this expression and the realization of a political project of global neo-colonial domination, put in place by Western societies with the aim of expanding capitalism of information to other regions of the world.

Chapter XIII
The Fabrication of Networked Socialities

Paolo Landri
Cnr and University of Naples, Italy

ABSTRACT

This chapter is dedicated to analyse the fabrication of networked socialities, that is to address the complex interweaving of technologies of information and communication and the manifold instantiations of sociality. Networked socialities are digital formations being produced out of the intertwining of social logics outside and inside digital spaces and society. Such contribution is organized as follows: first, it will present the theoretical frame necessary to grasp the fabrication of sociologies in our information age, drawing on some concepts elaborated by the social studies of science and technology, together with the studies of the global digital worlds. Then, it will highlight the analytical fruitfulness of this perspective by describing some digital formations, such as social network sites, virtual communities of practice, and electronic markets. Finally, it will discuss the effects and the implications of such fabrication as a re-configuration of social, the emerging post-social relationships as well as the increasing fragility of knowledge societies.

INTRODUCTION

The aim of this chapter is to analyse the *fabrication of networked socialities*, i.e. to address the complex interweaving of technologies of information and communication and the manifold instantiations of sociality. Networked socialities constitute emerging forms of society, and the materializations of its electronic constitution. Networked socialities are digital formations being produced out of the intertwining of social logics outside and inside digital spaces and society (Latham & Sassen, 2005b).

The role played by information and communication technologies in the transformation of contemporary societies has been widely acknowledged. Their diffusion contributes to re-design new scenarios and relationships among social and institutional agencies by affecting forms of government, regulative practices, as well as knowledge and organizational learning. However, most analyses of digital worlds are dominated by a focus on technical properties, drawing on simplistic accounts of the imbrications of technologies and society. The effect of such a stance is the application of technology to society, as if they were two separate worlds. On the other hand, sociological analyses are carried out by a macro approach that takes the 'social' (and 'society') for granted, infrequently investigating the practicing and the networking leading to the emergence of society and to the many forms of sociality.

In order to offer a more encompassing view of the electronic constitution of society, the present chapter adopts a perspective which looks at the mutual constitution of technology and society, arguing for the appropriateness of the analytical categories of the social studies of science and technologies (in particular, from the 'actor-network' theory) in addressing the imbrications of technology and society.

We will argue that technologies can be seen as destructive, reproductive as well as constitutive of forms of sociality, not relying on the essence, substance, or intrinsic logic of technology but on the situated fabrication of technology *and* society. In this sense, we will try to expand the analysis of the forms of sociality given by Latham and Sassen (2005a), by encompassing the dystopian effects of technologies, such as the destruction of sociality inherited by the sociology of industrial society, or the post-sociality forms of post-modern reflections on the re-shaping of knowledge societies.

This chapter will proceed as follows: first, it will present the theoretical frame necessary to grasp the fabrication of socio-logies in our information age, drawing on some concepts elaborated by the social studies of science and technology, together with the studies of the global digital worlds. Then, it will highlight the analyti-

cal fruitfulness of this perspective by describing some digital formations, such as social network sites, virtual communities of practice, and electronic markets. Finally, it will discuss the effects and the implications of such fabrication as a *re-configuration of social*, the *emerging post-social relationships* as well as the *increasing fragility of knowledge societies*.

THE MUTUAL CONSTITUTION OF TECHNOLOGY AND SOCIETY

Information and communication technologies represent a challenge to the vocabulary of social science. Actually, most sociological analyses tend to overemphasize the role of technologies, thus risking determinism, or the use of a repertoire of concepts ineffective to analyze their re-defining effect in shaping the social.1 In this respect, a more interesting approach can be devised by looking at the social studies of science and technology (Wajcman, 2002; 2006). This might help the analysis of the complex imbrications of technology and society, the understanding of the fabrication of sociality by the information and communication technologies, also in connection with the debate on knowledge societies. It will also include a set of issues related to how knowledge practices contribute to the making of the social.

By articulating this tradition of study with some conceptualizations recently drawn by organization studies, we will try to analyse the making of sociality in the area of internet. Sociality usually refers to ways of grouping, linking and mutuality among humans. In the case of Internet, it is the making of *networked socialities*, that is, the complex configurations of association among humans and non-humans embedded in the digital. In some respects, Internet can be considered as a laboratory for the making and the re-making of sociality. In order to grasp the fabrication of this sociality, however, we need to address the *materiality of sociality*, the *imbrications of the* social *'outside' Internet and the electronic space*, and the *mode of attachment* (the 'affiliative relationship', see Suchman, 2005) to what has been called 'virtuality'.

We will build on the work of Latham and Sassen (2005b) on the *digital formations* to be considered as instances of the networked sociality we would like to investigate. Their study unfolds in a bigger program on IT and on the manifold effects of the social realms in international cooperation, providing detailed case-studies of transboundary relationships. The construct of digital formation appears to define the object of their study, proving a useful tool to grasp the imbrications of technology and society in the case of cyberspace.

Digital formation refers directly to ICT facilitated networks emerging in the global space outside the frame of nation-states, namely in electronic markets, open source software, EU, China, NGOs, on-line discussions, economic configurations affecting the rescaling of traditional configurations of space globally and locally. Latham and Sassen (2005a) carefully suggest that their study focuses on social formations in their emerging stage, which evolve rapidly insofar as the dynamics of technology and society continue. The same concept can be useful to address the mutual constitutions of technology and society in Internet. A digital formation is 'a coherent configuration of organization, space and interaction' (p.10). Digital formations are constituted by the overlapping of three dimensions: organizing, interacting, and spatializing. By organization, Latham and Sassen (2005a) mean the ordering of the field of practice constituting the formation, in reference to the rules and the roles attributed to people, to machines as well as to the contents of the electronic space. By interaction, they consider the flows of exchange and communication among actors. Elsewhere, Lanzara and Morner (2005) have noted that the sociality on the web can be viewed as an interactive system where communication offers a chance for the unfolding of exchanges leading to the mutual shaping of technologies

and the societies accompanying them, mainly performed by technological artefacts. By space, finally, they mean 'the electronic staging of the substance (or content) and social relations at play in a digital formation' (p. 10). These dimensions overlap, and attribute a temporary stabilization—that is to say, coherence and (contingent) identity—to digital formations.

In a different vocabulary, we could say that these dimensions contribute to the momentary 'closure' of social digital formations, in respect to the possible Internet relations between space, interaction and organization. In this sense, the concept appears to be wide enough, albeit not completely abstract, to include forms of networked socialities – from the actual to the possible. A digital formation is a multi-layered entity (in the sense previously defined), which includes (but is not reducible to) social networks, virtual communities, electronic networks, electronic markets, etc..

At the same time, it is possible to re-read these constitutive concepts of digital formations through a vocabulary drawn from the 'actor network' theory, so as to suggest the dynamics (the making of) of temporary closures on the complexity of cyberspace. In this view, quoting Latour's insightful essay entitled 'Technology is society made durable' (1991), digital formations can be considered as social formations temporarily frozen in the complexities of digital space. Their contingent identity depends on the stabilization of the actor-networks, i.e. the fragile assemblages of humans and non-humans emerging from ongoing practices of *inscribing* (and performing) digital space, *translating* and *framing*, namely reaching an outcome in the practice with the subjects and the objects entering the fabrication of digital formations (Faraj, Know & Watts, 2004).

The notion of *inscription* is useful to understand the making of electronic space. According to Latham and Sassen, 'electronic space is composed of picto-textual social artefacts embodied in the electronic staging of texts, images, and graphics through hardware and software. A range of realized and potential relations and action is opened up to produce electronic space' (p.11). The definition points out at the *materialization of this sociality*, the matter composing it. Here, the concept of inscription sheds light on the fact that picto-textual artefacts are embodiments of the designers' configurations, of the actors' interests and visions of the digital artefacts, and of the different assumptions on what will be the use of a digital environment.

In addition, the practices of inscription involve a work of translation. In this case, the notion of translation refers to the strategies whereby an actor tries to interest other (human and non-human) actors in order to support the construction of a claim, a fact, a machine (Latour, 1987). The concept applies equally to the digital and to the non-digital, referring to machineries aiming at enrolling actors in digital formations, and visualizing the *complex imbrications among digital socialities and the 'outside' sociality*. Interactive technologies on Internet can be instrumental to organizational or institutional restructurings, or mobilized towards these aims, while the trajectories of the organizational or institutional change might support the making and stabilization of digital formations.

As a result of the ongoing negotiations and confrontations, and of the different strategies of translation, digital formations reach a temporary effect, that is, the *framing* implying an ordering of the practices regarding the roles, the rules (the legal limits of use, for example), the electronic space, as well as the boundaries within which the interactions among those who are involved in digital formations, exist. This explicitly refers to the model of digital formation that Latham and Sassen relate to the dimension of 'organization'. The term 'framing' implies the result of a structuration practice giving form and, accordingly, identity to a digital environment.

In order to complete our theoretical re-reading of digital formations, we would like to include the relevance of the issue of 'emotion', a neglected yet interesting problematic, within the fabric of the social on the web. Digital formations as forms of new social settings are *emotional arenas* to be analysed not as reductive modes of expression, but as places where emotions are presented in particular and proper ways. Here, in order to address this dimension, a relevant concept is given by 'affiliative relationships' (Suchman, 2005), and by the related conceptualization of 'attachment' (Latour, 2005; Hennion, 2004). These concepts show that social configurations present a specific architecture of association and disassociation so as to imply an attachment to the subjects and objects of the communities of practitioners, a relevant material repertoire, and the detachment from other—competing, or simply different—configurations of association.

The dynamics of attachment/detachment focus on the affective side of the configurations of association, and, accordingly, on the emotions accompanying (or as effects of) social digital settings. In recent times, this dimension is attracting attention in a growing number of contributions (Fineman, Maitlis, & Panteli, 2007; Gherardi, Nicolini, & Strati, 2007). In this area of research, a first wave described the digital environment as a site for an impoverished display of emotions, with a nostalgic tone, and a more than explicit preference to face-to face interactions. At the same time, it is also possible to find an enthusiastic, and in a way excessive, description of the novelty of emotions in the virtual world as a new revolutionary frontier (this is a revisionist view, see Fineman, 2006). In a more modest, yet middle-ground, point of view, the configurations of association producing and reproducing digital formations, 'offer creative opportunities for individuals to experiment with the construction and expression of feeling and to negotiate novel emotion protocols, some of which will become institutionalized for the medium' (Fineman et alii, 2007, p. 556). This position opens up an interesting agenda for research in the way digital formations, and more generally, virtuality, reconfigure the settings (cues, prompts, bodily dispositions etc.) that shape emotional displays.

FORMS OF NETWORKED SOCIALITIES

We will now see how the three dimensions of fabrication—materiality, imbrications and attachment—operate in practice, by presenting a set of networked socialities. Our focus is on digital formations to be considered as modes of inscription/interaction/and organization in the digital sphere. These modes do not constitute an exhaustive typology; they are rather diffused translations of 'social' mirroring and reshaping, constituting social formations - namely new socio-technical networks.2 We will see that these networked socialities are not limited to the discourse of the 'community on line', yet include a wider list of forms, such as electronic (or social) networks, virtual communities, and electronic markets.

We will start with the work by Latham and Sassen (2005a), expanded in a way to make it work on a more general level of argumentation. These forms do not present set-limits, but may easily move from one form (the on-line community, for instance) to another (a social network).

Social Network Sites

Social network sites (SNS) are popular and global phenomena attracting increasing interest in recent scholarly publications (Boyd & Ellison, 2007; Goldbeck, 2007). Participation is growing rapidly, and it concerns websites with more or less generalized audience targets, ranging from generic networking among lists of 'friends of friends' to networks with specific characteristics (religious orientation, or common professional interests). Worksites such as MyFace, Facebook, Cyworld, register millions of users and are being integrated in the everyday life of their users. In a recent analysis, Boyd and Ellison (2007) define SNS as 'web-based services that allow individuals to (1) construct a public or semi-public profile within a bounded system, (2) articulate a list of other users with whom they share a connection, and (3) view and traverse their list of connections and those made by others within the system'. In a way, SNS could be seen as technology web services able to connect and meet 'strangers'; in fact, they are more suited to allow exchange and communication from 'within', giving visibility to extended social networks. Frequently, they permit connections which could be otherwise difficult to sustain, for instance, due to the distance between those who are part of 'off-line' social networks.

SNS can present different technological affordances; yet, their materiality gathers common features. In order to participate to a SNS, the user is asked to define a visible profile, and display a list of friends. The user must provide information about his/her age, location, interests, plus a personal presentation. In some cases, it is possible to add photos to the profile, enhancing it through intervening media tools (images). Profiles can be more or less visible, depending on the rule of the website, and on the users' intention to make them visible. Thus, the site grants access by a reconfiguration of identity – the making of the profile being a way to highlight a portion of one's self, a particular set of characteristics of a 'life of screen' – which is, in the terms we borrow from the 'actor network' theory, an inscription of identity in the electronic space, the building block of the materiality of sociality.

The space is made of the profile, and also of the links between profiles. Once the users have gained access, they are asked to identify other users they have a relationship with. The connection is not necessarily a link of friendship – the term 'friend' is, in this case, misleading. The public display of connections allows users to cut across the connections themselves. SNS furnish instant services, blogs, the possibility of leaving messages. This digital formation has, of course, its history: it does not necessarily need this characteristic from its very beginning – some SNS start as messaging services; others are born as a communities, to then 'evolve' towards other features.

Recent research on SNS reveals 'unexpected' links between the sociality off-line and the social world on-line. Utopian views on cyberspace emphasize the role of Internet in reducing social inequalities by offering opportunities to widen the social capital, and in the unfolding of complex and expanded social networks. However, other studies on social network sites present a complex, not definite, landscape. In some respects, SNS seem to allow the cultivation of existing social networks, constituting less a tool to ease new socialities. Scholars have actually stressed that the re-construction of network socialities tends to reproduce differentiations and social inequalities. Hargittai's detailed study on well-known and frequented SNS, such as My Face, Facebook, Xanga and Friendster, underlines the importance of addressing the issue of the differentiation of users of SNS services. His analysis draws on a sample of mainly 18- and 19-year old students attending the University of Illinois, Chicago, a particular appropriate locus because of the ethnic diversities and social differences present in its campus. Moreover, in comparison with the older cohort of a population, the sample is chosen among those who are highly wired, so that the group can be considered appropriate to the effects of deep involvements in the on-line sociality. The research's findings suggest differential uses according to gender, race, ethnicity, and parental educational levels. Additionally, the research shows that it is possible to define different profiles for different SNS, that is to say, the commonalities which orient students to the use of a particular SNS (Facebook instead of MySpace, for example, or the other way around). One's off-line social network affects the choice of the site one enters, so that it is possible to say that SNS

people do not use the services in order to search for friends or new acquaintances, but to connect with already existing social networks.

However, other analyses provide different and fine-grained descriptions of SNS, which seem to account for the constitutive role of these services in sociality-making. The qualitative case-study of Cyworld, a South-Korean SNS, shows that the social network site represents a sort of intersected social world, where virtual, nonetheless lively, experiments of different socialities are performed (Kim & Yun, 2007). The electronic space of Cyworld, its pictorial elements, as well as its manifold tools, encourage the display of interpersonal and emotional communication by extending the core characteristics of the off-line communication in South-Korean culture. In turn, it contributes to reshape the forms of sociality off-line by introducing an individualistic attention to self-reflection and self-relationship, scarcely distributed in a dominantly collectivist culture.

Cyworld's case is interesting because it also furnishes descriptions of its emotional dynamics. SNS provide emotional arenas, that is, electronic venues where feelings and emotions are publicly displayed. This concerns questions of identity, as well as possibilities to represent an expressive space for emotions. Like others SNS, Cyworld inscribes identity, allowing, in a sense, an ongoing reshaping of digital selves. It represents an experiment of thought for a post-modern identity, where people can sense themselves as 'others'. In this way, it might be considered as a way to think of oneself reflexively, adjusting or acquiring a deeper sense of one's own identity.

Further, Cyworld is a place for posting messages, thus expressing emotions otherwise difficult to display and elaborate in off-line communications. Kim and Yun (2007) report the lack of emotional displays in off-line communication among Koreans. On the contrary, the accounts of the participants to Cyworld describe a richness of feelings and emotions when lived in the digital world, to be considered as a way to experiment and express affiliative relationships (sometimes, also in contrasted ways) with digital buddies.

In particular, the wide use of minihompies (template-based home page services, meant to expand and surf alternative minihompies) to be customized according to skin, music and miniroom (virtual rooms), furnishes alternative channels to overtly communicate feelings, possibly helping in relationship crises.

We have seen how SNS are ways to scale modelling social networks. They accomplish a doing and re-doing of sociality trough its specific materiality implying a complex re-articulation of the off-line sociality, leading to the reproduction or the transformation of both the off-line and the on-line sociality. Finally, we have outlined that SNS are not emotionally emptied, and that the digitalization can accompany emotional displays as well as providing a valuable support in emotional work. We will now draw the attention onto the most cited Internet formation: its virtual communities.

Virtual Communities of Practice

Both in virtual worlds and in Internet, 'community' is a recurrent and widespread term, so that it can undergo a notable conceptual 'pull'. Its word reference seems to point out at any digital formation, with Internet itself as a 'community' of networks of networks. By the same token, 'community' is at the core of many social sciences (philosophy, sociology, etc.). It is, moreover, possible to meet so many different perspectives on community that the term contributes to the growing ambiguity in the descriptive as well as the prescriptive usage of the concept.

Early scholars of Internet saw in on-line communities a brand new sociality, welcoming them as an alternative to the increasing individualization of everyday life (Rheingold, 1993). They would represent a route to re-gain social capital in our contemporary societies, or, to put it in other words,

to construct egocentric social networks in digital societies (Wellman & Giulia, 1997).

In general, the literature on virtual communities is extensive and tends to coincide with those contributions explicitly addressing the themes of the society on the web (Smith & Kollock, 1999; Jones, 1997; Werry & Mowbray, 2001; Keeble & Loader, 2001). Here, we will analyze these communities from a different (and, hopefully, more definite) view. Thus, 'virtual communities' is intended to mean that 'configurations of space, organization, and interaction sustain a common identity around shared goals and reciprocal relations among participants, and that such identity, goals, and reciprocity are an important and substantive aspect of each participant's life, professional or personal' (Latham & Sassen, 2005a, p. 13). We will approach the literature that considers the so called 'virtual communities of practice' in organizational settings, or in broader institutional environments (Dubé, Bourhis, & Jacob, 2005; Sproull, Dutton & Kiesler, 2007; Gherardi, 2007). By reducing the scope of its definition, we will try to analyse the specificity of such digital formation in respect to its possible typology by following the three dimensions – materiality/imbrications/emotions – we have so far identified. At the same time, the definition is a contextualization of basic sociological vocabulary, which nonetheless stretches the concept of 'community' as to include a distributed social formation.

In comparison to SNS, 'virtual communities' have a less specific socio-digital space. It has been noted that, for nearly 20 years (from '75 to '95), the sociomateriality of these communities has been growing characteristics of complexity and plurality. If in the past, most of them were mainly focused on e-mail groups and bulleting boards, now we have a wider range of possibilities, albeit in this case the electronic space seems to be defined by a 'constellation of site and electronic postings' (Latham & Sassen, 2005a). A well-know example of virtual community is the open source communities that, for some, represent a paradigmatic case of what should have been the novelty of Internet in terms of an alternative political economy.

These communities have been empowered, and, in a sense, created by Internet. In turn, their success suggests the possibility of reshaping off-line societies by imagining other modes of 'open source economies' (Weber, 2005). Their core is the production of software, considered to be free and public - that is, with the double features of being public and non-proprietary. The source code is the object of work, collaboratively and continuously done and re-done within these distributed communities, voluntarily working in a relatively unstructured way, and without monetary compensation. Those communities operate in inscribing the rule of action in a source code, to be performed by sets or clusters of computers (Lanzara & Morler, 2005). This does not unfold in a social vacuum; namely virtual communities can contribute to the reshaping of the social off-line logic, as in the case of the open source communities which overtly contribute to the transformation of the rule of software intellectual property (Weber, 2005). They can mutually support the constitution of technology and society. In this respect, the relationships between NGOs and Internet are particularly interesting (Bach & Stark, 2004 e 2005). NGOs and interactive technology can be regarded as 'co-evolving actants'; both appear to be isomorphic in privileging the concept of 'network' as operational logic. Thus, the combination between NGOs with Internet seems to define organizational forms that are capable of redefining the global political space, and to contribute to the growing success of these organizations in the '80 and particularly during the '90.

However, this did not constitute a straightforward outcome of a simplistic and additive logic between organizational forms (fashionable networked) and interactive technologies. Rather, it appeared to be an emergent effect of the hybridization of NGOs organizational strategies - from

autarky to a collaboration with state agencies and market operators – and an appreciation of the combinatory and multiplicative logic of Internet ('link, search and interact') which supported their transformation into knowledge communities. Again, in this case, the overlapping between off-line and on-line socialities bridge continuity, defining this digital formation as a form of sociality made, expanded and re-made through Internet, yet embedded in organizational and social logics outside cyberspace.

Moreover, these communities are emotional arenas - configurations triggering passions, and conveying emotions. To illustrate this aspect, we will draw on some researches on 'empathic communities', and on the analyses of passion and community. The term 'emphatic communities' was introduced by Preece (1999) to shows how the success of on-line communities relies on a delicate balance between the emotional side and its informational content. In that research, a focus on self-help medical communities revealed that computer-mediated communication is helpful to activate and hold communities together. This role implies a shared cognitive interest, and a less instrumental aspect: on-line communities are helpful in providing information and news, yet they provide support, sustain emotional work, and help patients by establishing empathic relationships drawn on feelings.

These results suggest the relevance of the emotional side in holding on-line communities together, and in managing the knowledge-base of virtual communities of practice. Here, the most frequent reference in addressing those themes regards the above mentioned open source communities, which, in some respects, are viewed as a 'model' for highlighting the intertwining of emotion and virtuality (Kollock, 1999).

The intrinsic motivation of contributing to the development of the source code – namely the possibility of fostering the writing of appropriate lines of programming for its own sake – triggers the engagement of the software designer in the collective endeavour of the open source movement. By the same token, this involvement is accompanied by joy and pleasure, and by the effort of spending time and energy in the on-line collaborating to the refinement of the code. The emotional engagement is not limited to the open source communities, but it seems to accompany the virtual communities of practice, and, in particular, the communities of practitioners in the field of software making.

In some cases, the engagement has been described in terms of *flow states* (Kaiser, Muller-Seitz, Pereira Lopes & Pina Cuhna, 2007). The flow is a state of immersion that describes the pleasure of being involved in everyday activities or innovative activities (Csikszentmilhalyi, 1997). In this case, the experience of flow is associated with high freedom in terms of work; with the possibility of affecting the code writing; with the reciprocal social exchange engendering feelings of belonging and identity. The challenging nature of the tasks is particularly important; the experience of satisfaction and pleasure in having contributed, via discursive practices of the community, to the accumulation and the creation of knowledge.

Electronic Markets

Electronic markets represent a relevant case in the illustration of major novelties carried by the processes of globalization. They embed the process and the outcome of the processes of global market integration. In terms of digital formations, they allow complex intersections among the different macro and micro actors who participate in the global market of capital. In a way, they can be considered as 'global microstructures', that is, local configurations of organization, interaction and electronic space within a global reach.

A global market of capital is not a recent invention, of course; it has existed long before our information society. Yet, the digitization of finance is a notable novelty, since it implies a 'jump in orders of magnitude and the extent of worldwide interconnectedness' (Sassen, 2005). The materiality of this networked sociality is made possible by the use of complex software increasing the level of liquidity, and allowing the 'liquefying' of wealth usually regarded as 'nonliquid'. The multiplicative effects of the combination between the use of computers and the novelty of the software, draw on the possibility of using sophisticated algorithms without implying a specific knowledge and expertise in mathematics or in the subtleties of the design of software applications.

Moreover, the use of—not public like Internet, but with private access—chains of networks assures interconnectivity and simultaneity, decentralizing the access which, in turn, sustains the development of a growing number of transactions, the distance between the financial means and the assets, the number of participants who all over the world produce the emergent effects of global space with their dynamics, partly imbricated with national strategies (ibidem, 2005). In their detailed analysis of foreign exchange markets, Knorr Cetina and Bruegger (2000; 2002) present a clear description and analysis of a significant instance of the global market of capital. They show that, in practice, the screens are the objectual loci of this re-newed sociality developing through an engagement of traders and markets on the screen. The screens do not represent the 'reality' of the market outside; they are constitutive of it. That is to say that they integrate complex and dispersed networks of transaction between traders involved in businesses relationships. They make and re-make an assemblage of differently aligned activities of brokers, dealers, bookkeepers, analysts, researchers, newsagents, etc.:

In a sense, the screen is a building site on which a whole economic and epistemological world is erected (Knorr Cetina & Bruegger, 2002, pp. 165-167).

The empirical findings on the social embeddedness of these global microstructures (*social imbrications*) are particularly interesting. While the affordances of new interactive technologies as well as the de-regulation of the industries facilitated the geographical dispersal of the global capital market with the inclusion of new centers, and also the integration of the system at the international level, there has been the consolidation of the share of exchanges in a few financial centers. In addition, the tendency to concentration is clear within single countries, with a growing relevance of specific centers. Thus, it is registered the strength of some leading financial markets, together with the role of some global cities (London, Tokyo, New York, Paris, Hong Kong and Frankfurt) with a major share in the worldwide market of capital.

This trend seems to depend on the relevance of coordination and control in an era where, partly because of the effect of distributed technology of ICTs, the dispersed geography of firms and financial markets makes the central functions even more complicated. That is to say that it requires an innovative environment, and some high skills to be performed – accounting, technology, legal services, design, and most specialized corporate services. Comparatively speaking, global cities have a repertoire of expertise and competencies that can make the innovation conveyed by the massive use of financial digitization, even more efficient and appropriate.

Electronic markets draw on banks and traders located locally, and on important financial centers. These centers abound in knowledge resources for interpreting the turbulence of markets and the ambiguity of transaction by providing local interpretations of information. They have, in a way, the 'tacit knowledge' for the complexity to execute international deals. In other words, in these centers, there is the possibility to mobilize the appropriate knowledge to handle difficult and unpredictable financial environments:

In brief, financial centers provide the social connectivity that allows a firm or market to maximize

the benefits of its technical connectivity (Sassen, 2005).

The redesign of the geography of the capital markets and the growing intersections and alliances among financial centers, if it guarantees a space for globalization, it also produces the de-nationalization of financial centers, namely the detachment from national logics and the integration to the rationality of the global market of capitals. Furthermore, the analyses on the micro-conditions, and on the social imbrications at the micro level, reveal the emotional dynamics of electronic markets (Knorr-Cetina & Bruegger, 2002; Abolafia, 1999). The global microstructures are supported by bonding forces implying a notable engagement of traders and the objects of their work, namely the market-on-the-screen. The descriptions of everyday practices highlight the deep involvement that traders experiment in their work, their efforts and continuous involvement as a sort of obsessive fixation on the market-on-the-screen even beyond working hours.

The links between the material infrastructures of the setting, and the engagements of the traders, resemble the flow state (see the previous example of the virtual communities of software practitioners). It is a situation of immersion and undifferentiation from the immanence of the becoming of agency; hence the bonding forces, the attachment developed in incompleteness, the liquidity of the financial markets. They appear stable for the moment of price fixation, but, most frequently, they present themselves as always changing objects, 'incomplete objects' which reveal their characteristics on the screen on time. Financial markets are for traders the objects of a never-ending project of knowledge. It is the lack of completeness which accounts for the binding force that develops between traders and the market. This force has an emotional side: it involves a sense of excitement – traders are market makers for currencies, committed not only to the monetary reward, but also to the enjoyment of winning. Traders remains literally attached to their worksites for long periods of time. They continue to follow the unfolding of the markets at home while watching the news on the financial market on CNN, or on other specialized TV channels. They also experiment negative feelings, such as fear, anxiety, terror, expressed in a vocabulary that sometimes describes them in terms of physical assaults.

The excitement and the loss depending on the unfortunate results of dealing, require the careful emotion work of chief traders. These are aimed to downplay the excessive emphasis in case of luckily gains, or to provide a sense of confidence for traders experimenting losses or failures in gaining positions on the markets. The binding force, in this case, implies a major relationship between the market as an object and the trader as a subject. In Knorr Cetina's words, this performs a kind a post-social relationship where the central link is not between subjects, but between the mutual incompleteness of objects and subjects (we will return on this point in the next and final section of the chapter).

RESHAPING THE SOCIAL

Having addressed the fabrication of socialities through the three lens of *materiality*, *mutual constitution* and *emotions*, we will now focus on some effects of networked sociality. We will reflect on the reconfiguration of the social accompanying these technologies, and on the emergence of the qualification of specific types of socio-technical links of the sociality acknowledged through the notion of post-social relationships. We will complement our analysis by highlighting an associated risk in terms of the increasing surveillance and control as well as the growing fragility of society.

The Reconfiguration of the Social

As we have seen, the fabrication of networked socialities represents an experiment in the *reconfiguration of the social*. It is a sort of laboratory for the making and the remaking of the social through digitations. Here, the new technology of information and communication does not simply reflect upon, but tries to constitute and partly stabilize forms of sociality derivative or transformative of the society. This reconfiguration is not virtual, in the sense of being potential; it has its specific materialization, its electronic space and the respective socio-technical infrastructures. The reconfiguration has to do with the *scale modelling*, and it supports (or promises) epistemic advantages for (actual) potential users (Latour, 1998). Social network sites are venues where it is possible to observe a reconfiguration of identity, the inscription of social relationships (the list of 'friends'), the visualization of global social networks.

Virtual communities of practices are representation of communities of practices, their performation in electronic space – a venue where the practices performing the community and the knowledge accompanying them, are represented and made accessible. Similarly, the electronic markets allow the integration of otherwise dispersed networks of exchanges and dealers – via sophisticated software by providing a materialization of the process of globalization of capital markets.

This process affects the older hierarchies of scale. It supports a trend of destabilization and erosion of the old stable agglomerates of sociality, stabilized mainly around the geography of nation-states. This destabilization does not imply the definite demise of older hierarchies of scale; rather, it highlights, and makes visible, the multiscalar characteristic of the social space and, in particular, of the forms of networked socialities. The networked socialities instantiate the global space, make direct connections of local to global dynamics possible, as in the case of the electronic markets, without following the nested hierarchies of scale, thus allowing the multiplication of lateral and horizontal connections, like in the case of the virtual communities of practices.

Of course, the new technologies are not the main cause of these processes; yet they have facilitated them, and contributed to consistently shape them (Sassen, 2005). With some notable exceptions, these processes challenge the vocabularies of social sciences, usually derived and canonized by the stabilization of nation-states and by integration (Beck, 2006).

Post-Social Relationships?

The erosion of (old) forms of sociality asks for a deep reflection on the texture of the sociality emerging in the mutual fabric of interactive technology (Wittel, 2001). The concept of post-social relationship seems to be able to grasp some characteristics of this emergent sociality. It implies an engagement bringing the object-centred social relationship to the forefront. Knorr Cetina (1997; 2007) introduced the notion of post-social relationship in the analyses of the dynamics of the increasing structuration of contemporary societies as knowledge societies.

Most critical theories focus on the negative, and dystopian, effects of the recent transformations. These seem to signify an increasing destruction of social settings and of identities, and a growing individualization in the life-course of individuals. Yet, it fails to recognize the transformative and stabilizing effects of these changes.

In order to address this aspect, we should probably refine the way of conceiving sociality, usually understood in reference to humans with human relationships, by taking into account the relevance of the non-human side (objects, artefacts, tools, technologies) in the social fabric (Latour, 2005). This post-social perspective helps to visualize how the modern emancipation of selves from previous social belongings (communities, social

classes) has been accompanied by an increasing *objectualization of social life.*

Object-worlds contribute to stabilize identities, and to experiment new forms of sociality by complementing those usually studied by social scientists (Knorr Cetina, 1997). Here, we can see that the theme of the post-social relationship comes from the conceptualization of the contemporary societies as knowledge societies, from the recognition of the reshaping and constituting effects of diffusion of the culture of science and technology, and from the acknowledgement of the importance of knowledge processes in the social world (Stehr, 1999; 2001).

The forms of networked socialities are relevant empirical instances of these current transformations. They involve an objectualization of social worlds: they concern, in these cases, the social networks for SNS, the virtual communities, as well as the markets. The redistribution of the human and the non-human in the fabric of sociality does not imply the lack of emotion involvement. The object-centred environments we have analysed are *emotional arenas*. The SNS called Cyworld, for example, has a considerable success probably because it is a venue for displaying emotions and feelings publicly. In this case, the design of the site elaborates an object-world that accompanies the attachment, helping the expressive side otherwise channelled in different ways in the off-line social life. In addition, we have seen how virtual communities of practitioners display an attachment to the objects of activity which can trigger a flow state, namely a situation of immersion in the practices and a deep involvement conveyed by emotions and passions for the challenging work the communities are confronted with.

This attitude is not confined to these communities; it concerns other virtual communities whose 'aliveness' and rhythms are sustained by the mobilization of emotions. In some cases, the communities of self-help for example, they draw on emphatic relationships as well as on the services furnished to the participants and the users of their activities. A closer look at the dynamics of post-social relationships reveals how it develops on a specific objectuality. The forms of network socialities draws on an *ever-changing object* (epistemic objects) - the electronic space has the property of incompleteness, and engages participants in a never completed process of being materially defined. This incompleteness seems to provide an account of the binding forces of leading processes of post-social relationships. In following Knorr-Cetina's Lacanian reading, this sociality draws less on positive links and more on lack, with a corresponding 'structure of wanting' which orientates the processes of subjectivation.

The Fragility of Society

The analyses of these forms, and the reflection on the objectualization of sociality, introduce the theme of the risks resulting from the electronic constitution of society. This issue can be addressed from different angles: in a sense, the traceability of the sociality can multiply the possibility of growing *surveillance and control*, as some notable scholars have remarked with insistence (Lyon, 2001); on the other hand, this objectualization could also reveal the *fragility of modern societies* (Stehr, 2001).

Traceability sustains the increasing accountability of the human conduct as well as the 'dreams' of those who intend to augment the possibility of steering at a distance. SNS are maintained and partly support by the dynamics of marketing. Participants provide their personal information as well as information about their interests and tastes. They represent a focused and 'free' database for companies interested in investing on advertisement for target audience. Further, if the privacy policy of these sites and the use of the large amounts of information they have, is often unclear, it can represent a repertoire useful to mobilize the surveillance or the control of particular kinds of users. Similarly, the virtual

communities of practice are articulated around common knowledge interests and the communication of information, embodying the possibility of knowledge management via a solely technological instrumentation.

Here, the forms of networked socialities appear to be more a way of controlling rather than enabling social agencies. On the contrary, a possible alternative reading focuses on the increase in the capacity to act (knowledge) supported by interactive technology. The argument of knowledge society suggests the possibility for self-made social relationships, like those fabricated in networked socialities. The chance of making, re-making and even destroying social relationships visualize the construction of social realities on unpredictable scales. The increase in the capacity of act, however, occurs in unequal ways, so that, while traditional authorities lose their capacity for governing social action, some individuals and small groups are gaining a disproportioned advantage in the possibility of acting. The emergent effect of this tendency is a growing vulnerability of society. That is to say that the diffusion of networked socialities might reveal the objectualization of the social, and the fragility of the architecture of our modern societies (Stehr, 2001).

CONCLUSIVE REMARKS

This chapter has addressed the theme of the electronic constitution of society through the 'lens' of the categories of the social studies of science and technology. This approach is relevant to the analyses of the contribution of information and communication technologies to the making of sociality. The essay has focused on varied forms of networked socialities (social network sites, virtual communities of practice, electronic markets) in order to highlight the complexities of the imbrications among organization/interaction/electronic space, and, in particular, the type of socio-technical assemblage that they contribute to perform via the materialization, the intermeshing of the social and the digital, as well as the emotional involvement.

These networked socialities have been presented as widespread exemplars of social on Internet. However, they do not constitute an exhaustive typology of all possible socio-digital assemblages; further research is needed to cover other emerging alternative and existing forms, and to refine the reasoning here presented. Our analysis has had the aim of showing the enacting role of such technologies, qualifying some of their formal features (reconfiguration, objectualizations, fragility) in terms of sociality.

REFERENCES

Abolafia, M. (1998). Markets as Cultures: An Ethnographic Approach. In Callon, M. (Ed.), *The Laws of the Market* (pp. 69-85). Oxford: Blackwell Publishers.

Bach, J., & Stark, D. (2004). Link, Search, Interact: The Co-evolution of NGOs and the Interactive Technology. *Theory Culture and Society, 21*(3), 101-117.

Bach, J., & Stark, D. (2005). Recombinant Technology and New Geographies of Association. In R. Latham & S. Sassen (Eds.), *The Digital Formations: IT and New Architectures in the Global Realm*. Princeton: Princeton University Press.

Beck, U. (2006). *Cosmopolitan vision*. Cambridge: Polity Press.

Boyd, D. M., & Ellison, N. B. (2007). Social network sites: Definition, history, and scholarship. *Journal of Computer-Mediated Communication, 13*(1), article 11. Retrieved February 10, 2008, from http://jcmc.indiana.edu/vol13/issue1/boyd.ellison.html

Castells, M. (1996). *The Rise of the Network Society: The Information Age: Economy, Society and Culture*. Cambridge, MA: Blackwell Publishers, Inc.

Csikszentmilhalyi, M. (1997). *Finding Flow. The Psychology of Engagement with Everyday Life*. New York, NY: Basic Books.

DiMaggio, P., & Hargittai, E. (2002, August). *The new digital inequality: Social stratification among Internet users*. Paper presented at the American Sociological Association Annual Meeting, Chicago.

Dubé, L., Bourhis, A., & Jacob, R. (2003). *Towards a Typology of Virtual Communities of Practice* (Cahier du GReSI No. 03-13). Montréal, Canada: HEC Montréal, Groupe de Recerche en Systémes d'Information.

Faraj, S. Kwon, D., & Watts, S. (2004). Contested artefact: technology sensemaking, actor networks, and the shaping of web browser. *Information Technology and People, 17*(2), 186-209.

Fineman, S. (2006). Emotion and organizing (pp. 675–700). In C. Hardy, S. Clegg, T. Lawrence, & W. Nord (Eds.), *Handbook of organization studies*. London: Sage.

Fineman, S., Maitlis S., & Panteli, N. (2007). Virtuality and Emotion. *Human Relations, 60*(4), 555–560.

Gherardi, S. (2008). Breve storia di un concetto in viaggio: dalla comunità di pratica alle pratiche di una comunità. *Studi Organizzativi, 10*(1), 49-72.

Gherardi, S., Nicolini, D., & Strati, A. (2007). The Passion for Knowing. *Organizations, 14*(3), 315-330.

Halcli, A., & Webster, F. (2000). Inequality and Mobilization in The Information Age. *The European Journal of Social Theory, 3*(1), 67-81.

Hargittai, E. (2007). Whose space? Differences among users and non-users of social network sites.

Journal of Computer-Mediated Communication, *13*(1), 14. Retrieved February 10, 2008 from http:// jcmc.indiana.edu/vol13/issue1/hargittai.html.

Hennion, A. (2004). Une sociologie des attachments. D'une sociologie de la culture à une pragmatique de l'amateur. *Sociètès, 85*(3), 9-24.

Golbeck, J. (2007). The dynamics of Web-based social networks: Membership, Relationships, and Change. *First Monday, 12*(11). Retrieved 5 November 2007 from http://www.uic.edu/htbin/cgi-wrap/bin/ojs/index.php/fm/article/view/2023.

Jones, S. G. (1995). *Cybersociety. Computer-Mediated Communication and Community.* London: Sage.

Kaiser, S., Müller-Seitz, G., Pereira Lopes, M., & Pina e Cuhna, M. (2007). Weblog as a Trigger to Elicit Passion for Knowledge. *Organizations, 14*(3), 373-390.

Keeble, L., & Loader, B.D. (2001). *Community Informatics. Shaping Computer-Mediated Relations.* London: Routledge.

Kim, K.-H., & Yun, H. (2007). Cying for me, Cying for us: Relational dialectics in a Korean social network site. *Journal of Computer-Mediated Communication, 13*(1), 15. Retrieved February 10 2008 from http://jcmc.indiana.edu/vol13/issue1/kim.yun.html.

Knorr Cetina, K. D. (1997). Sociality with Objects: Social Relations in Postsocial Knowledge. *Theory Culture and Society, 14*(4), 1-30.

Knorr Cetina, K. D. (2007). Culture in global knowledge societies: knowledge cultures and epistemic cultures. *Interdisciplinary Science Reviews, 32*(4), 361-375.

Knorr Cetina, K. D., & Bruegger, U. (2000). The Market as an Object of Attachment: Exploring Postsocial Relations in Financial Markets. *Canadian Journal of Sociology, 25*(2), 141-168.

Knorr Cetina, K., & Bruegger, U. (2002). Traders' engagement with market: a Post-social Relationship. *Theory Culture Society, 19*(5-6), 161-185.

Kollock, P., & Smith, M. (Eds) (1999). *Communities in Cyberspace.* London: Routledge.

Kollock, P. (1999). The economics of online cooperation: Gifts and public goods in cyberspace (pp. 220-239). In M. A. Smith & P. Kollock (Eds.), *Communities in Cyberspace* London: Routledge.

Lanzara G. F., & Morner, M. (2005). Artifact Rule! How organizing happens in an open source project (pp. 67-90). In B. Czarniawska & T. Hernes (Eds.), *Actor Network Theory and Organizing.* Copenhagen: Liber & Copenhagen Business School Press.

Latham, R., & Sassen, S. (2005a). Introduction. Digital Formations: Constructing an Object of Study (pp. 1-34). In R. Latham, & S. Sassen (Eds.), *The Digital Formations: IT and New Architectures in the Global Realm.* Princeton: Princeton University Press.

Latham, R., & Sassen, S. (Eds) (2005b). *The Digital Formations: IT and New Architectures in the Global Realm.* Princeton: Princeton University Press.

Latour, B. (1987). *Science In Action: How to Follow Scientists and Engineers through Society.* Cambridge, MA: Harvard University Press.

Latour, B. (1991). Technology is society made durable. In Law (Ed.), *A Sociology of Monsters: Essays on Power, Technology and Domination,* (pp. 103-131). London: Routledge.

Latour, B. (1998). *Thought Experiments in Social Science: from the Social Contract to Virtual Society.* Annual Public Lecture at 1st Conference on Virtual Society?, Brunel University, UK. Retrieved 08 February 2008 from www.artefaktum. hu/it/Latour.html

Latour, B. (2005). *Reassembling the Social*. Oxford: Oxford University Press.

Lyon, D. (2001). *Surveillance Society: Monitoring Everyday Life*. Milton Keynes: Open University Press.

Nardi, B, Schiano, D., & Gumbrecht (2004, November). *Blogging as Social Activity, or, Would You Let 900 Million People Read Your Diary?* Paper presented at CSCW 2004, Chicago, Illinois, USA.

Nardi, B., & Harris, J. (2006, November). *Strangers and Friends: Collaborative Play in World of Warcraft*. Paper presented at CSCW 2006, Banff, Alberta, Canada.

Peddibothla, N. B., & Subramani, M. R. (2007). Contributing to Public Document Repositories: A Critical Mass Perspective. *Organization Studies, 28*(3), 327-348.

Preece, J. (1999). Empathic communities: Balancing emotional and factual communication. *Interacting with Computers, 12*(1), 63-77.

Rheingold, H. (1993). *The Virtual Community: Homesteading on the Electronic Frontier*. Reading, MA: Addison Wesley

Sassen, S. (2005). Electronic Markets and Activist Networks: The Weight of Social Logics in Digital Formations (pp.54-88). In R. Latham & S. Sassen (Eds.), *The Digital Formations: IT and New Architectures in the Global Realm*. Princeton: Princeton University Press.

Sproull, L., Dutton, W., & Kiesler, S. (2007). Introduction to the Special Issue: On Line Communities. *Organization Studies, 28*(3), 277-283.

Stehr, N. (1999). *Knowledge Societies*. London: Sage.

Stehr, N. (2000). Deciphering Information Technologies. Modern Societies as Networks. *European Journal of Social Theory, 3*(1), 83-94.

Stehr, N. (2001*). The Fragility of Modern Societies. Knowledge and Risk in the Information Age*. London: Sage.

Suchman, L. (2005). Affiliative Objects. *Organization, 12*(3), 379-399.

Wajcman, J. (2002). Addressing Technological Change: The Challenge to Social Theory. *Current Sociology, 50*(3), 347–363.

Wajcman, J. (2006). New connections: social studies of science and technology and studies of work. *Work, Employment and Society, 20*(4), 773–786.

Weber, S. (2005). The Political Economy of Open Source Software and Why it Matters (pp. 178-212). In Latham, R., & Sassen, S. (2005b). *The Digital Formations: IT and New Architectures in the Global Realm*. Princeton: Princeton University Press.

Wellman, B. (2001). Computer Networks As Social Networks. *Science, 293*(14), 2031-2034.

Wellman, B., & Giulia, M. (1999). Net Surfer do not ride alone. Virtual Communities as Communities. In Kollock and Smith, *Communities and Cyberspace*. New York: Routledge.

Werry, C., & Mowbray, M. (2001). *Online Communities: Commerce Community Action, and the Virtual University*. Upper Saddle River, N.J.: Prentice Hall.

Wittel, A. (2001). Towards a Network Sociality. *Theory, Culture and Society, 18*(6), 51-76.

ENDNOTES

[1]   Castells' depiction of the information age draws on post-industrialism, albeit with a stance away from technological determinism. However, his analysis is not entirely successful; the role of technology in shap-

ing society is rather simplistic (Wajcman, 2002; Stehr 2000). Moreover, Castells argues in favour of the emancipatory role of information and communication technologies, elaborating a metanarrative of the information age where ICTs should play a positive effect on social inequalities. Halcli and Webster (2000) suggest that this claim is not sufficiently supported by empirical facts.

2   I have not included in the study: the weblogs (or blogs), the networks of on-line games (Nardi, et alii 2004; Nardi, & Harris, 2006), the public document repositories (Peddibothla, & Subramani, 2007), which represent further instances of digital formations.

Chapter XIV
Virtual Nations

William Sims Bainbridge
National Science Foundation, USA

ABSTRACT

Virtual worlds are computer environments in which large numbers of human beings may interact, do useful work for each other, and build enduring social connections. For example, in World of Warcraft an estimated nine million subscribers form short-term action-oriented groups and long-term guilds, employing a variety of software tools to manage division of labor, spatial distributions, activity planning, individual reputations, and channels of communication, to accomplish a variety of often complex goals. A broader system of essentially permanent allegiances, comparable to current national governments and major corporations, frames the volatile forming and dissolving of small and medium-sized cooperative groups. New social technologies have a clear potential to supplement and render more flexible the existing structures of government, but they may also represent a significantly new departure in human social organization. The chapter will describe the diversity of information technology tools used to support social cooperation in virtual worlds, and then explain how they could be adapted to mediate in new ways between government and its citizens.

Virtual worlds are computer-generated environments having some similarity to the physical world, in which humans are represented by avatars or other surrogates, and where people may interact socially and economically. They have a very real potential to enhance government operations and offer new possibilities for popular involvement in public decision-making. However, several lines of argument suggest this potential may be modest. Current virtual worlds have serious technical limitations, some of which may be inescapable. Some technical limitations might be overcome only by concerted investment in research and development, raising the question where the necessary funds will be found. Other technologies compete with virtual worlds, notably the existing World Wide Web and teleconferencing systems. Perhaps even more serious, governments themselves are mired in inefficiency, and seem incapable of handling many of the challenges facing them. This means both that governments may be too slow to adopt new technologies that might help them fulfill their missions, and that the problems governments face might be too severe for solution by any means.

Government use of information technology, and government regulation of it, are however changing rapidly. In her recent book, *Change of State* (2006), Sandra Braman surveyed much of the recent socio-legal and socio-technical literature on the topic and came to two conclusions. First, she notes that electronic communication media are merging, rendering existing regulatory systems obsolete. For example, electronic mass media (radio, TV) were regulated differently from private electronic communication (telegraph, telephone), yet today there are many hybrid technologies and all of them coexist on the Internet. Second, she feels that new information technologies are giving governments greater control over their citizens, and this may be especially true when leaders like George W. Bush exploit the excuse of the so-called War on Terror in a cynical attempt to increase their power. However, it could equally well be argued that some of the new technologies give greater power to the people, notably political blogs, YouTube exposés, and low-cost political email spamming. Virtual worlds highlight both of these major issues, because they are the ultimate merging of all media into one, and because they provide environments where people may not merely talk about but fully act out radical, utopian, and revolutionary dreams.

These concerns suggest that this essay will need to juggle multiple perspectives. Much of its coherence will be provided by the two premiere examples of contemporary virtual worlds *Second Life* (SL) and *World of Warcraft* (WoW). Both were created by California companies, Linden Lab in San Francisco, and Blizzard Entertainment in Irvine. However both are international in scope. North Americans are outnumbered by Europeans in SL, and WoW is popular in China and Korea as well as Europe (Reuters 2007). Indeed, Blizzard

is owned by Vivendi, a French media company, and recently merged with Activision to become the world's leading pure electronic game company (Kennedy 2007). However, the author of this essay will emphasize American perspectives. One reason is that he has spent decades studying the culture and society of the United States, and thus can draw more confidently upon analyses of this nation. A second reason is more important: At this time this essay was written, the United States had moved a considerable distance toward fascism, and there is reason for concern that any movement back toward a more liberal form of government might be only temporary. This perspective, among others, suggests that we need to consider the revolutionary potential of virtual worlds, as well as how they might comfortably support conventional democratic institutions.

At best, however, we can only outline socio-technical possibilities and identify some of the questions that would need to be answered before major applications of innovations from virtual worlds could transform government operations. That caveat allows us to be a bit speculative on some topics, but does not release us from the responsibility to base our comments on the best available current knowledge. We must begin with a description of today's virtual world technology, based on research the author has carried out in our two very influential examples.

TWO VIRTUAL WORLDS

Virtual worlds, such as *Second Life* and *World of Warcraft*, offer models of future computer-organized virtual groups that could become extremely important for digital government. Virtual worlds are computer environments in which large numbers of human beings may interact, do useful work for each other, and build enduring social connections. For example, in *World of Warcraft* ten million subscribers form short-term action-oriented groups called *parties* and long-term groups called *guilds*, employing a variety of software tools to manage division of labor, spatial distributions, activity planning, individual reputations, and channels of communication, to accomplish a variety of often complex goals (Nardi & Harris, 2006; Williams *et al.*, 2006; Ducheneaut, *et al.*, 2006, 2007). A broader system of essentially permanent allegiances, comparable to current national governments and major corporations, frames the volatile forming and dissolving of small and medium-sized cooperative groups. Developed for online virtual worlds, these social technologies have a clear potential to supplement and render more flexible the existing structures of government, but they may also represent a significantly new departure in human social organization.

Before we can analyze these possibilities, we need a clear picture of what today's virtual worlds actually are like. Both SL and WoW run on ordinary desktop or laptop computers, although fast processors and graphics cards enhance the experience. They use the conventional computer screen for display, and today's virtual worlds are not to be confused with *virtual reality* (VR). For decades, engineers, scientists, and science fiction writers have imagined physically immersive VR environments, that surround the user with three-dimensional images mimicking a dynamic physical environment at high fidelity. Two distinct approaches are commonly used. First, a computer-generated scene may be projected on the walls, floor and ceiling of a room, often called a *cave*, perhaps adjusting to the actions of the user and simulating such visual phenomena as shadows and movement. Second, the user may wear a *head-mounted display* that presents different images to the two eyes, thus achieving stereoscopic illusion of depth. Today's virtual worlds employ neither of these methods (Schroeder and Bailenson, 2008).

However, the experience of dwelling for long in a virtual world reveals that the conventional view of immersive environments may be wrong.

Despite the lack of expensive VR equipment, these worlds can be extremely immersive in a psychological sense (Castronova, 2005; Taylor, 2006; Boellstorff, 2008). I find WoW especially so.

First of all, a number of features of human vision harmonize well with existing technology. My own computer setup places a wide computer monitor about 18 inches from my eyes, so it fills most of my vision. Humans can see details only near the center of their field of view, and peripheral vision detects nothing more than gross movements, neither detail nor even color. We use several methods to perceive in three dimensions, and binocular vision is only one of them. In WoW, distant mountains are hazy, just as in the real world, simulating the distance effect of the opacity of the atmosphere. Almost all virtual worlds display distant objects smaller than near objects, and show objects growing in angular size in a realistic manner as the viewer subjectively approaches them. A natural consequence of this is that straight lines provide the perspective depth cues that Renaissance painters labored so hard to master. Thus, rapid movement of the person through the scene, for example running through a forest in WoW, correctly shows objects flowing past in three dimensions.

A second insight concerning the immersive quality of virtual worlds is that action inside them can become so meaningful to participants, that *emotions* make the environment feel real. In *World of Warcraft*, one undertakes a number of quests and other goal-oriented activities that give the world purpose. Interacting with other players and the dangerous environment, one feels anger, fear, surprise, anticipation, pride, shame, and even sometimes gratitude. Thus, the world is psychologically impressive, therefore immersive.

The third point on realism is that these are *persistent* environments. Suppose you are standing in *Second Life*, talking with two other people. As often happens, events in the physical world intrude, for example you need to go to the bathroom. You return to your computer to find that one of the people has left, and the other fills you in on the end of their conversation. While you were away, SL persisted, and events took place that you did not observe.

A fourth point is that virtual worlds are *multimodal* and thus multi-sensory. Especially in the case of WoW, sounds are integrated to some extent with actions that can be seen. In mid-2007, both SL and WoW added integrated voice communications, but even before that users could run auxiliary voice channels, using special services like Ventrillo and TeamSpeak, or conference calls in an online phone system like Skype. Both music and sound effects have long been available in both virtual worlds.

None-the-less, it is clear that current virtual worlds are far more modest than future ones might be. For example, one can well imagine an advanced-technology input system that made avatars look and act realistically like their owners, using computer vision and cheap transponders, like RFID tags, attached to clothing to detect limb movement. As Jason Leigh and Maxine Brown quoted me:

Today's virtual worlds contrast sharply with the concept of total immersive VR that has long been popular with science fiction writers but has proven so difficult for computer scientists to achieve in the real world. Second Life and World of Warcraft images are restricted to the screen of an ordinary computer monitor, rather than filling the walls of a VR cave or binocular head-mounted display. On the one hand, this may suggest that people really do not need visually perfect VR. On the other hand, today's virtual worlds may be preparing millions of people to demand full VR in the future. (Leigh and Brown, 2008, p. 84).

Most virtual worlds, including SL and WoW, employ special client software, rather than web browsers. Thus while they communicate over Internet, they are not part of the World Wide Web. However, objects in *Second Life* can provide hyperlinks to web pages, and both SL and WoW are supported by massive, diverse web-based information systems.

Second Life describes itself as "a 3D online digital world imagined and created by its residents" (cf. Rymaszewski *et al.*, 2008).[1] Linden Lab sells virtual land to subscribers and provides tools for creating three-dimensional objects, plus a scripting language that gives them the power to animate these objects and control such things as automatic text generation and links to information resources. Linden Lab manages an internal currency system, which is based on "Linden dollars." On January 13, 2008, the exchange rate to buy Linden dollars was L\$265 = US\$1. Aside from a few legal restrictions, such as a recent prohibition against gambling casinos, few rules constrain the behavior of residents.

At the risk of oversimplifying, *Second Life* residents can be categorized as follows. Many are tourists and curiosity seekers, who visit SL once or twice, then never return. Some are representatives of corporations, government agencies, and educational institutions, who have created public relations displays in SL, in hopes that visitors will find them interesting. A few are members of organizations like IBM that conduct seminars or other group meetings inside the environment. Others are entrepreneurs or innovators who hope they can create profitable businesses in SL, dealing in virtual real estate, products like virtual clothing, or services. Some are attracted to the SL "red light district" where they explore novel sexual orientations and seek partners. A fair number belong to social groups that meet in SL, often organized around special interests like astrology, environmentalism, or *Star Trek*. Finally, perhaps the most important groups are artists, computer programmers, or design students who create the virtual spaces and objects that give SL its vitality.

My experience with *Second Life* began at the August 2006 conference of the World Transhumanist Association in Helsinki, Finland, where my presentation was carried in an auditorium of the virtual world, as well as one at the university. Remote participants from around the world could hear my words and watch two screens on a virtual wall, one showing me and the other my PowerPoints. Or, they could ignore me altogether, and chat with the other remote attendees in SL. The government agency for which I work, the National Science Foundation, is considering what presence it ought to have in virtual worlds, and I have been leading an effort to prototype possibilities. Over the summer of 2007, I worked with Stephanie Nieves, a college student intern who created demonstration displays based on NSF-funded research projects, and with NSF colleague Mary Lou Maher, who has considerable experience directing design students in both *Second Life* and the earlier virtual environment, *Active Worlds*. In Figure 1, Stephanie is showing Mary Lou and me how a combination lock can

be added to a door, then programmed to open it when the right combination is keyed in. We are in Mary Lou's virtual design studio, and everything visible in the image including our clothing was created by an SL user.

World of Warcraft, with ten million subscribers, is the most successful massively multi-player online role playing game or "MMORPG" (Davis, 2005; Lummis and Kern, 2006, 2007; Kern *et al.*, 2006). Central to the game are approximately 5,000 quests, pre-designed adventures of varying lengths and difficulties that earn the player valuable rewards if completed. But it is far more than a game; it is also a drama, an allegory, and a realm of the real world. From the standpoint of commerce, it is a billion-dollar harbinger of a new mode of communications. From the standpoint of the humanities, it is a total work of art, as prophesied by Richard Wagner's 1849 treatise, *The Artwork of the Future* (Wagner, 1893; cf. Newman,

Figure 1. A design team in Second Life

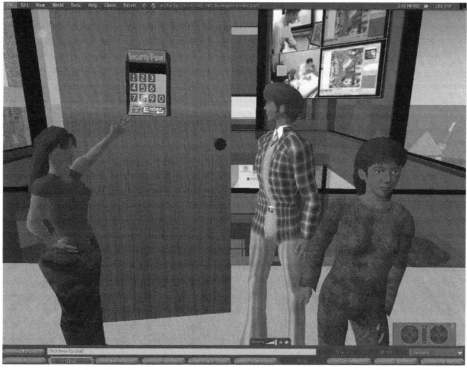

1924). From the standpoint of the social sciences, it is a magnificent laboratory for understanding social and economic relations, cultural change, and the technological transformation of human personalities. However, it cannot be modified in the myriad ways in which *Second Life* can be, so it is much better adapted for observational rather than experimental research.

For over a year I have carried out an ethnographic research project inside WoW, rather as a cultural anthropologist might study an exotic society, with some methodologies rather more like those of a cultural historian. This research project was carried out by running 21 characters through WoW, like research assistants or native informants, using two computers and two subscriptions so they could interact with each other. They covered both factions (Alliance and Horde) and all ten races: Human, Night Elf, Dwarf, Gnome, Draenei, Orc, Troll, Tauren, Undead, and Blood Elf. The classes are: priest, shaman, mage, druid, warlock, rogue, paladin, hunter, and warrior. To explore the diversity of supernatural cultures, the team includes six priests, two shamans, and two druids, plus at least one of each of the others. At least one practiced each of the professions: mining, herbalism, skinning, alchemy, enchanting, jewelcrafting, leatherworking, tailoring, blacksmithing, engineering, fishing, cooking, and first aid. A total of more than 2,100 hours of ethnographic work took two of the characters to the top 70 level of experience, the others to lower levels that allowed the team to explore all corners of the world, and generated a vast trove of data notably in nearly 20,000 "screen shot" pictures.

SOCIAL COMPUTING APPLICATIONS

Acrimonious political debates about the alleged incompetence, rigidity, or dogmatism of particular government administrations may obscure the important point that the public expects its government to accomplish unprecedented tasks, some of which may be impossible and many of which cannot be completed by following pre-set plans. More flexible structures may be needed, both for setting goals and for reaching them. Numerous information technology tools developed for the virtual worlds could be adapted to manage the work of constantly changing groups and networks of individuals for the purposes of government. In a very real sense, they are a "third way" alternative to representative democracy and market economies.

In representative democracy, citizens vote for candidates and parties whose policies they tend to favor, and then legislatures and executives establish programs to carry out the public's wishes. Unfortunately, the practicality of the goals or the adequacy of the means are often uncertain, and governments experience great difficulty in fine-tuning their programs or abandoning them altogether. Market economies are very good at finding trade-offs between competing values or investments, and in conducting efficient exchange between individuals or groups, but they are criticized for failing to take account of public goods and other "externalities." Like a market, the groups in virtual worlds bring together individuals for their mutual benefit, giving people considerable freedom concerning which goals they will seek and which paths they will follow. Yet like government agencies and contractors, the goals they select from are largely set by a higher authority, and thus they are capable of producing public goods.

Already, many government agencies recognize the training potential of electronic virtual environments, notably the armed forces (Prensky, 2001). The BBN Technologies Corporation, which was instrumental decades ago in the creation of Internet, has been active in adapting electronic games for training, and the company's website lists fully 28 commercial games that have been used for this purpose by the US military.[2] For example, *Falcon 4.0* has been adapted by the US

Air Force as an F-16 combat flight simulator, and *Sub Command* is used similarly by the US Navy. Interestingly, the Air Force has also made use of *Starcraft*, the science-fiction twin to *Warcraft* from which *World of Warcraft* was developed. *Starcraft* is a strategy game, rather than a virtual world, but *Falcon 4.0* and *Sub Command* put the user in command of a fairly realistic war vehicle inside a complex environment and thus would qualify as virtual worlds, especially since many of the newer examples connect multiple players.

Another application that is gaining serious attention is modeling social processes, such as the spread of infectious diseases through a population. Rather than simply assuming the ways humans would behave, and programming them into a multi-agent system or other pure computer simulation, researchers can create or observe situations in virtual worlds that mimic epidemiological processes from the real world, integrating real human beings into a multi-agent system. A team led by Yasmin Kafai at the University of California, Los Angeles, has already used the children's virtual world *Whyville* in an experimental study of reactions to a measles-like epidemic affecting the avatars (Kafai *et al.*, 2007).

Perhaps the most famous example was accidental, a plague that spread throughout *World of Warcraft* on September 15, 2005. The game employs many biological metaphors in motivating action, and characters often contract short-duration diseases from the "non-player characters" (simple artificial intelligence beasts and humanoids) that are common across this virtual world. But in this case, a contagious disease was added to an isolated region of the world, on the assumption that sufficient safeguards existed to prevent it from spreading from player to player. This proved to be false, and by one estimate four million characters were infected. In the prestigious medical journal *The Lancet Infections Diseases*, Eric Lofgren and Nina Fefferman (2008) argued that this WoW plague actually told us something important about real epidemics that had not been

factored into current computer models: Some people might move toward the source of infection, rather than flee from it. When *Science News* interviewed me about this idea, I responded:

I tend to think it's more realistic than we acknowledge, that there would be motivations for people to go to the disaster. If you believe, as I do, that the federal government can't succeed in containing it, you would rush to the place where they were giving immunizations, knowing that the smallpox was going to get everywhere pretty soon. It goes well beyond mere curiosity seeking. (Vastag, 2007, p 265).

Quite beyond the issue of government planning in the area of public health, many social programs could potentially be tested before implementation. The *Boston Globe* newspaper has quoted economist Edward Castronova as saying, "Down the road, you might have a situation where every government maintains a whole bunch of virtual worlds, trying out variations on its policies to see how they work" (Johnson, 2007). This would especially be the case for new groupware systems designed to facilitate the flexible completion of large numbers of small government-sponsored projects.

Let us imagine a possible future example, allowing the US Environmental Protection Agency to monitor water quality across a wide geographic territory, such as every pond and stream in the nation. The data will be instantly updated and analyzed by information systems, and the main challenge is to obtain a very large number of field measurements of water quality widely distributed over time and space. You want to volunteer — or play the game for prestige points in your community — so you log into an online site, let's call it *World of Watercraft* (WoWater). You key in your general location, perhaps your postal zip code, and select the *looking for group* software modeled on that in WoW. In a few moment, WoWater has linked you to four other volunteers who happen to be near different points along a nearby river. For your entire group to "win," you must each go to several locations along the river and collect data at specified points in time.

Technically, the measurements are made simply by plugging a probe into your cellphone and dipping the probe into the water. The cellphone automatically records the exact time and location, and sends a chemical analysis of the water to the WoWater database. Once all the geographically dispersed members of your expedition have completed all the measurements, the Honor section of the WoWater website automatically updates your personal pages to display your success and award you the points you earned in this very serious game.

If you appreciated the help of the four other members of your expedition, you may add them to your friends list so you can invite them to go on a variety of environmentalist quests in the future, or even invite them to join your persistent guild. Such guilds may organize their own group activities, such as carrying out campaigns against egregious polluters in your environment, both online and in the material world. The social utilities in *World of Warcraft*, developed on the basis of extensive experience in other games, are already well designed to coordinate small, ad-hoc groups undertaking somewhat complex tasks, as well as mobilizing the membership of persistent social groups on either a periodic or emergency basis.

A *WORLD OF WARCRAFT* CASE STUDY

An example of how the WoW groupware operates to bring people together for a common goal is the experience my character Catullus had on the night of January 27-28, 2008. Catullus was a Blood Elf priest in the Horde faction, and thus an enemy of the competing Alliance faction. Having reached experience level 65, out of the total of 70 levels,

he was solo questing in the north central part of Nagrand, fourth most difficult of the 52 virtual-geographic zones of *World of Warcraft*. He had joined the largest guild, Alea Iacta Est (AIE, "The Die Is Cast"), which had fully 1,186 members of level 10 or above. He received a private text message from Bunks, a fellow member of his guild, saying "Hey, want to do the Ring of Blood quests here in Nagrand? Lots of experience and great rewards." The social modules of the WoW software interface allowed Bunks to immediately discover which other members of his guild are currently online and what zones they are in, specifically to facilitate cooperation. After a brief exchange of information, Catullus agreed, and a memo box opened on his screen, formally asking him if he wanted to join a party led by Bunks. He clicked the box to agree. This immediately gave him access to a new chat channel, just for members of Bunks's questing party. Gold dots appeared on his nested set of maps, showing where the other members were in virtual space, and mousing over each dot revealed the character's name. Face icons and status bar graphs opened along the left edge of the computer screen, and Catullus could mouse over the icons or use the groupware to get more information about the other members.

Bunks was listed as a level 66 shaman, which is a rather good specialization for fighting in an enclosed space like the Ring of Blood. The second member was Norser, also a member of the AIE guild, who was a powerful spell-caster, a level 70 warlock who could summon a supernatural minion. The third member of the party was a level 65 mage named Borza who belonged to a different guild, Tears of Draenor. Catullus sized up his partners and asked who was going to tank. According to standard MMORPG theory, there are primarily three roles in battle. A *tank* attacks the enemy and stands toe-to-toe in melee combat, aggravating the enemy and drawing all the enemy's fire. A tank needs heavy armor, but need not be a powerful killer, because that role is played by the *DPS* (Damage Per Second) members of the team, for example a hunter who stands at a distance, untouched by the enemy but shooting bullets or arrows at him. The third role is *healer*, whose main job is using magical spells to counteract any damage done to the tank. The ideal tank is the class of character called warriors, but this team of four had none. For DPS, the team had three members who could deal damage at the moderate distances in the enclosed battle ring. Catullus was a priest, the ideal class to serve as healer, so his primary role was to heal the tank, if they could recruit one. After about ten minutes, they were joined by Damurota, a level 65 warrior of the Warmaul Hill guild, and the team was complete.

The six quests of the Ring of Blood were a series of melee combats, in which a huge and powerful opponent would enter the ring, towering over the team, and attack Damurota. Catullus could see a bar graph representing Damurota's health, and could tell when it was dropping far enough to require him to cast a healing spell, often several spells in rapid succession. On occasion, the opponent's aggression would be directed at a different team member, so when Catullus saw another partner's health bar dropping, he could select the correct person for healing. This did not always succeed, so one or more of the team would "die" and need to be resurrected after the combat ceased; one time all five died, but death is only a temporary setback in this virtual world (Klastrup, 2006). In one of the combats, all four other players were killed, and Catullus needed to switch to offensive mode, healing himself as he employed a magic wand to kill the enemy at the last possible moment. Frankly, despite the quality of the groupware, it would have been quite useless if the players had not been both highly experienced cooperative fighters, and familiar with the standard roles and functions each combatant can be expected to perform. After the team achieved its final victory, Catullus returned to his solo questing, only to be interrupted again suddenly.

As he usually does, Catullus was reading the AIE guild text chat when not thoroughly occupied with his own individual battles. At 10:49, a fellow member named Moobie posted this message: "I just got off my Alliance spy; they are planning a raid on Undercity tonight, 50 people." What Moobie meant was that he had just been running a different character in the Alliance faction, leaning that members of the Alliance were organizing a major raid on one of the Horde cities. Very quickly, AIE organized those of its members who were online to travel to Undercity to defend it. When Catullus got there, he ran through the central trade district, helping the others kill Alliance characters who had broken in. Once all the invaders had been exterminated, Catullus went to the front entrance of the city, where small numbers of Alliance characters repeatedly attacked over the next hour. When that activity had died down, AIE spontaneously decided to organize its own raid on an Alliance city, at first mentioning Darnassus, then deciding on Ironforge.

The groupware can handle teams with forty members, but AIE's counterattack recruited more than this, so two raid groups were organized. Catullus joined the group organized by Astuss, a level 70 rogue who had an extensive experience fighting members of the Alliance, notably a remarkable 8,890 lifetime honorable kills. Figure 2 shows the data displayed on the main page of the raid section of the social groupware system incorporated in the *World of Warcraft* user interface, listing all forty members. Notice that the names of the forty members are arranged in eight groups of five, identifying the character's experience level and functional specialization. Each of the eight groups can function like a five-man questing party; for example, Catullus will see constantly updated information about the location and condition of the four other members of Group 3: Cyleidor, Balerius, Stigg, and Eldacar. The mark to the left of Astuss's name indicates he is the group leader. He has delegated fully a dozen assistant leaders, also marked, giving them the power to invite new members and giving them access to a special leadership chat channel. All forty members will be able to communicate over the general raid chat channel, and all members of the AIE guild will also be able to communicate on the guild channel.

Astuss, but not Catullus, has the power to move someone from one group to another, simply by clicking his mouse on their name and moving their name where he wants it. If this were a well-planned attack, or one of the teams competing in one of the battleground arenas in WoW, he would have set up each group as a self-sufficient unit, with tank, DPS and healer. However, this mass attack on a city is bound to be chaotic, so he has made no attempt to do so. There is no discernable pattern to the distribution of the following classes of characters totaling forty: druid (1), hunter (8), mage (4), paladin (4), priest (7), rogue (5), shaman (3), warlock (7), and warrior (1). Similarly, Astuss could have arranged the players in some logical order by experience level, either distributing the lower-level players evenly among the 26 level 70 players, or concentrating them in weaker groups that would not be at the forefront of the attack. Along the right side of the in-WoW control utility are tabs for the different classes or specialties of

Figure 2. The Raid Membership Window from World of Warcraft

	Group 1				Group 2		
X	Astuss	70	Rogue		Teniven	27	Paladin
	Eiji	70	Mage	x	Helloise	70	Priest
	Tirielian	70	Paladin		Moobie	48	Hunter
x	Bobosha...	70	Priest	x	Xanar	70	Hunter
x	Vothos	70	Mage		Bovne	70	Druid
	Group 3				Group 4		
	Cyleidor	69	Priest		Izeila	21	Hunter
	Balerius	65	Hunter		Soulock	70	Warlock
x	Stigg	70	Hunter		Deaderin	70	Mage
x	Eldacar	70	Paladin		Uchihavi...	66	Rogue
	Catullus	65	Priest		Larathydo	64	Mage
	Group 5				Group 6		
	Sunnypark	70	Priest		Meatpu...	55	Warlock
	Qhelas	48	Warlock	x	Shadowt...	70	Rogue
	Bokrasuo	52	Shaman		Vixxca	70	Hunter
	Pelell	70	Rogue	x	Nural	70	Priest
	Voodoo...	70	Shaman	x	Wurmw...	70	Warrior
	Group 7				Group 8		
	Venusia	70	Priest		Lollock	27	Warlock
x	Akathia	70	Paladin		Posonby	70	Hunter
	Varlope	70	Shaman	x	Thyatira	70	Warlock
	Hollus	70	Hunter	x	Ashayo	70	Warlock
	Madam...	63	Rogue		Alergan	70	Warlock

the characters, allowing the leader to recruit the needed skills and assign the appropriate division of labor. Given that this raid was a rapid response to an attack by the other faction, there was no time to recruit a well-balanced team. For example, there is only one warrior, when one might have wanted to have a warrior in each of the eight subgroups.

Figure 3 shows a typical display just before the attack began, that gives all the information and tools available to Catullus. Most participants had rendezvoused at Kargath, a Horde outpost in a zone adjacent to the territory dominated by Ironforge. Once many participants reached that point, they rode north on their mounts (not only horses but also giant birds and trained dinosaurs), and the picture shows them just after they have crossed the border into Loch Modan, part of the Ironforge Territory. The picture is dark simply because the time is evening, as indicated also by

the tiny half moon in the far upper left corner; clicking on it would reveal the exact time. The icon in the upper left displays the current situation of Catullus himself: level 65, member of Group 3 in the raid, and with full bar graphs for health and for the mana he will use to cast healing spells. Below that are smaller but identical displays for the four other members.

At this point, Izeila, a level 21 hunter, was in the group, along with her trained beast represented by an even smaller icon below hers, but Astuss shortly moved her to Group 4, and moved Cyleidor into group 3. To the right of the icon representing Catullus is one for Hollus, a member of the raid but not the same group. Part of the preparation for an attack is the sharing of resources, and Catullus has just given Hollus two strengthening buffs that only a priest can give, represented by the fourth and sixth of the tiny square icons under the bar graphs for Hollus; these will last for several

Figure 3. The user interface during a major raid in World of Warcraft

minutes, well into the battle. Catullus had clicked his mouse on several of the characters around him, looking for ones who needed protective buffs, most recently displaying the information for Hollus, a display that will vanish again in a moment. The special buffs and other conditions applying to Catullus himself are visible as the six small square icons along the right top edge. Note the box saying "Lieutenant General Eiji" near the bottom right; this indicates the character that happens to be under the mouse cursor at the moment, and clicking the mouse would display Eiji's information in the place where the information for Hollus now stands.

The circle in the upper right corner is a simple zoomable map, with dots and arrows representing the locations of fellow raid members. Clicking the map would fill the screen first with a map of the entire local zone, displaying locations of raid members, then maps of adjacent zones or the whole world. The text near the lower left corner of the image is the most recent messages of three separate chat channels: local defense, the guild, and the raid, using different colors to make it easy to focus on the raid chat. To make sure that the most important messages are noticed, the raid leader can display a short command in huge letters on the middle of the screen, announced with the sound of a claxon, but does this only at crucial moments in the battle itself. The square icons all across the bottom and right side are things Catullus can do, such as casting a spell, checking resources in one of the five supply containers he is carrying, or opening some of the groupware modules such as the raid census shown in Figure 2. It should be abundantly clear that this interface is extremely complex, and only very experienced users can handle it effectively in the heat of battle.

After this picture was taken, the raid rushed forward, battled its way through a lightly defended tunnel at North Gate Pass, and assembled again just outside the gate of Ironforge, where a few members of the Alliance were ready to defend their city. The raid was able to blast through the complex gate area before being stopped and all its members slaughtered just inside the Ironforge commercial area between the auction house and the bank. Again, death is only a temporary setback in *World of Warcraft*, so the invaders prepared to resurrect themselves simultaneously when a countdown reached zero. Only at this point did the raid leader announce the special target of the raid, presumably in order to reduce the possibility that a spy would tell the Alliance defenders where the Horde attackers were actually headed.

The Alliance defenders may have felt the most likely target was the throne room that lay just the other side of a narrow passage, where King Magni Bronzebeard could be assassinated. But this would have been a mistake, because the real target was the Deeprun Tram, the unique subway train that connects Ironforge, the capital of the Dwarves, to Stormwind, the capital of their allies, the Humans. After great effort, Catullus and about twenty other members of the Horde raid were able to reach their goal and turn off the battle flags that allow Alliance players to attack them. Then they could ride back and forth between the two enemy cities, perfectly safe and able to taunt members of the Alliance at every stop.

Many readers may feel that this fanciful evening of adventure is frivolous and has nothing to do with modern government operations. I believe there are several reasons why this would be a mistake, of which three deserve mention here. First, the same software could be used to organize a rapid community response to a real disaster. I happened to grow up in a Connecticut town that had a volunteer fire department. When a house caught fire, for example during the middle of the night when my own parents' home did in fact burn, a call to the fire station caused a tremendously loud horn to blare out a geographic code telling the volunteers they needed to rush to a particular part of town. Whether because of the primitive nature of the whole system, or simple bad luck, my parents and sister were killed in the fire, although their house itself was saved.

In an era when many citizens carry Internet-connected cellphones or similar devices, we will need emergency response systems to galvanize the community for action, whether or not these systems are directly inspired by the ones currently operating in virtual worlds.

Second, virtual worlds can be valuable sites for research in the social, behavioral, and economic sciences, as well as in human-centered computer science. The very differences between *Second Life* and *World of Warcraft* let them exemplify somewhat different future scientific applications for government-sponsored research:

In terms of scientific research methodologies, one can do interviews and ethnographic research in both environments, but other methods would work better in one than the other. SL is especially well designed to mount formal experiments in social psychology or cognitive science, because the researcher can construct a facility comparable to a real-world laboratory and recruit research subjects. WoW may be better for nonintrusive statistical methodologies examining social networks and economic systems, because it naturally generates a vast trove of diverse but standardized data about social and economic interactions. Both allow users to create new software modules to extract data (Bainbridge, 2007, p. 472).

Third, people who have invested extensive time in WoW or similar virtual worlds, develop a host of skills that potentially transform their ability to handle other information and communications systems. Obviously, practice using the complex user interface under emotionally exciting conditions would help the person react in a well-organized fashion when using a comparable interface to deal with real-world challenges. More broadly, I think virtual worlds teach people to employ information technology tools when thinking through problems, to plan ahead, and to cooperate on short notice with equally well-trained strangers. I have often observed unruly players, apparently pre-teen boys, learn how to play responsible roles in teams, questing in *World of Warcraft*. Thus, the social lessons taught in virtual worlds, as well as the technical lessons, can be priceless.

CRITICAL AND UTOPIAN APPLICATIONS

To this point we have described the diversity of information technology tools used to support social cooperation in virtual worlds, then explained how they could be adapted to mediate in new ways between government and its citizens. However, this assumes that conventional systems of governance are beneficial and need only to be improved in some minor ways to meet the challenges of the future in a manner that is both just and effective. An alternate set of assumptions would hold that the current system is unjust or doomed or both. One need not accept these radical views to find their implications interesting, and we must conclude this essay with the possibility that virtual worlds are more subversive than supportive of the current system.

Interviewed on the influential television program, *Meet the Press*, late in 2007, presidential candidate Ron Paul expressed concerns felt by many Americans that their nation was decaying into some form of imperialism or fascism: "We're not moving toward Hitler-type fascism, but we're moving toward a softer fascism: Loss of civil liberties, corporations running the show, big government in bed with big business."[3] These words echoed those of President Dwight David Eisenhower, uttered on the same television channel nearly forty-seven years earlier: "In the councils of government, we must guard against the acquisition of unwarranted influence, whether sought or unsought, by the military-industrial complex. The potential for the disastrous rise of misplaced power exists and will persist."[4] Eisenhower then described the technological revolution that could make this form of soft fascism all the more likely:

Today, the solitary inventor, tinkering in his shop, has been overshadowed by task forces of scientists in laboratories and testing fields. In the same fashion, the free university, historically the fountainhead of free ideas and scientific discovery, has experienced a revolution in the conduct of research. Partly because of the huge costs involved, a government contract becomes virtually a substitute for intellectual curiosity. For every old blackboard there are now hundreds of new electronic computers.

The prospect of domination of the nation's scholars by Federal employment, project allocations, and the power of money is ever present – and is gravely to be regarded.

Yet, in holding scientific research and discovery in respect, as we should, we must also be alert to the equal and opposite danger that public policy could itself become the captive of a scientific-technological elite.

Some will react to the words of these two Republicans as if they were passages from Lenin's angry little book, *Imperialism, the Highest Stage of Capitalism*, dismissing them for being disloyal, unpleasant, or simplistic (Lenin, 1916). For Dr. Paul, a clear sign of America's fascism is the Iraq War, which he considers to have been unnecessary and thus profoundly immoral. Paul assumes that by the middle of the twentieth century, educated and free people agreed that an aggressive war is a collective form of murder, never permissible. Opinions differ still about the decision by the Bush administration to go to war, the degree of dishonesty versus incompetence in the case it made for war, and who should take the blame now for at least 100,000 deaths and perhaps many more. Yet one reading of events is that Americans were by-and-large happy to send their military forces to kill the citizens of other nations, until their own people began to be killed in some numbers and military success proved difficult to achieve. Dr. Paul's rhetorical reference to Hitler raises the specter of Nazism, and part of the mythology about the Nazi era holds that the German nation was mesmerized by a few evil individuals. This myth is preferable to the earlier myth that the Germans were by nature a primitive and violent people, but it may be no more true (Viereck, 1941; Schoenbaum, 1966; Allen, 1984). Rather, all human groups are both by nature and by necessity ready to kill members of competing groups, and any attempt to establish a world-wide pacifist mentality will be doomed to failure. This is where virtual worlds come in, both as allegories and utopias.

In the most intellectually developed religion of *World of Warcraft*, the doctrine of the Holy Light promulgated from the Cathedral in Stormwind, there are three cardinal virtues: respect, tenacity, and compassion. The most difficult of these is compassion, because whenever we act to help someone we rob them of some of their autonomy. Thus, WoW's vision of compassion is more Buddhist than Christian. The goal is not to save people from suffering, but to help them learn from it. Tenacity is one of the things they should learn, never to allow even death to deter them from accomplishing their goals. Respect is a militaristic virtue connected with chivalry: It is permissible to kill someone to obtain his resources, so long as you do so with respect. This lesson America has not learned, because it routinely slanders its opponents, whether at the moment they are Germans or Islamists.

Indeed, *World of Warcraft* overflows with anti-imperialist rhetoric. On one level, each player is supposed to be loyal to one of the factions, either Horde or Alliance, but on another level the faction leadership often proves to be incompetent, self-serving, and capricious. Ironically, many of the quests turn out to be futile. In the midst of dire attacks from enemies, one quest sends the player on a hunt for hors d'oeuvre for an aristocrat's garden party; another sends the player to a distant land for scientific soil samples that wind up getting dumped by a government flunky on a mud pile.

Many other quests are effective yet anti-establishment. The player meets "deforesters" for the Venture Trading Company who are clear-cutting a forest in the Charred Vale, and gets the opportunity to chop them down in retaliation. Indeed, WoW incorporates a tremendous amount of environmentalist propaganda against the military-industrial complex that has done great harm with polluting industries, resource exhaustion, and weapons of mass destruction. The expedition of a big game hunter, named Hemet Nesingwary, sends the player slaughtering dozens of animals, a clear anagram reference to Ernest Hemingway who paradoxically extolled macho virtues while claiming to oppose fascism.

Some WoW quests draw upon politically radical movies or similar objects of popular culture. In tropical Stranglethorn Vale, the player is supposed to assassinate Colonel Kurzen, who is clearly based on Colonel Kurtz, played by Marlon Brando, in *Apocalypse Now*, a movie satirizing American neo-colonialism in Vietnam, based in turn on Joseph Conrad's "*Heart of Darkness*" that critiques European Colonialism in Africa. While Toshley's Station appears to be named after a location in *Star Wars*, it really represents the outpost attacked by huge bugs in the anti-fascist satire, *Starship Troopers*, complete with a military officer named Razak Ironsides after the commander in that movie, Jean Rasczak played by Michael Ironside. Two non-player characters, Klaatu and Barada, echoed the most famous phrase of alien language in science fiction, "Klaatu barada nikto" from the anti-war film *The Day the Earth Stood Still*; the absence of a character named Nikto reflects the disaster that occurred in the fantasy satire, *Army of Darkness*, when the hero could not remember this entire phrase correctly. When a warlock's imp attacks in *World of Warcraft*, it often shouts,

"Can't we all just get along?" This phrase was famously spoken by the American President in *Mars Attacks!*, played by Jack Nicholson, and searching the web for the phrase indicates that it became prominent in American popular culture after Rodney King repeatedly spoke a variant of it after his wanton 1991 beating by racist Los Angeles police was captured on videotape and provoked riots.

If *World of Warcraft* is fundamentally subversive, *Second Life* is utopian. In principle, except for the land sold and taxed by Linden Lab, this virtual world is entirely created by its residents. Some rules have crept in over time, such as bans on child pornography, gambling casinos, and most recently unregistered financial institutions. Said to have been inspired by the imaginary Metaverse in Neal Stephenson's (1992) cyberpunk novel, *Snow Crash*, like *World of Warcraft* it is tied to a west-coast American counterculture, but it seeks to empower ordinary people more than to critique the elite.

The logical extension of these principles would be full cultural and legal independence, in which virtual worlds seceded from the (perhaps) dysfunctional nations of the mundane world. Already, social theorists have contemplated non-spatial government that represents the interests of online communities that are not limited to any particular patch of dirt (Tonn and Feldman, 1995). My own research has turned up many indications that the subculture to which virtual worlds belong has departed from conventional culture in many ways. Notably, participants are much less likely to be guided by religious belief, and more likely to prefer the suspension of disbelief associated with science fiction and fantasy (Bainbridge and Bainbridge, 2007). So, we can expect that virtual worlds will prototype many social innovations that might then diffuse to offline governance, while often preaching sedition. The question then becomes how much this revolution is real, rather than virtual.

CONCLUSION

Second Life and *World of Warcraft* may be the two most influential current virtual worlds, but they only hint at the diversity of such environments that may exist in the future. As users become more accustomed to the technical and social characteristics of virtual worlds, the worlds themselves will evolve still further, posing new challenges and opportunities for users. This technology is at a crucial stage in its development, possibly the breakthrough point after which it becomes an important medium of human communication. At the same time, arguably the wider world is entering a new phase in its history, possible a dark period marked by chaotic competition for scarce resources. Conceivably, many people will derive a sense of wealth and status from the possessions and accomplishments that belong to them in virtual worlds, thereby burdening the natural environment far less than in the twentieth century that was marked by resource-wasting conspicuous consumption in the economically advanced nations. Or, virtual worlds could become military training grounds for governments and insurgents alike, thereby causing even more bloody conflict in the so-called "real world." Rather than take a "wait and see" attitude, many people may need to become personally involved in creating new virtual worlds that will be better, rather than worse, for humanity.

REFERENCES

Allen, W. S. (1984). *The Nazi seizure of power: The experience of a single German town, 1922-1945*. New York: F. Watts.

Bainbridge, W. S. (2007). The scientific research potential of virtual worlds. *Science 317*, 472.

Bainbridge, W. S., & Bainbridge, W. A. (2007). Electronic game research methodologies: Study-ing religious implications. *Review of Religious Research, 49*, 35-53.

Boellstorff, T. (2008). *Coming of age in Second Life*. Princeton, NJ: Princeton University Press.

Braman, S. (2006). *Change of state: Information, policy, and power*. Cambridge, MA: MIT Press.

Castronova, E. (2005). Synthetic worlds: The business and culture of online games. Chicago: University of Chicago Press.

Davis, H. L. (Ed.). (2005). *World of Warcraft Atlas*. Indianapolis, IN: BradyGames.

Ducheneaut, N., Yee, N., Nickell, E., & Moore, R. J. (2006). Building an MMO with mass appeal: A look at gameplay in *World of Warcraft*. *Games and Culture, 1*, 281-317.

Ducheneaut, N., Yee, N., Nickell, E., & Moore, R. J. (2007). The life and death of online gaming communities: A look at guilds in *World of Warcraft*. In *Proceedings of CHI2007* (pp. 839-848). New York: Association for Computing Machinery.

Johnson, C. Y. (2007). Online gamers become guinea pigs: Epidemics uncorked in virtual worlds. *Boston Globe*, August 25. Retrieved June 6, 2008, from http://www.boston.com/news/local/articles/2007/08/25/online_gamers_become_guinea_pigs/

Kafai, Y. B., Feldon, D., Fields, D., Giang, M., & Quintero, M. (2007). Life in the times of whypox: A virtual epidemic as a community event. In C. Steinfield, B. Pentland, M. Ackerman & N. Contractor (Eds.), *Communities and Technologies* (pp. 171-190). New York: Springer.

Kennedy, S. (2007). Vivendi and Activision to create Activision Blizzard: Two companies merge to become the most profitable game publisher. 1up.com, December 2. Retrieved June 6, 2008, from http://www.1up.com/do/newsStory?cId=3164668

Klastrup, L. (2006). Death matters: Understanding gameworld experiences. In *Proceedings of the International Conference on Advances in Computer Entertainment Technology (ACE)* (2006) June 14-16, Hollywood, California. New York: Association for Computing Machinery.

Leigh, J., & Brown, M. D. (2008). Cyber-commons: Merging real and virtual worlds. *Communications of the ACM, 51* (January), 82-85.

Lenin, V. I. (1916). *Imperialism, the Highest Stage of Capitalism.* Retrieved June 6, 2008, from http://www.marxists.org/archive/lenin/works/1916/imp-hsc/.

Lofgren, E. T., & Fefferman, N. H. (2007). The untapped potential of virtual game worlds to shed light on real world epidemics. *The Lancet Infections Diseases, 7,* 625-629.

Lummis, M., & Kern, E. (2006). *World of Warcraft master guide.* Indianapolis, IN: BradyGAMES/DK Publishing.

Lummis, M., & Kern, E. (2007). *World of Warcraft: The Burning Crusade.edition.* Indianapolis, IN: BradyGAMES/DK Publishing.

Nardi, B., & Harris, J. (2006). Strangers and friends: Collaborative play in *World of Warcraft.* In *Proceedings of CSCW 2006* (pp. 1–10). New York: Association for Computing Machinery.

Newman, E. (1924). *Wagner as man and artist.* New York: Knopf.

Prensky, M. (2001). *Digital game-based learning.* New York: McGraw-Hill.

Rymaszewski, M., Au, W. J., Ondrejka, C., Platel, R., Van Gorden, S., Cézanne, J., Cézanne, P., Batstone-Cunningham, B., Krotoski, A., Trollop, C., & Rossignol, J. (2008). *Second Life: The official guide.* Hoboken, NJ: Wiley.

Reuters, A. (2007). Europe takes lead in Second Life users, Reuters Second Life News Center, February 9, Retrieved June 6, 2008, from http://secondlife.reuters.com/stories/2007/02/09/europe-takes-lead-in-second-life-users/.

Schoenbaum, D. (1966). *Hitler's social revolution.* Garden City, NY: Doubleday.

Schroeder, R., & Bailenson, J. (2008 in press). Research uses of multi-user virtual environments. In R. Lee, N. Fielding, & G. Blank (Eds.), *The handbook of Internet research.* London: Sage.

Stephenson, N. (1992). *Snow crash.* New York: Bantam Books.

Taylor, T. L. (2006). *Play between worlds: Exploring online game culture.* Cambridge, MA: MIT Press.

Tonn, B., & Feldman, D. (1995). Non-spatial government. *Futures, 27,* 11-36.

Vastag, B. (2007). Virtual worlds, real science: Epidemiologists, social scientists flock to online world. *Science News, 172* (October 27), 264-265.

Viereck, P. (1941). *Metapolitics: From the romantics to Hitler.* New York: A. A. Knopf

Wagner, R. (1893). *The art-work of the future.* In *Richard Wagner's prose works* (vol. 1, pp. 69-213). London: K. Paul, Trench, Trübner.

Williams, D., Ducheneaut, N., Xiong, L., Zhang, Y., Yee, N., & Nickell, N. (2006). From tree house to barracks: The social life of guilds in *World of Warcraft. Games and Culture, 1,* 338-361;

ENDNOTES

[1] http://secondlife.com/
[2] http://seriousgames.bbn.com/behaviorauthoring/militaryuse.html
[3] Dr. Ronald Paul, interviewed by Tim Russert, Meet the Press, NBC, December 23, 2007.

4   Dwight David Eisenhower, "Eisenhower's Farewell Address to the Nation," January 17, 1961, http://mcadams.posc.mu.edu/ike.htm.

Chapter XV
A Research Agenda for the Future

Francesco Amoretti
University of Salerno, USA

This volume does not constitute yet another account of the blessings of ICTs. Nor does it add new criticism to the old, nurturing fears about the future. The goal of this book is to provide an overview and reinterpretation of the main issues on digital information technology in world politics, relating them to the processes of transformation of the current historical system.

Inspired by the Braudelian concept of the multiplicity of time—and space—diachronic and synchronic and of the close-knitted unity of the phenomenon under investigation, i.e. the capitalist world-economy, an interpretative key is developed in an approach which could substantially enable advancement in this field of study both in theoretical and methodological terms.

Despite the limited number of cases and issues investigated, the contributions to this volume show that the diffusion of new technologies engender transformations that go beyond declared political objectives. Often this is understood as an expression of the "unintentional consequences" of social action. However, this is not the case. What appears as "unintentional consequences"—socio-cultural tensions and contradictions—is instead, *constitutive* of the capitalist system in its historical development.

It is my sincere hope that the volume will stimulate further research in the area of ICTs and political and socio-cultural development in the contemporary world. In this final section I identify specific issues I consider worthy of further research study in this field.

At the *longue durée* level, which appears the most complicated and unexplored, an extremely relevant issue is *public/private* relationships: not so much in terms of relationships between institutions and citizens but rather in terms of relationships between institutions that implicate standing and decision-making, institutions that allocate resources, and elaborate collective symbols etc... In the literature on ICTs the net distinction still exists between state institutions—at a central, transnational and sub-national level and, on the contrary, private institutions—the market and/or social market. A position yet to be fully defined but which, above all in the immense development of private digital networks, is founded on unquestionable empirical grounds. As Saskia Sassen maintains in *The Impact of the Internet on Sovereignty: Unfounded and Real Worries*: "My argument is that economic globalization and technology have brought with them significant transformations in the authority of national states. Especially important here is the growth of the new non-state centred governance mechanisms which have transformed the meaning of national territorial sovereignty independently from whatever impact the Internet has so far had, and further, the formation of partly digitalized global financial markets which can deploy considerable power against the will of national states".

In theory, capitalists operate via the market and wish governments to stay out of market operations. In practice, "the will of national states" is crucial to their market success in many different ways.

In the VI Charter of *Civilisation matérielle, économie et capitalisme (XV-XVIII siècle). Le temps du monde*, Fernand Braudel recalls how the State and Capital—or at least specific capital, that of large firms and monopolistic businesses – co-exist happily, and how furthermore, this good relationship cuts across the centuries of the modern age up to the present time. The tight-knit interaction underlying this scenario never falters, but the grounds on which the relationship is reinforced or consolidated changes in time, as do the tools adopted—not only the juridical ones—and the solutions pursued. The development of digital networks does not affect this relationship. Nor does it split up so-called globalisation, favoured (Castells) by this very development. ICTs at a certain stage find their place in history in contributing to the redefinition of that relationship. Pertinently re-constructing the solutions given in historic terms—and the ideological

representations that have accompanied such development—would undoubtedly contribute to throwing a different light on the *issues* emerging and/or that have emerged in this phase, such as *outsourcing, partnership* etc….

Another area I suggest for further research at the *longue durée* level is the *Individualism/Community* alternative. Positioned in current debate as an alternative between two concepts of cyberspace or better, of the experiential practice it favours, in effect, the *Individualism/Community* alternative does not originate with the diffusion of digital networks; it spreads its roots in the juxtaposing between principles - modalities of institutional and organisational practice—and their respective cultural models of political legitimisation—which are right from the origin of society and the capitalist economies, solutions hoped for and/or pursued by means of political tools. Language on the other hand, is classical albeit enriched with the neologisms of the cyberpunk culture. Even in this case, a more careful reconstruction of ideological production that historically, can be traced to individualism and to the community factor as pillars of the liberal tradition, can contribute to clarifying the actual terms of the alternative, if it is alternative we intend.

At the *conjuncture* level, the research agenda is extraordinarily varied. We see, albeit with different degrees of sensitivity and from different interest perspectives, how the contributions presented have made reference to the geo-political dynamics of the capitalist world-system in order to understand the origin and significance of the policies of diffusion of digital networks. How, more precisely, the *reference* to the neo-liberal paradigm and to globalisation have provided the coordinates – ideological coordinates - of the programme policies within which to collocate interest and to direct investments for cyberspace worldwide.

Over the last thirty years, technological innovation has been one of the principal drivers for the re-definition of geo-political power relationships on a world-wide scale. This interpretative key deserves to be used further, to analyse scenarios absent from this work, but nevertheless equally significant. The experience of the Indian continent or the Latin-American countries come to mind: how have they developed e-democracy and e-government policies, with what characteristics and with what results? For these countries, what is the significance of the electronic constitution?

We still know too little to rest on our laurels. The *conjuncture* level requires further commitment on the part of scholars and researchers. I cannot avoid however, pointing out that while the necessity is underlined of broadening research and extending the field to other countries and continents, the conjuncture phase shows signs of profound crisis. Indeed in truth, we are out of one phase and about to enter another i.e. a new conjuncture crisis seems to have begun.

In an article written in 2005 on Social Forces, Immanuel Wallerstein posed the question: *After Developmentalism and Globalization, What?*, indicating the currency crisis in East and Southeast Asia in 1997, the slide downward of the World Trade Organization from Seattle to Cancun, al-Qaeda and September 11, the crisis of the United States and China's economic growth as some of the signs of a new world (dis)order.

We would do well to ask with the euphoria of the 90s dampened, what strategies and narrations are beginning to take hold? The crisis and the rejection of the ideology of globalization has/have paved the way for different dynamics in historical terms, specifically in some cases, by redefining strategies of re-organisation on a more markedly macro-regional as opposed to a more state-national basis. All this crucially affects the core of the electronic constitution. A case in point is the *Carta Iberoamericana De Gobierno Electronico,* approved at the IX Iberian/Spanish-American Conference of the Ministries of Public Administration and State Reform held at Pucon in Chile from 31 May to 1 June 2007.

These developments will need to be followed more closely. In the same way, the implications of this *conjuncture* will need analysing in the context of the world economic crisis, in terms of eventual repercussions on the policies of diffusion of ICTs and on its politico-institutional value in different countries and geo-political areas. The crucial issue of the relation between new technologies and democratic development, if we examine some specific instances – that of China but certainly also that of many African and Arab States – has certainly not been resolved. Authoritarian trends appear reinforced by recourse to digital networks, and forms of control hitherto not pursued are tentatively gaining ground. Even in Western countries where a solid representative democracy reigns supreme, on balance it would seem that there are (few) *luci* and (many) shadows in the use of digital networks. Or, at least, based on empirical evidence, we do not see as yet any fundamental change in how the democratic institution operates thanks to the new ICTs. (Gibson, Rommele, and Ward, 2004).

This problematical consideration leads us to the third level of analysis, that on which scholars and researchers are concentrating most of their efforts, i.e. the short term level or *evenementielle.* In other words, the historical time context of events: the most problematical and threatening Often, too often, events are those filling the newspapers, their headlines screeching out from their pages. It is as though every tiny change is paving the way for a revolution: in different cultures, in institutional practices, in organisational models. There are numerous examples. The new technologies and the virtual reality they have generated, have become a hunting ground for those who are tracking down the latest scoop. The evolution of the World Wide Web and the Internet in particular, lends itself perfectly to this scope. From civic networks to the e-mail, to Second Life and youTube: cyberspace has expanded under our eyes becoming a kind of supermarket in which everybody—individuals, groups and institutions—find whatever they need.

However, the short term is above all, that in which innovations are produced and from which they spread: those that take off with difficulty or do not even take off at all and those that on the contrary, propagate at different rates eventually forming stable structures. Pertinent examples are the transnational movement phenomenon or collective mobility processes favoured or made possible by digital networks above all starting from the 90s, or in other words, since the Internet has sustained *counter-hegemonic* discourses, challenging the established system of domination and legitimising and publicizing political claims by the powerless and marginalized. The literature on these events is plentiful. What is needed however, is an analysis of the reasons and outcomes of this innovation: why it occurred in that period and what still exists of the experience.

In conclusion, a wider overview is reserved for the articulated universe of digital cultures, i.e. the re-organisation of knowledge via digital networks.We still know far too little of the social practices concerning virtual reality and their implications on the lives of individuals and we still know far too little about how the new technologies are changing the institutional and organisational balance of the production and transmission of knowledge. For instance, the increasing diffusion of immense data banks of magazines and books on demand; to the transformation of university systems worldwide, not the least to be favoured by recourse to new methodologies in didactics (e-learning); or to imposing programmes financed by the international institutions such as the World Bank, aimed at creating mega-portals of knowledge, such as the *Development Gateway*.

I began this book inspired by his work, with a quotation from Braudel. I would like to end it recalling yet again that

«History is the sum of all possible histories, a combination of tecniques and points of view from yesterday, today and tomorrow. It would be a mistake to go simply in a single direction, choosing one rather than the other path. [...] Either tracing events back to 1558 or to the year 1958, the purpose, for those who wish to comprehend the world, is always that of attempting to define a hierarchy of power, thought and specific movements, then to re-assemble the same. [...] Every 'contingency' includes different movements of origin and pace: the present time can be traced back to yesterday and at the same time, to a more distant past and even remote past»[1].

If this book even on a small scale, has been inspired by Braudel's thought, or stimulates others to do so, then it will have achieved its purpose.

ENDNOTE

[1]   «Pour moi, l'histoire est la somme de toutes les histoires possibles, - une collection de métiers et de points de vue, d'hier, d'aujourd'hui, de demain. Le seule erreur, à mon avis, serait de choisir l'une de ces histoires à l'exclusion des autres. [...] Qu'on se place en 1558 ou en l'an de grâce 1958, il s'agit, pour qui veut saisir le monde, de définir una hiérarchie de forces, de courants, de mouvements particuliers, puis de ressaisir une constellation d'ensemble. [...] Caque "actualité" rassemble des muvements d'origine, de rythme différent: le temps d'aujourd'hui date à la fois d'hier, d'avant-hier, de jadis». From F. Braudel (1958), "Histoire et Science Sociale: La Longue Durée", Annales E.S.C., Vol. 13, No 4, pp. 725-753.

Compilation of References

1999: The government online year. (1999, January 3). *People's Daily* (p. 4).

Abdulai, A. I. (2005). Of visions, development plans and resource mobilization in S Africa: The case of Ghana vision 2020. *Africa Insight, 35*(1), 28-35.

Abolafia, M. (1998). Markets as cultures: An ethnographic approach. In Callon, M. (Ed.), *The laws of the market* (pp. 69-85). Oxford: Blackwell Publishers.

Abrahamsen, R. (2000). *Disciplining democracy: Development discourses and good governance in Africa.* London: Zed Books.

Adams, M. (2005). The management of chieftaincy records in Ghana: An overview. *African Journal of Library Archives and Information Science, 15, 1,* 67-73

Adanu, T. S. A. (2006). Planning and implementation of the University of Ghana library automation project. *African Journal of Library Archives and Information Science, 16*(2), 101-108.

Adcock, R., & Collier, D. (2001). Measurement validity: A shared standard for qualitative and quantitative research. *The American Political Science Review, 95*(3), 529-546.

Adjei, E. (2004). Retention of medical records in Ghanaian teaching hospitals: Some international perspectives. *African Journal of Library Archives & Information Science, 14*(1), 37-52.

Adjei, E., & Ayernor E. T. (2005). Automated medical record tracking system for the Ridge hospital, Ghana Part 1: Systems development and design. *African Journal of Library Archives and Information Science, 15*(1), 1-14.

Agre, P. E. (1999, Octomber). *Growing a democratic culture.* Paper presented at the Media in Transition Conference. Cambridge: Massachusetts Institute of Technology.

Akrich, M. (1992). The description of technical objects. In W. E. Bijker, & J. Law (Eds.), *Shaping technology/building society* (pp. 205-224). Cambridge, MA: MIT Press.

Akussah, H. (2006). The state of document deterioration in the national archives of Ghana. *African Journal of Library Archives and Information Science, 16*(1), 1-8.

Alabau, A. (2004). *The European Union and its eGovernment development policy. Following the Lisbon strategy objectives.* University of Valencia: Valencia.

Albright, K. S. (2005). Global measures of development and the information society. *New Library World, 106*(7/8), 320-331.

Alhassan, A. (2004). *Development communication policy and economic fundamentalism in Ghana.* Finland: Tampare University Press.

Alhassan, A. (2005). Market valorization in broadcasting policy in Ghana: Abandoning the quest for media democratization. *Media, Culture & Society, 27*(2), 211-228.

Alhassan, A. (2007). Broken promises in Ghana's telecom sector. *Media Development, 3,* 45.

Allen, W. S. (1984). *The Nazi seizure of power: The experience of a single German town, 1922-1945.* New York: F. Watts.

American Chamber of Commerce in the People's Republic of China (2002). *2002 White Paper: American Business in China.* Section on the Information Technology. Retrieved February 29, 2004, from http://www.amcham-china.org.cn/publications/white/2002/en-20.htm

Amnesty International. (2002, November 26). *China: Internet Users at Risk of Arbitrary Detention, Torture and Even Execution.* Retrieved January 5, 2003, from http://web.amnesty.org/library/index/ENGASA170562002

Amnesty International. (2002, November 26). *State Control of the Internet in China* (pp. 3-4). Retrieved June 5, 2008, from http://www.amnesty.org/en/library/asset/ASA17/007/2002/en/dom-ASA170072002en.pdf

Amnesty International. (2006, July). *Undermining Freedom of Expression in China: The Role of Yahoo!, Microsoft and Google* (p. 14). Retrieved January 7, 2008, from http://irrepressible.info/static/pdf/FOE-in-china-2006-lores.pdf

Amoretti, F. (2006). Benchmarking electronic democracy. In A. V. Anttiroiko, & M. Malkia (Eds.), *Encyclopedia of digital government* (3 Voll.), Hershey, PA: Information Science Reference.

Amoretti, F. (2006). La Rivoluzione Digitale e i processi di costituzionalizzazion euopei. L'e-democracy tra ideologia e pratiche istituzionali. *Comunicazione Politica, 7*(1), 49-74.

Amoretti, F. (2006). *The digital revolution and Europe's constitutional process. E-democracy between ideological ad institutional practices.* Paper presented at the VII Congresso Espanol De Ciencia Politica Y De La Administration, Grupo De Trabajo 9 Communicacion Politica.

Amoretti, F. (2007). International organizations ICTs policies: E-democracy and e-government for political development. *Review of Policy Research, 24*(4), 331-344.

Amsden, A. H. (1989). *Asia's next giant: South Korea and late industrialization.* New York: Oxford University Press.

Andreson, C. (2006). *The long tail: Why the future of business is selling less of more.* New York: Hyperion.

Ansell, C., & Gingrich, J. (2003). Reforming the Administrative State. In B.E. Cain, R.J. Dalton, & S.E. Scarrow (Eds.), *Democracy Transformed? Expanding Political opportunities in Advanced Industrial Democracies* (pp. 164-191). Oxford: Oxford University Press.

Appadurai, A. (1990). Disjuncture and difference in the global cultural economy. *Theory, Culture and Society,* (7), 295-310.

Armatte, E. (2002). Informatique et libertés: de big brother à little sisters ? *Terminal, n°88,* 11-21.

Arndt, H. W. (1987). *Economic development. The history of an idea.* IL: The University of Chicago Press.

Arrighi, G., & Silver, B. J. (1999). *Chaos and governance in the modern world system.* Minneapolis: University of. Minnesota Press.

Bach, J., & Stark, D. (2004). Link, Search, Interact: The Co-evolution of NGOs and the Interactive Technology. *Theory Culture and Society, 21*(3), 101-117.

Bach, J., & Stark, D. (2005). Recombinant Technology and New Geographies of Association. In R. Latham & S. Sassen (Eds.), *The Digital Formations: IT and New Architectures in the Global Realm.* Princeton: Princeton University Press.

Badu, E. E. (2004). Academic library development in Ghana: Top managers' perspectives. *African Journal of Archives and Information Science, 14*(2), 93-107.

Baehr, P., & Richter, M. (Eds.). (2004). *Dictatorship in History and Theory: Bonapartism, Caesarism, and Totalitarianism.* Cambridge: Cambridge University Press.

Bainbridge, W. S. (2007). The scientific research potential of virtual worlds. *Science 317*, 472.

Bainbridge, W. S., & Bainbridge, W. A. (2007). Electronic game research methodologies: Studying religious implications. *Review of Religious Research, 49*, 35-53.

Bairoch, P., & Kozul Wright, R. (1996). *Globalization Myths: Some historical reflections on integration, industrialization and growth in the world economy* (Discussion Papers, No. 113). Geneva: UNCTAD.

Baldersheim, H. (2006, May*). The future of the periphery in information society. Or: Stein Rokkan meets Manuel Castells.* Paper presented at the conference on *"Towards a New Nordic Regionalism?",* Sogn og Fjordane University College, Balestrand.

Baptista, M. (2005). e-Government and state reform: Policy dilemmas for Europe. *Electronic Journal of e-Government, 3*(4), 167-174.

Barber, B. R. (1984). *Strong democracy: Participatory politics for a new age.* Berkeley: University of California Press.

Barlow, J. P. (9ᵗʰ February 1996). *A cyberspace independence declaration.*

Barney, D. (2004) *The network society.* Cambridge: Polity Press.

Barro, R. J., & Sala-i-Martin, X. (1991). Convergence across states and regions. *Brookings Papers on Economic Activity, 1.*

Barzilai-Nahon, K. (2006). Gaps and bits: Conceptualizing measurements for digital divide/s. *The Information Society, 22*, 269-278.

Battelle, J. (2005). *The search. How Google and its rivals rewrote the rules of business and transformed our culture.* New York: Portfolio.

Beck, U. (2006). *Cosmopolitan vision.* Cambridge: Polity Press.

Beech, H. (2002, July 22). Living it up in the illicit Internet underground. *Time, 160*(4), 4.

Beijing Internet Cafes ordered to stop operation for rectification. (2002, June 17). *People's Daily.* Retrieved June 18, 2002, from http://english.peopledaily.com.cn/200206/17/eng20020617_97950.shtml

Benkler, Y. (2006). *The wealth of the networks. How social production transfirms markets and liberty.* New Haven and London: Yale University Press.

Bertolini, R., Dawson Sakyi O., Anyimadu A., & Asem P. (2001). *Telecommunication use in Ghana: Research from the Southern Volta Region.* University of Ghana-Center for Development Research, Bonn University, Working Paper.

Bhuiyan, A. J. M. S. A. (2008). Peripheral View: Conceptualizing the Information Society as a Postcolonial Subject. *The International Communication Gazette, 70*(2), 99-116.

Bianco, C., Lugones, G., & Peirano, F. (2003, February). *A methodological proposal for measuring the transition to Knowledge Society in Latin American countries.* Paper presented at the Second Workshop on Indicators for the Information Society, Lisbon, Portugal.

Bignami, F. (2004). Three Generations of Participation Rights before the European Commission. *Law and Contemporary Problems, 68*, 61-83.

Bijker, W. E. (2006). Why and how technology matters. In R. E. Goodin & C. Tilly (Eds.), *The Oxford Handbook of Contextual Political Analysis* (pp. 681-706). Oxford: Oxford University Press.

Bingham, L. B., Nabatchi, T., & O'Leary, R. (2005). The new governance: Practices and processes for stakeholder and citizen participation in the work of the government. *Public Administration Review, 65*(5), 547-558.

Binns, T., Kyei P., Nel, E., & Porter G. (2005). *Africa Insight, 35*(4), 21-31.

Blumler, J. G., & Coleman, S. (2001). Realising democracy online: A civic commons in cyberspace. *IPPR Citizens Online Research Publication*, (2), March 2001. Retrieved from http://www.citizensonline.org.uk/site/media/documents/925_Realising%20Democracy%20Online.pdf. Last access 31st May 2008.

Boellstorff, T. (2008). *Coming of age in Second Life.* Princeton, NJ: Princeton University Press.

Boldrin, M., & Canova, F. (2001). Inequality and convergence in Europe's regions: Reconsidering European Regional Policies. *Economic Policy*, April.

Bonner, P., Delius, P., & Posel, D. (1993). The shaping of apartheid: Contradiction, continuity and popular struggle. In P. Bonner, P. Delius, & D. Posel (Eds.), *Apartheid's genesis 1935-1962* (pp. 1-41). Johannesburg: Wits University Press.

Börzel, T., & Risse, T. (2000). When Europe Hits Home. Europeanization and Domestic Change. *European Integration Online Papers, 4*(15). Retrieved June 20, 2007, from http://eiop.or.at/

Bourdieu, P. (1993). La démission de l'Etat. In P. Bourdieu (Eds.), *La misère du monde* (pp. 219-228). Paris: Seuil.

Boyd, D. M., & Ellison, N. B. (2007). Social network sites: Definition, history, and scholarship. *Journal of Computer-Mediated Communication, 13*(1), article 11. Retrieved February 10, 2008, from http://jcmc.indiana.edu/vol13/issue1/boyd.ellison.html

Brah, K. (2001). Ghana goes for IT lead. *African Business*, July/August, p.25.

Braimah, I., & King, R.S. (2006). Reducing the vulnerability of the youth in terms of employment in Ghana through the ICT sector. *International Journal of Education and Development using ICT, 2*(3), 23-32.

Braman, S. (2006). *Change of state: Information, policy, and power.* Cambridge, MA: MIT Press.

Brandon, B., & Calitz, R. (2002). Online rulemaking and other tools for strengthening our civil infrastructure. *Administrative Law Review, 54*(4), 1421-1478.

Bratton, M., Gyimah-Boadi, E., & Mattes, R. (2005). *Public opinion, democracy and market reforming Africa.* Cambridge, U.K.: Cambridge University Press.

Brazil (2001). *Information society in Brazil: Green book.* Brasilia: Ministry of Science and Technology.

Bridges.org (2002). *Government efficiency vs. citizens' rights: The debate about electronic public records comes to developing countries.* Cape Town: Bridges.org.

Breckenridge, K. (2002a, May). *From hubris to chaos: The making of the bewysburo and the end of documentary government.* Paper presented at the Wits Interdisciplinary Research Seminar, WISER, University of the Witwatersrand, Johannesburg.

Breckenridge, K. (2002b, October). *Biometric government in the new South Africa.* Paper presented at the The State We Are In Seminar Series, WISER, University of the Witwatersrand, Johannesburg.

Brewer, G. A., Neubauer, B. J., & Geiselhart, K. (2006). Designing and Implementing E-Government Systems. Critical Implications for Public Administration and Democracy. *Administration and Society, 38*(4), 472-499.

Brewer, G. A., Neubauer, B. J., & Geiselhart, K. (2006). Designing and Implementing E-Government Systems. Critical Implications for Public Administration and Democracy. *Administration and Society, 38*(4), 472-499.

Bridges.org (2004). *Straight from the Source: Perspectives from the African Free and Open Source Software Movement.* Retrieved December 3, 2007, from http://www.bridges.org/files/active/1/straight_from_the_source_may04.pdf

Briggs, J. (2001). Geeks in do-good rampage. *Fortune, 144*(1), 182.

Brümmer, S. (2002). From dompas to smart card. Johannesburg: *Mail & Guardian*, 22 February.

Brümmer, S. (2002). Buthelezi and Masetlha at it again. Johannesburg: *Mail & Guardian*, 22 March.

Burgelman, J. C., & Clements, B. (2003). A New Paradigm for eGovernment Services. *The European Science and Technology Observatory*. Retrieved March 25, 2008, from http://www.jrc.es/home/report/english/articles/vol78/ICT1E786.htm

Buthelezi, M. (1997). *Department of Home Affairs budget vote 1997/1998*. Cape Town: National Assembly – Republic of South Africa, 17 April.

Buthelezi, M. (2002). *Parliamentary media briefing*. Cape Town, GCIS – Republic of South Africa, 11 February.

Buthelezi, M. (2003). *Home Affairs budget speech*. Cape Town, National Assembly – Republic of South Africa, 19 May.

Calenda, D. (2007). La Società dell'Informazione nelle regioni del Mezzogiorno. Sfide, visioni, comportamenti. *Polis*, XXI, April, (pp. 65-94).

Calise, M. (2002). Corporate authority in a long-term comparative perspective. Differences in institutional change between Europe and the United States. *Rechtstheorie, Beiheft 20*, 307-324.

Calleja, D. (2002). Heart of geekness. *Canadian Business*, April 29, (pp. 65-71).

Cammaerts, B. (2008). Civil society participation in multi-stakeholder processes: In between realism and utopia. In D. Kidd, C. Ropdriguez, & L. Stein (Eds.), *Making our media: Mapping global initiatives toward a democratic public sphere*. Cresskill: Hampton Press.

Cammaerts, B., & Padovani, C. (2006, July). *Theoretical reflections on multi-stakeholderism in global policy processes: the WSIS as a learning space*. Paper presented at IAMCR conference, Cairo, Egypt.

Capgemini (2006). *Online Availability of Public Services: How Europe is Progressing*. Brussels.

Caral, J. (2004). Lessons from ICANN: Is Self-Regulation of the Internet Fundamentally Flawed. *International Journal of Law and Information Technology*, 12(1), 2.

Carmel, E. (1997). American Hegemony in Packaged Software Trade and the Culture of Software. *The Information Society*, 13(1), 125-142.

Cassese, S. (2006, November). Four Features of the European Administrative Space. Paper presented at the Connex thematic conference "*Towards a European Administrative Space*", Birbeck College, London.

Cassese, S. (2006). *Oltre lo Stato*. Roma-Bari: Laterza.

Castells, M. (1996). The rise of the network society. In M. Castells, *The information age: Economy, society and culture, 1*. Malden, Oxford: Blackwell.

Castells, M. (1997). *The rise of the network society Volume 2: The power of identity*. Oxford: Blackwell.

Castells, M. (2000). End of millenniun. In M. Castells, *The information age: Economy, society and culture, 3*. Oxford: Blackwell.

Castells, M. (2000). *The information age: Economy, society and culture, vol.1 The rise of the network society*, Oxford: Blackwell.

Castells, M. (2001). *The Internet galaxy*. (p. 1). Oxford: Oxford University Press.

Castronova, E. (2005). Synthetic worlds: The business and culture of online games. IL: University of Chicago Press.

Centeno, C., van Bavel, R., & Burgelman, J. C. (2005). A prospective view of e-government in the European Union. *The Electronic Journal of e-Government, 3*(2), 59-66.

Center for public service innovation. (2003). *Government unplugged – mobile and wireless technologies in the public service*, Pretoria: Center for public service innovation.

Chadwick, A. (2003). Bringing e-democracy back in: Why it matters for future research on e-governance. *Social Science Computer Review, 21*(4), 443-455.

Chadwick, A., & May, C. (2003). Interaction between states and citizens in the age of the Internet: "E-government" in the United States, Britain and the European Union. *Governance: International Journal of Policy, Administration and Institutions, 16*(2), 271-300.

Chandler, A. D., & Cortada, J. W. (Eds.). (2000). *A nation transformed by information: How information has shaped the United States from colonial times to the present.* New York: Oxford University Press.

Chang, H. (2002). *Kicking away the ladder: Development strategy in historical perspective.* London: Anthem Press.

Chase, M. S., & Mulvenon, J. C. (2002). *You've got dissent! Chinese dissident use of the Internet and Beijing's counter-strategies.* Santa Monica, CA: RAND.

Cheneau-Loquay, A. (2005). Comment les nouvelles technologies de l'information et de la communication sont-elles compatibles avec l'économie informelle en Afrique. *Annuaire Français de Relations Internationales, 5*, 345-375.

China closes over 3,300 Internet cafes in six months (2002, December 27). *People's Daily.* Retrieved April 20, 2008, from http://english.peopledaily.com.cn/200212/27/eng20021227_109161.shtml

China orders unlicensed Internet cafes closed nationwide. (2002, June 29). *People's Daily.* Retrieved April 30, 2008, from http://english.peopledaily.com.cn/200206/29/eng20020629_98774.shtml

China seeks to build boundary on Internet (2003, April 1). *People's Daily.* Retrieved April 17, 2008, from http://english.peopledaily.com.cn/200304/01/eng20030401_114386.shtml

China's eleventh five year plan (2006-2010)—Information industry, Ministry of Information Industry. (2007, March 1). Retrieved January 14, 2008, from http://www.mii.gov.cn/art/2007/03/01/ art_3986_1936.html

China's Web surveillance slows access even as government promotes Internet use. (2003, March 5). *The Associated Press.* Retrieved April 2, 2008, from http://www.clearwisdom.net/emh/articles/2003/3/7/zip.html#34

Chokshi, M., Carter, C., Gupta, D., Martin, T., & Robert, A. (1995). *Computers and the apartheid regime in South Africa.* Retrieved October 28, 2003, from http://www-cs-students.stanford.edu/~cale/cs201/

Chung, Y. (2002). *Anatomy of the decision-making process in China: (De)Concentration of the Internet industry.* (p. 55). Unpublished master thesis, Seoul National University, South Korea.

Ciborra, C. (2005). Interpreting e-government and development: Efficiency, transparency or governance at a distance? *Information Technology & People, 18*(3), 260-279.

Closson, R. B., Mavima, P., & Siabi-Mensah, K. (2002). The shifting development paradigm from state centeredness to decentralization: What are the implications for adult education? *Convergence, 35*(1), 28-42.

Coene, Y., & Gasser, R. (2007). *Joint operability workshop report "Towards a single information space for environment in Europe".* Frascati.

Coglianese, C. (2004). Information Technologies and Regulatory Policies: New Directions of Digital Government Research. *Social Science Computer Review, 22*(1), 85-91.

Coleman, S. (2005). *Direct representation. Towards a conversational democracy.* Institute for Public Policy Research.

Coleman, S., & Gøtze, J. (2001). *Bowling together Online public engagement in policy deliberation.* London: Hansard Society.

Commission of the European Communities (1993). *Growth, competitiveness, employment: The challenges and ways forward into the 21st century - White Paper.* Parts A and B. COM (93) 700 final/A and B, 5 December 1993, Bulletin of the European Communities, Supplement 6/93.

Commission of the European Communities (1994). *Report on Europe and the global information society: Recommendations of the high-level group on the information society to the Corfu European Council.* Bulletin of the European Union, Supplement No. 2/94.

Commission of the European Communities (1999). *eEurope – An information society for all.* Communication from the Commission of 8 December 1999, COM(1999)687 final.

Commission of the European Communities (2001). *eEurope 2002 – Impact and priorities.* Communication from the Commission to the Spring European Council in Stockholm, 23-24 March 2001, COM(2001)140 final.

Commission of the European Communities (2001). *Unity, solidarity, diversity for Europe, its people and its territory - Second Report on Economic and Social Cohesion.* Luxembourg: OOPEC.

Commission of the European Communities (2002). *eEurope 2005 action plan – An information society for all.* Communication from the Commission of 28 May 2002, COM(2002) 263 final.

Commission of the European Communities (2004). *Internet – Eurobarometre spécial 194*, Vague 59.2 – European Opinion Research Group EEIG.

Commission of the European Communities (2004). *A new partnership for cohesion: Convergence, competitiveness, cooperation. Third Report on Economic and Social Cohesion.* Luxembourg: OOPEC.

Commission of the European Communities (2005). *i2010 – A European information society for growth and employment.* Communication from the Commission of 1 June 2005, COM(2005)229 final.

Commission of the European Communities (2006). *Information society and the regions: Linking European policies.* DG Information Society and Media, Luxembourg, OOPEC.

Conseil De L'Union Européenne (2000). *List of eEurope benchmarking indicators.* 13493/00, Limite, ECO 338, Annex, Bruxelles, le 20 novembre 2000.

Cordella, A. (2007). E-government: towards the e-bureaucratic form? *Journal of Information Technology, 22*(3), 265–274. Retrieved January 15, 2008 from http://www.palgravejournals.com/jit/journal/v22/n3/full/2000105a.html#bib53

Council of Ministers (2006). *Ministerial declaration on the occasion of the Ministerial Conference "ICT for an inclusive society".* 11 June 2006, Riga (Latvia).

Council of the European Union, & Commission of the European Communities (2000, June 19- 20). *eEurope 2002: an Information Society for All.* Action Plan prepared by the Council and the European Commission for the Feira European Council.

Coyle, M. (2001, October 1). September Attacks Prompt Sharp Debate on Scope of Surveillance Law. *Law.com.* Retrieved December 18, 2003, from http://www.law.com/cgi-bin/nwlink.cgi?ACG=ZZZ1NUCI6SC

Crozier, M., Huntington, S. P., & Watanuki, J. (1975). *The crisis of democracy. Report on the governability of democracies to the Trilateral Commission.* New York: New York University Press.

Crush, J. (1992). Power and surveillance on the South African goldmines. *Journal of Southern African Studies, 18*(4), 825-844.

Csikszentmilhalyi, M. (1997). *Finding flow. The psychology of engagement with everyday life.* New York: Basic Books.

Cullen, R., & Choy, P. D. W. (1999). The Internet in China. *Columbia Journal of Asian Law, 13*(1), 116.

Cyber police to guard all Shenzhen Websites. (2006, January 5). *Shanghai Daily.* Retrieved November 2, 2007, from http://www1.china.org.cn/english/government/154200.htm

Dai, X. (2000). *The digital revolution and governance.* (pp. 151-2). Aldershot: Ashgate.

Dai, X. (2002). Towards a digital economy with Chinese characteristics? *New Media and Society, 4*(2), 144.

Daly, J. A. (2004, September). *What is wrong with ICT expenditures as a measure of ICT penetration.* Paper presented at the ASIS&T/EC and AoIR workshop on Measuring the Information Society, University of Sussex, Brighton, UK.

Damm, J., & Thomas, S. (Eds.). (2006). *Chinese Cyberspace: Technological Changes and Political Effects*. London and New York: Routledge.

Davies, S. G. (1997). Re-engineering the right to privacy : How privacy has been transformed from a right to a commodity. In P. E. Agre & M. Rotenberg (Eds.), *Technology and privacy: The new landscape* (pp. 143-165). Cambridge MA: MIT Press.

Davis, H. L. (Ed.). (2005). *World of Warcraft Atlas*. Indianapolis, IN: BradyGames.

Dawson, M., & Bellamy Foster, J. (1998). Virtual capitalism. In R. W. McChesney, E. Meiksins Wood, & J. Bellamy Foster (Ed.), *Capitalism and the information age*. New York: Monthly Review Press.

De Kock, E. (2005). Data protection in South Africa. *De Rebus*, December.

Deng, K., & Wu, C. (2002, June 21). Wangba he Tade Shengcun zhi Dao (Internet Cafés and Their Ways of Existence). *Nanfang Zhoumo* (Southern Weekend). Retrieved June 4, 2008, from http://www.people.com.cn/GB/it/48/297/20020621/758428.html

Department of Home Affairs (2001). *Strategic Plan 2002/2003 to 2004/2005*, Document n°2/6/8/P, Pretoria : Department of Home Affairs.

Desrosières, A. (1993). *La politique des grands nombres: Histoire de la raison statistique*. Paris: La Découverte.

Desrosières, A. (1997). Du singulier au général: L'argument statistique entre la science et l'Etat. In B. Conein & L. Thévenot (Eds.), *Cognition et information en société* (pp. 267-282). Paris: Ecole des Hautes Etudes en Sciences Sociales.

Deutsch, K. W. (1963). *The nerves of government*. New York: Free Press.

DG Infortmation Society and Media, eGovernment unit (2007). *EU: Study on Interoperability at Local and Regional Level*. Brussels.

Diffie, W., & Landau, S. (1998). *Privacy on the Line: The Politics of Wiretapping and Encryption*. Cambridge, MA. & London: MIT Press.

DiMaggio, P., & Hargittai, E. (2001). *From the 'Digital Divide' to 'Digital Inequality': Studying Internet use as penetration increases*. Princeton University, Center for Arts and Cultural Policy Studies, Working Paper, 15. Retrieved September 23, 2007 from: https://www.princeton.edu/arts pol/work-pap15.html

DiMaggio, P., & Hargittai, E. (2002, August). *The new digital inequality: Social stratification among Internet users*. Paper presented at the American Sociological Association Annual Meeting, Chicago.

DiNicolo, D. (2002). The African beam. *Canadian Business, April 29*, 73.

Donkor, M. (2002). Educating girls and women for the nation: Gender and educational reform in Ghana. *International Education, 32*(1), 72-85.

Dubé, L., Bourhis, A., & Jacob, R. (2003). *Towards a Typology of Virtual Communities of Practice* (Cahier du GReSI No. 03-13). Montréal, Canada: HEC Montréal, Groupe de Recerche en Systémes d'Information.

Dubow, S. (1995). *Scientific racism in modern South Africa*. Cambridge: Cambridge University Press.

Ducatel, K., Webster, J., & Herrmann, W. (Ed.) (2000). *The information society in Europe, work and life in an age of globalization*. New York-Oxford: Rowman & Littlefield.

Ducheneaut, N., Yee, N., Nickell, E., & Moore, R. J. (2006). Building an MMO with mass appeal: A look at gameplay in *World of Warcraft*. *Games and Culture, 1*, 281-317.

Ducheneaut, N., Yee, N., Nickell, E., & Moore, R. J. (2007). The life and death of online gaming communities: A look at guilds in *World of Warcraft*. In *Proceedings of CHI2007* (pp. 839-848). New York: Association for Computing Machinery.

Dyer-Witheford, N. (2002). E-capital and the many-headed hydra. In G. Elmer (Ed.), *Critical perspectives on the Internet,* (pp. 129-64). Lanham, MD: Rowman and Littlefield.

Ebbers, W. E., & van Dijk, J. A. M. G. (2007). Resistance and support to electronic government, building a model of innovation. *Government Information Quarterly, 24*(3), 554-575.

e-Business W@tch (2005). *A Guide to ICT Usage Indicators. Definitions, sources, data collection.* Special Report, Bonn-Brussels: European Commission, & Enterprise & Industry Directorate General.

Eells, R. (1962). *The Government of Corporation.* New York: The Free Press of Glencoe.

Elijah, O., & Ogunlade, I. (2006). Analysis of the uses of information technology for gender empowerment and sustainable poverty alleviation. *International Journal of Education and Development using ICT, 2*(3), 45-69.

Engelbrecht, L. (2007). Five in court for marriage fraud. Johannesburg: *ITWeb*, 10 December.

Engelbrecht, L. (2007). Blank cheque for Home Affairs information technology. Johannesburg: *ITWeb*, 5 September.

Engelbrecht, L. (2007). HANIS gets Rands 30m 'refresh. Johannesburg: *ITWeb*, 6 September.

Engelbrecht, L. (2007). Getting Home Affairs cup-ready. Johannesburg: *ITWeb*, 5 November.

Entsua-Mensah, C. (2005). Revitalizing the indigenous agricultural marketing system in Ghana through the e-commerce project: A performance appraisal. *IAALD Quarterly Bulletin, 50*(3), 141 -147

ESCAP (1990). Restructuring the developing economies of Asia and the pacific in the 1990s. New York: United Nations.

ESPON (2006a). *Territory matters for competitiveness and cohesion – Facets of regional diversity and potentials in Europe – ESPON Synthesis Report III.* Results by autumn 2006. Retrieved May 13, 2008 from: http://www.espon.lu

EU Commission (2005). *I2010 – A European information society for growth and employment.* COM 229 final {SEC(2005) 717}.

Eu task force report (2006), *Fostering the competitiveness of Europe's ICT industry.* Retrieved April 15, 2006, from ec.europa.eu/enterprise/ict/policy/doc/icttf_report.pdf

European Commission (1994). *Europe's way to the information Ssociety – An action plan.* Retrieved February 10, 2008, from http://europa.eu.int/ISPO/docs/htmlgenerated/i_COM(94)347final.html

European Commission (1999). *eEurope: an Information Society for All.* Communication on a Commission Initiative for the Special European Council of Lisbon, 23 and 24 March 2000.

European Commission (2000). *eEurope 2002. Action Plan.* Brussels.

European Commission (2003). *Linking up Europe: the importance of interoperability of e-Government services.* Brussels.

European Commission (2003). *The Role of eGovernment for the Europe's Future.* Brussels.

European Commission (2005). *i2010 – A European Information Society for growth and employment.* Brussels.

European Commission (2006). *i2010 – eGovernment Action Plan: Accelerating eGovernment in Europe for the Benefit of All.* Brussels.

European Commission (2006). *Interoperability for Pan-European eGovernment Services.* Brussels.

European Commission (2007). *i2010 – Annual Information Society Report.* Brussels.

Evans, D., & Yen, D. C. (2005). E-government: An analysis for implementation: Framework for understanding cultural and social impact. *Government Information Quarterly, 22,* 354-373.

Evans, D., & Yen, D. C. (2006). E-government: Evolving relationship of citizens and government, domestic and international development. *Government Information Quarterly, 23,* 207-235.

Evans, I. (1997). *Bureaucracy and race: Native administration in South Africa.* Berkeley, CA: University of California Press.

Executive Office of the President (2001). *A blueprint for new beginnings*. Washington DC.

Executive Office of the President (2001). *The President's management agenda*. Washington DC.

Executive Office of the President, (2002). *E-government strategy. Implementing the President's management agenda for e-government*. Washington DC.

Faraj, S. Kwon, D., & Watts, S. (2004). Contested artefact: Technology sensemaking, actor networks, and the shaping of Web browser. *Information Technology and People*, *17*(2), 186-209.

Fauvelle-Aymar, F. X. (2006). *Histoire de l'Afrique du Sud*. Paris: Seuil.

Fineman, S. (2006). Emotion and organizing (pp. 675–700). In C. Hardy, S. Clegg, T. Lawrence, & W. Nord (Eds.), *Handbook of organization studies*. London: Sage.

Fineman, S., Maitlis S., & Panteli, N. (2007). Virtuality and Emotion. *Human Relations*, *60*(4), 555–560.

Fire prompts tight control on Internet cafes in China. (2002, June 18). *People's Daily*. Retrieved April 18, 2008, from http://english.peopledaily.com.cn/200206/18/eng20020618_98048.shtml

First, R., Steele, J., & Gurney, C. (1973). *The South African connection: Western investment in apartheid*. Harmondsworth: Penguin.

Ford, P. Why China Shut Down 18,401 Websites. (2007, September 25). *The Christian Science Monitor*. Retrieved January 9, 2008, from http://www.csmonitor.com/2007/0925/p01s06-woap.htm

Forman, M. (2003). *Memorandum for the chief information officers. Procedures for requesting funds from the e-government fund*. Executive Office of the President, Office of Management and Budget, Washington.

Fortsakis, T. (2005). Principles Governing Good Administration. *European Law Review*, *11*(2), 207-217.

Foucault, M. (1995). *Surveiller et punir: Naissance de la prison*, Paris: Gallimard.

Foucault, M. (1997). Cours du 14 janvier 1976. In F. Ewald, A. Fontana & M. Bertani Ed.), *Il faut défendre la société* (pp. 21-36). Paris: Gallimard-Seuil.

Fountain, J. E. (2001). *Building the virtual state: Information technology and institutional change*. Washington DC: Brookings Institution Press.

Fountain, J. E. (2007). Bureaucratic reform and e-government in the United States: An institutional perspective. National Center for Digital Government Working Paper, No. 07-006, to be printed in A. Chadwick & P. N. Howard (Eds.), *The handbook of Internet politics*. New York: Routledge.

Free and Open Source Software Foundation for Africa (FOSSFA) (2004). *FOSSFA action plan 2004–2006*, Nairobi: FOSSFA.

Frick, M. M. (2005). *Parliaments in the digital age*. Geneve: E-Democracy Centre.

Fuchs, C. (2008). *Internet and society: Social theory in the information age*. New York: Routledge.

Fukuyama, F. (2004). *State-building: Governance and world order in the 21 Century*. Ithaca: Cornell University Press.

Garson, D. G. (2003). Toward an information technology research agenda for public administration. In D. G. Garson (Ed.), *Public Information Technology: Policy and Management Issues*. Hershey, PA: IGI Global Publishing.

Garson, G. D. (2004). The promise of digital government. In A. Pavlichev, & D. Garson, (Eds.), *Digital government: principles and best practices* (pp. 2-15). Hershey, PA: IGI Global Publishing.

Garson, G. D. (2005). Patriotic information systems: Evaluating Bush administration information policy. *Social Science Computer Review*, *23*(4), 395-400.

Germino, A. C., Parent, M., & Vandebeek, C. (2005). Building citizen trust through e-government. *Government Information Quarterly*, *22*, 720-736.

Gherardi, S. (2008). Breve storia di un concetto in viaggio: dalla comunità di pratica alle pratiche di una comunità. *Studi Organizzativi*, *10*(1), 49-72.

Gherardi, S., Nicolini, D., & Strati, A. (2007). The passion for knowing. *Organizations, 14*(3), 315-330.

Ghosh, R. A. (2003). Licence fees and GDP per capita: The case for open source in developing countries. *First Monday, 8*(12). Retrieved November 18, 2007, from http://firstmonday.org/issues/issue8_12/ghosh/index.html

Gigaba, M. (2004). *Home affairs department budget vote 2004/2005.* Cape Town: National Assembly – Republic of South Africa, 11 June.

Gilder, B. (2003). *Media briefing.* Johannesburg: Department of Home Affairs – Republic of South Africa, 5 November.

Gil-Garcia, J. R., & Luna-Reyes, L. F. (2003). Towards a definition of electronic government: A comparative review. In A. Mendez-Vilas, et al. (Eds.), *Techno-legal aspects of information society and new economy: An overview.* Extremadura, Spain: Formatex Information Society Series.

Gil-Garcia, J. R., & Pardo, T. (2005). E-government success factors: Mapping practical tools to theoretical foundations. *Government Information Quarterly,* (pp. 187-216).

Gillet, S. E., & Kapor, M. (1996). The self-governing Internet: Coordination by design. In B. Kahin & J. Keller (Eds.), *Coordination of Internet.* MIT Press.

Gillett, S. E., Lehr, W., Wroclawski, J. T., Clark, D. D. (2001, October). *A taxonomy of Internet appliances.* Sloan Working Paper 4186-01 eBusiness@MIT Working Paper 112.

Gilliom, J. (2001). *Overseers of the poor: Surveillance, resistance and the limits of privacy.* Chicago, IL: University of Chicago Press.

Golbeck, J. (2007). The dynamics of Web-based social networks: Membership, Relationships, and Change. *First Monday, 12*(11). Retrieved 5 November 2007 from http://www.uic.edu/htbin/cgiwrap/bin/ojs/index.php/fm/article/view/2023.

Goldsmith, J., & Wu, T. (2006). *Who controls the Internet? Illusions of a borderless world.* New York: Oxford University Press.

Goudos, S. K., Peristeras, V., & Tarabanis, K. (2007). *Mapping citizen profiles to public administration services using ontology implementations of the governance enterprise architecture (GEA) models.* Proceedings of the 40th Hawaii International Conference on System Sciences, Maui (Usa): IEEE Press.

Government Reform Minority Office (2004). *Secrecy in the Bush administration.* U.S. House of Representatives, Washington.

Gramsci, A. (1997). *Le opere. La prima antologia di tutti gli scritti.* Ed. Antonio A. Santucci. Roma: Editori Riuniti.

Guilaume, M. (1978). *Eloge du désordre.* Paris: Gallimard.

Guillaume, P., Péjout, N., & Wa Kabwe Segatti, A. (2004). *L'Afrique du Sud dix ans après: Transition accomplie?.* Johannesburg – Paris: Institut Français d'Afrique du Sud – Karthala.

Guseh, J. S., & Oritsejafor, E. (2005). Democracy and economic growth in Africa: The cases of Ghana and South Africa. *Journal of Third World Studies, 22*(2), 121-137

Habermas, J. (1986). *Between facts and norms. Contributions to a discourse theory of law and democracy.* Cambridge: MIT Press.

Hachigian, N. (2002). Telecom taxonomy: How are the one party states of East Asia controlling the political impact of the Internet? In J. Zhang & M. Woesler (Eds.). *China's Digital Dream: The Impact of the Internet on Chinese Society* (pp. 35-67). Bochum: The University Press Bochum.

Hachigian, N. (2002). The Internet and power in one-party East Asian states. *The Washington Quarterly, 25*(3), 41-58.

Hagen, M. (2003). *A typology of electronic democracy.*

Hagen, M. (2004). Electronic government in the United States. In Eifert, M. & Püschel, J.O. (Eds), *National Electronic Government: Comparing Governance Structures in Multilayer administrations* (211-240). London: Routledge.

Hague, B. N., & Loader, B. D. (Eds.). (1999). *Digital democracy. Discourse and decision making in the information age*. London & New York: Routledge.

Halcli, A., & Webster, F. (2000). Inequality and mobilization in the information age. *The European Journal of Social Theory, 3*(1), 67-81.

Hamilton, A. (1791). *Report of the Secretary of the Treasury of the United States, on the subject of manufactures*. Presented to the House of Representatives, December 5, 1791, Philadelphia, PA: Childs and Swaine.

Hargittai, E. (2007). Whose space? Differences among users and non-users of social network sites. *Journal of Computer-Mediated Communication, 13*(1), 14. Retrieved February 10, 2008 from http://jcmc.indiana.edu/vol13/issue1/hargittai.html.

Harlow, C. (2005). Law and public administration: convergence and symbiosis. *International Review of Administrative Sciences, 7*(2), 279-294.

Hart, K., (1973). Informal income opportunities and urban employment in Ghana. *The Journal of Modern African Studies, 11*(1), 61-89.

Haruna, P. F. (1999). *An empirical analysis of motivation and leadership among career public administrators: The case of Ghana*. Unpublished doctoral dissertation, University of Akron, Ohio.

Haruna, P. F. (2003). Reforming Ghana's public service: Issues and experiences in comparative perspective. *Public Administration Review, 63*(3), 343-354.

Harvey, D. (1990). *The condition of postmodernity*. Basil Blackwell.

Harvey, D. (2003). *The new imperialism*. Oxford: Oxford University Press.

Harvey, D. (2005). *A brief history of neoliberalism*. Oxford: Oxford University Press.

Harwit, E. (1998). China's telecommunications industry: Development patterns and policies. *Pacific Affairs, 71*(2), 175-194.

He, Q. (2004). *Zhongguo zhengfu ruhe kongzhi meiti* (How the Chinese government controls the media). (pp. 117-48). Retrieved January 25, 2008, from http://www.ir2008.org/PDF/initiatives/Internet/Media-Control_Chinese.pdf

Heeks, R. (2002). Reinventing government in the information age. In Heeks, R. (Ed.), *Reinventing Government in the Information Age. International Practice in IT-Enabled Public Sector Reform* (9-21). London: Routledge.

Heeks, R. (2003). Most eGovernment-for-development projects fail: How can risks be reduced? *iGovernment Working Paper Series*. Paper no. 14. Retrieved March, 15, 2008, from unpan1.un.org/intradoc/groups/public/documents/NISPAcee/UNPAN015488.pdf

Heeks, R. B. (1999). Software strategies in developing countries. *Communications of the ACM, 42*(6), 15-20.

Heeks, R. B. (2002). E-government in Africa: Promise and practice. *Information Polity, 7*(2/3), 97-114.

Heeks, R. B. (2002). Information systems and developing countries: Failure, success and local improvisations. *The Information Society, 18*(2), 101-112.

Held, D. (2004). *Global covenant: The social alternative to the Washington consensus*. Cambridge: Polity Press.

Held, D., & McGrew, A. (2003). *The global transformations reader*. Cambridge: Polity Press.

Held, D., McGrew, A., Goldblatt, D., & Perraton, J. (1999). *Global transformations: Politics, economics, and culture*. Cambridge: Polity Press.

Hemmati, M. (2002). *Multi-stakeholder processes for governance and sustainablility: Beyond deadlock and conflict*. London: Earthscan

Hennion, A. (2004). Une sociologie des attachments. D'une sociologie de la culture à une pragmatique de l'amateur. *Sociètès, 85*(3), 9-24.

Heritier, A. (2001). Differential Europe: National administrative responses to community policy. In M.G. Cowles, J. Caporaso & T. Risse (eds). *Transforming Europe.*

Europeanization and domestic change (pp. 257-294). Ithaca, NY: Cornell University Press.

Hess, J. B. (2003). Imaging architecture 11, "treasure storehouses" and constructions of Asante regional hegemony. *Africa Today, 50*(1), 27-48.

Hill, D. T., & and Sen, K. (2005). *The Internet in Indonesia's new democracy*. London and New York: Routledge.

Hinson, R. E. (2005). Internet adoption among Ghana's SME nontraditional exporters. *African Insight, 35*(1), 20-27.

Hockings, B. (2006). Multistakeholder diplomacy: Forms, functions and frustrations. In J. Kurbalija & E. Katrandjiev (Eds.), *Multistakeholder Diplomacy. Challenges and Opportunities* (pp. 13-32). Diplo Foundation.

Hoffmann, J. (2006). *Internet governance: A regulative idea in flux*. Retrieved from http://duplox.wz-berlin.de/people/jeanette/texte/Internet%20Governance%20english%20version.pdf. Last access March 1, 2008

Hofmann, H. C. (2006, November). Mapping the European administrative space. Paper presented at the Connex thematic conference *"Towards a European Administrative Space"*, Birbeck College, London.

Holland, S. (1980). *Uncommon market: Capital, class and power in the European community*. New York: St. Martin's Press

Homburg, V. (2004). E-government and Npm: A perfect marriage?. *Proceedings of the 6th International Conference on Electronic Commerce* (547–555). New York: International Center for Electronic Commerce Publisher.

Hooghe, L. (1996). Introduction: Reconciling EU-wide policy and national diversity. In L. Hooghe (Ed.), *Cohesion policy and European integration. Building multi-level governance*, Oxford: Clarendon Press.

Hooghe, L. (1996). Building a Europe with the regions. The changing role of the European commission. In L. Hooghe (Ed.), *Cohesion policy and European integration. Building multi-level governance*. Oxford: Clarendon Press.

Hu, A. (2002). *Zhongguo Zhanlue Gouxiang* (Strategy of China). (p. 15). Hangzhou: Zhejiang Renmin Chubanshe.

Hughes, C. R., & Wacker, G. (Eds.). (2003). *China and the Internet: Politics and the digital leap forward*. London and New York: Routledge Curzon.

Hulianwang Xinwen Xinxi Fuwu Guanli Guiding (Provisions on the Administration of Internet News Information Services). Retrieved March 12, 2008, from http://www.cnnic.cn/html/Dir/2005/09/27/3184.htm

Hulianwang Xinxi Fuwu Guanli Banfa (2000, October 1). Retrieved February 12, 2004, from http://past.people.com.cn/GB/channel5/28/20001001/257566.html

Human rights in China. (2008, January 2). Press Advisory: HRIC Launches 2008 Take Action Website and Calls on China To Release Shi Tao. Retrieved January 7, 2008, from http://hrichina.org/public/contents/press?revision%5fid=46424&item%5fid=46414

Hung, C. F. (2003). Public discourse and "virtual" political participation in the PRC: The Impact of the Internet. *Issues & Studies, 39*(4), 1-38.

Hung, C. F. (2005). The interaction between Internet entrepreneurs and the Chinese authorities: Possible implications for civil society. *Issues & Studies, 41*(3), 145-80.

Hung, C. F. (2006). The politics of cyber participation in the PRC: The implications of contingency for the awareness of citizens' rights. *Issues & Studies, 42*(4), 142-5.

Hurst, J. W. (1977). *Law and social order in the United States*. Ithaca, NY: Cornell University Press.

Hyder, S. (2005). The information society: Measurements biased by capitalism and its intent to control-dependent societies - a critical perspective. *The International Information & Library Review, 37*, pp. 25-27.

i2010 High Level Group (2006). *i2010 Benchmarking Framework*. Retrieved May 4, 2008, from http://

ec.europa.eu/information_society/eeurope/i2010/docs/i2010_high_level_group/i2010_benchmarking_framework.doc

i2010 High Level Group (2006). *The challenge of convergence.* Brussels.

Idabc (2004). *European interoperability framework for Pan-European e-government services.* Brussels.

Idabc (2005). *The impact of e-government on competitiveness, growth and jobs.* Brussels.

Idabc (2005). *eGovernment on the member states of the European Union.* Brussels.

Idabc (2006). Bringing government closer to the people. *Synergy. The Idabc quarterly,* Brussels.

Ifinedo, P. (2005). Measuring Africa's e-readiness in the global networked economy: A nine-country analysis. *International Journal of Education and Development using ICT, 1*(1), 53-71.

Independent Electoral Commission. (2004). *Independent Electoral Commission on reports about giving out details from voters' roll.* Pretoria: Independent Electoral Commission, 15 June.

International Labour Organisation. (1972). *Incomes and equality: A strategy for increasing productive employment in Kenya,* Geneva: International Labour Organisation.

International working group on Administrative Burdens (2004). *The Standard Cost Model; a framework for defining and quantifying administrative burdens for businesses.* Retrieved May 15, 2008, from www.compliancecosts.com

Isenberg, D. (1997, August). *Rise of the stupid network.* Computer Telephony.

ITU (2005). *World Summit on the Information Society. Outcome Documents. Geneva 2003-Tunis 2005.* Retrieved January 21, 2008, from http://www.itu.int/wsis/outcome/booklet.pdf

ITU (2006). *World Telecommunication/ICT Development Report 2006: Measuring ICT for social and economic de-*velopment. Retrieved January 13, 2008, from http://www.itu.int/ITUD/ICT/publications/wtdr_06/index.html

ITU (2007). *World Telecommunication Indicators database,* Retrieved January 20, 2008, from www.itu.int/ITUD/ICT/publications/world/world.html

ITU (2007). *World Information Society Report 2007. Beyond WSIS.* Geneva: ITU.

ITU (International Telecommunication Union). (2007). *Telecommunication Indicators Handbook.* Retrieved April 14, 2008, from http://www.itu.int/ITU-D/ICT/handbook.html

Jacobs, S. J., & Herselman, M. E. (2005). An ICT-hub model for rural communities. *International Journal of Education and Development using ICT, 1*(3), 57-93.

Jaeger, P. T. (2005). Deliberative democracy and the conceptual foundations of electronic government. *Government Information Quarterly, 22,* 702-719.

Japan (2004). *Information and communication in Japan - White Paper.* Ministry of Public Management, Home Affairs, Posts and Telecommunications. Retrieved May 11, 2008 from: http://www.soumu.go.jp/joho_tsusin/eng/whitepaper.html

Jedlowski, P. (1994). *Il Sapere dell'Esperienza.* Milano: il Saggiatore.

Johnson, C. Y. (2007). Online gamers become guinea pigs: Epidemics uncorked in virtual worlds. *Boston Globe,* August 25. Retrieved June 6, 2008, from http://www.boston.com/news/local/articles/2007/08/25/online_gamers_become_guinea_pigs/

Jones, S. G. (1995). *Cybersociety. Computer-Mediated Communication and Community.* London: Sage.

Juma, C. (2005). The way to wealth. *New Scientist, 185*(21), 15-21.

Kafai, Y. B., Feldon, D., Fields, D., Giang, M., & Quintero, M. (2007). Life in the times of whypox: A virtual epidemic as a community event. In C. Steinfield, B. Pentland, M. Ackerman & N. Contractor (Eds.), *Communities and Technologies* (pp. 171-190). New York: Springer.

Kahler, M. (2002). The State of the State in World Politics. In I. Katznelson and H. Milner (Eds), *Political Science: the State of the Discipline*. New York-London: Norton.

Kahn, M. H. (2002). State failure in developing countries and strategies of institutional reform. *Proceedings of the ABCDE Conference*, Oslo, 24-26 June.

Kaiser, S., Müller-Seitz, G., Pereira Lopes, M., & Pina e Cuhna, M. (2007). Weblog as a trigger to elicit passion for knowledge. *Organizations*, *14*(3), 373-390.

Kalathil, S. (2001). The Internet and Asia: Broad band or broad bans? *Foreign Service Journal*, *78*(2). Retrieved April 18, 2001, from http://www.ceip.org/files/Publications/internet_asia.asp?p=5&from=pubdate

Kalathil, S. (2003). Dot com for dictators. *Foreign Policy*, *135*, 42-49.

Kalathil, S., & Boas, T. C. (2001). The Internet and state control in authoritarian regimes: China, Cuba, and the counterrevolution. *First Monday*, *6*(8). Retrieved June 8, 2008, from http://www.firstmonday.org/issues/issue6_8/kalathil/

Kalathil, S., & Boas, T. C. (2003). *Open networks, closed regimes: The impact of the Internet on authoritarian rule*. Washington, DC: Carnegie Endowment for International Peace.

Kalir, E., & Maxwell, E.E. (2002). *Rethinking boundaries in Cyberspace. A report of the Aspen Institute*. Washington.

Kamarc, E. C., & Nye, J. S. (Eds.) (2002). *Governance. com. Democracy in the information age*. Washington: Brookings Institution Press.

Kankpeyeng, B. W., & DeCorse, C. R. (2004). Ghana's vanishing past: development, antiquities, and the destruction of the archaeological record. *African Archaeological Review*, *21*(2), 89-128.

Katchanovski, I., & La Porte, T. (2005). Cyberdemocracy or Potemkin e-villages? Electronic governments in OECD and post-vommunist vountries. *International Journal of Public Administration*, *28*(7), 665-681.

Katz, H., & Anheier, H. (2006). Global vonnectedness: The structure of transnational NGO getworks. In M. Glasius, M. Kaldor, & H. Anheier (Eds.), Global Civil Society 2005/6 (pp. 240-65). London: Sage.

Keck, M. E., &. Sikkink, K. (1998). *Activists beyond borders: advocacy networks in international politics*. New York: Cornell University Press.

Keeble, L., & Loader, B.D. (2001). *Community informatics. Shaping computer-mediated relations*. London: Routledge.

Keinwachter, W. (2007). *The power of ideas: Internet governance in a global multi-stakeholder environment*. Berlin: Marketing fur Deutschland GmbH.

Kenis, P., & Schneider, V. (1991), Policy network and policy analysis: Scrutinizing a new analytical toolbox . In B. Marin & R. Mayntz (Eds.), *Policy networks. Empirical evidence and Theoretical considerations* (pp. 25-62). Boulder: Westview Press

Kenis, P., & Schneider, V. (1991). Policy network and policy analysis: Scrutinizing a new analytical toolbox, (pp. 25-62). In B. Marin & R. Mayntz (Eds.), *Policy networks. Empirical evidence and theoretical considerations*. Boulder, CO: Westview Press.

Kennedy, S. (2007). Vivendi and Activision to create Activision Blizzard: Two companies merge to become the most profitable game publisher. 1up.com, December 2. Retrieved June 6, 2008, from http://www.1up.com/do/newsStory?cId=3164668

Khulumani Support Group. (2004). Complaint. South District Court, New York State, 15 December.

Kim, J. (1998). Universal service and Internet commercialization: Chasing two rabbits at the same time. *Telecommunication Policy*, *22*(4-5), 281-288.

Kim, K.-H., & Yun, H. (2007). Cying for me, Cying for us: Relational dialectics in a Korean social network site. *Journal of Computer-Mediated Communication*, *13*(1), 15. Retrieved February 10 2008 from http://jcmc.indiana.edu/vol13/issue1/kim.yun.html.

Klastrup, L. (2006). Death matters: Understanding gameworld experiences. In *Proceedings of the International Conference on Advances in Computer Entertainment Technology (ACE)* (2006) June 14-16, Hollywood, California. New York: Association for Computing Machinery.

Kleinwächter, W. (2004). Beyond ICANN vs. ITU?How WSIS tries to enter the new territory of Internet governance. *Gazette, 66* (3-4), 233-251

Kleinwächter, W. (2007). The history of Internet governance. In C. Mőeller & A. Amoroux (Eds.), *Governing the Internet* (pp.41-66). OSCE Representative on Freedom of Media.

Knight, R. (1986). *US computers in South Africa*. Retrieved October 26, 2003, from http://richardknight.homestead.com/files/uscomputers.htm

Knoke, D., Pappi, F. U., Broadbent, J., & Tsujinaka, Y. (1996). *Comparing policy networks*. Cambridge: Cambridge University Press.

Knorr Cetina, K. D. (1997). Sociality with Objects: Social Relations in Postsocial Knowledge. *Theory Culture and Society, 14*(4), 1-30.

Knorr Cetina, K. D. (2007). Culture in global knowledge societies: Knowledge cultures and epistemic cultures. *Interdisciplinary Science Reviews, 32*(4), 361-375.

Knorr Cetina, K. D., & Bruegger, U. (2000). The market as an object of attachment: Exploring postsocial relations in financial markets. *Canadian Journal of Sociology, 25*(2), 141-168.

Knorr Cetina, K., & Bruegger, U. (2002). Traders' engagement with market: A post-social relationship. *Theory Culture Society, 19*(5-6), 161-185.

Kolko, G. (1963). *The triumph of conservatism: A reinterpretation of American history, 1900-1916*. New York: The Free Press of Glencoe.

Kollock, P. (1999). The economics of online cooperation: Gifts and public goods in cyberspace (pp. 220-239). In M. A. Smith & P. Kollock (Eds.), *Communities in Cyberspace* London: Routledge.

Kollock, P., & Smith, M. (Eds) (1999). *Communities in cyberspace*. London: Routledge.

Kooiman, J. (2003). *Governing as governance*. London: Sage Publications.

Kratochwil, F. (1989). *Rules, norms and decisions: On the conditions of practical and legal reasoning in international relations and domestic affaires*. Cambridge: University Press.

Krugman, P. (1991). *Geography and trade*. Leuven: Leuven University Press.

Kshetri, N. (2004). Economics of Linux adoption in developing countries. *IEEE Software, 21*(1), 74-81.

Kubicek, H., Dutton, W. H., & Williams, R. (Eds.). (1997). *The social shaping of information superhighways*. Frankfurt: Campus Verlag.

Kumar, K. (1995). *From post-industrial to post-modern society. New theories of the contemporary world*. Oxford: Blackwell Publishing.

Kwapong, O. A. T. (2007). Problems of policy formulation and implementation: The case of ICT use in rural women's empowerment in Ghana. *International Journal of Education and Development using ICT, 3*(2), 1-21.

Lall, S., Navaratti, G. B., Teitel, S., & Wiggnaraja, G. (1994). *Ghana under structural adjustment*. New York: St Martin's Press

Lam, W. (2008). Stability trumps reform at China's parliamentary session. *China Brief, 8*(6), 2-4. http://www.jamestown.org/terrorism/news/uploads/cb_008_006a.pdf

Lambinon, I. (2003). *Briefing at the Home Affairs portoflio committee*. Cape Town, National Assembly – Republic of South Africa, 18 March.

Lanzara G. F., & Morner, M. (2005). Artifact rule! How organizing happens in an open source project (pp. 67-90). In B. Czarniawska & T. Hernes (Eds.), *Actor network theory and organizing*. Copenhagen: Liber & Copenhagen Business School Press.

Lascoumes, P., & Le Gales, P. (Ed.). (2004). *Gouverner par les instruments*. Paris: Presses de Sciences Po.

Latham, R., & Sassen, S. (Eds) (2005). *The Digital Formations: IT and New Architectures in the Global Realm*. Princeton: Princeton University Press.

Latour, B. (1987). *Science in action. How to follow scientists and engineers through society*. Cambridge, MA: Harvard University Press.

Latour, B. (1991). Technology is society made durable. In Law (Ed.), *A sociology of monsters: Essays on power, technology and domination*, (pp. 103-131). London: Routledge.

Latour, B. (1998). *Thought experiments in social science: From the social contract to virtual society*. Annual Public Lecture at 1st Conference on Virtual Society?, Brunel University, UK. Retrieved 08 February 2008 from www. artefaktum.hu/it/Latour.html

Latour, B. (2005). *Reassembling the social*. Oxford: Oxford University Press.

Laumann, E. O., & Knoke, D. (1987). *The organizational state*. Madison: University Winsconsin Press.

Lautier, B. (2004). *L'économie informelle dans le tiers monde*. Paris: La Découverte.

Layne, K., & Lee, J. (2001). Developing fully functional e-government: A four stage model. *Government Information Quarterly, 18*, 122-136.

Lazuly, P. (2004). Il mondo secondo Google. In AA. VV., *Il pensiero unico al tempo della rete* (pp. 44-46). Numero speciale fuori serie di *Le monde diplomatique/il manifesto*.

Leigh, J., & Brown, M. D. (2008). Cyber-commons: Merging real and virtual worlds. *Communications of the ACM, 51* (January), 82-85.

Lenin, V. I. (1916). *Imperialism, the highest stage of capitalism*. Retrieved June 6, 2008, from http://www. marxists.org/archive/lenin/works/1916/imp-hsc/.

Lentini, O. (2003). *Saperi sociali, ricerca sociale, 1500-2000*, Milan: Franco Angeli.

Leonardi, R. (2005). *Cohesion policy in the European Union: The building of Europe*. London: Palgrave Macmillan.

Lessig, L. (1999). *Code and other Laws of Cyberspace*. New York: Basic Books.

Lessig, L. (2006). *Code. Version 2.0*. New York: Basic Books.

Li, L., & Yu, L. (2005, August 18). 14 Buwei Lianhe Jinghua Hulianwang. (Fourteen Departments United to 'Purify' the Internet) *Nanfang Weekend*. Retrieved June 6, 2008, from http://www.nanfangdaily.com.cn/zm/20050818/xw/szxw1/200508180019.asp

Lips, M. (2001). Designing electronic government around the world. Policy developments in the USA, Singapore, and Australia. In J. E. J. Prins (Ed.), *Designing e-government: On the crossroads of technological innovation and institutional change* (pp. 199-216). Boston: Kluwer Law International.

Lisbon European Council (2000) *Presidency conclusions*.

List, F. (1885 [1841]), *The national system of political economy*. London: Longmans, Green and Company.

Lofgren, E. T., & Fefferman, N. H. (2007). The untapped potential of virtual game worlds to shed light on real world epidemics. *The Lancet Infections Diseases, 7*, 625-629.

Lu, X. (Ed.). (2002). *Zhongguo Xinxihua* (China's Informatization). (p. 53). Beijing: Electronics Industry Publisher.

Lummis, M., & Kern, E. (2006). *World of Warcraft master guide*. Indianapolis, IN: BradyGAMES/DK Publishing.

Lummis, M., & Kern, E. (2007). *World of Warcraft: The Burning Crusade* edition. Indianapolis, IN: BradyGAMES/DK Publishing.

Lusoli, W. (2006). Of windows, triangles and loops: the political economy of the e-democracy discourse. *Comunicazione politica, 7*(1), 27-48.

Lynch, D. C. (1999). *After the propaganda state: Media, politics, and "thought work" in reformed China*. (p. 173). Stanford: Stanford University Press.

Lyon, D. (1994). *The electronic eye: The rise of surveillance society*. Cambridge: Polity Press.

Lyon, D. (2001). *Surveillance society: Monitoring everyday life*. Milton Keynes: Open University Press.

Lyon, D. (2002). Everyday surveillance: Personal data and social classifications. *Information, Communication & Society, 5*(2), 242-257.

Lyon, D. (2003). Introduction. In D. Lyon (Ed.), *Surveillance as social sorting: Privacy, risk and digital discrimination* (pp. 1-9). London: Routledge.

Machlup, F. (1962). *The production and distribution of knowledge in the United States*. Princeton: Princeton University Press.

Mangesi, K. (2007). *ICT in education in Ghana, Survey of ICT and Education in Africa: Ghana Country Report*. www.Infodev.org/ict4edu-Africa

Mansell, R., & When, U. (Eds.). (1998). *Knowledge societies: Information technology for sustainable development. Report for the United Nations Commission on science and technology for development*. New York: Oxford University Press.

Mapisa-Nqakula, N. (2004). *Home Affairs department budget vote 2004/2005*. Cape Town: National Assembly – Republic of South Africa, 11 June.

Mapisa-Nqakula, N. (2007). DHA budget speech for budget vote 2007. Cape Town: National Assembly – Republic of South Africa, 7 June.

Martin, B. (2005). Information society revisited: From vision to reality. *Journal of Information Science, 31*(1), 4-12.

Martin, D.-C. (1998). Le poids du nom: Culture populaire et constructions identitaires chez les 'métis' du Cap. *Critique Internationale*, n°1, 73-100.

Martin, S., & Robinson, J. P. (2004). The income digital divide: An international perspective. *IT&Society, 1*(7).

Retrieved April 3, 2008 from: http://www.itandsociety.org

Masuda, Y. (1981). *Information society as post-industrial society*. Bethesda, MD: World Future Society.

Mattelart, A. (2003). *The information society: An introduction*. London, Thousand Oaks, CA, & New Delhi: Sage.

May, C. (2006). Escaping the TRIPs' Trap: The political economy of free and open software in Africa. *Political Studies, 54*(1), 123-146.

May, C. (2006), The FLOSS alternative: TRIPs, non-proprietary software and development. *Knowledge, Technology, & Policy, 18*(4), 142-163.

Mayur, R., & Daviss, B. (1998). The technology of hope: Tools to empower the world's poorest peoples. *Futurist, 32*(7), 46-51.

McCaughey, M., & Ayers, M. D. (Eds.). (2003). *Cyberactivism: Online activism in theory and practice*. London and New York: Routledge.

McChesney, R. (2000). *Rich media, poor democracy: Communication politics in dubious times*. New York: New Press.

McMichael, P. (2000). States and governance in the era of "globalization". In J. D. Schmidt, *Globalization and Social Change* (pp. 181-198). London: Routledge.

McMicheal, P. (2004). *Development and social change. A global perspective*. London: Sage.

Menou, M. J., & Taylor, R. D. (2006). A «Grand Challenge»: Measuring information societies. *The Information Society, 22*, 261-267.

Mensah, J. V. (2005). Problems of district medium-term development plan implementation in Ghana. *International Development Planning Review, 27*(2), 245-270.

Midelfart-Knarvik, K. H., & Overman, H. G. (2002). Delocation and European integration. Is structural spending justified? *Economic policy*, October.

Milievic, I., & K. Gareis (2003). *Disparities in ICT take-up and usage between EU regions.* Presentation at NESIS workshop in Milan. Retrieved September 23, 2007 from: http://www.biser-eu.com

Miller, A. S. (1976). *The modern corporate state: Private governments and the America constitution.* Westport: Greenwood.

Minnaar, C.-L. (2004). Putting maps to work. Johannesburg: *ITWeb*, 12 March.

Mooney, P. (2004, April). China's "Big Mamas" in a Quandary. *YaleGlobal.* Retrieved June 8, 2008, from http://yaleglobal.yale.edu/display.article?id=3676

Morris-Suzuki, T. (1986). Capitalism in the computer age. *New Left Review, 160,* 81-91.

Mosco, V. (1998). Myth-ing links: Power and community on the information highway. *Information Society, 14*(1), 57-62.

Mosco, V. (2004). *The digital sublime. Myth, power, and cyberspace.* Cambridge, MA: MIT Press.

MPRA (Munich Personal RePEc Archive) (2008). *Mapping the ICT in EU regions: Location, employment, factors of attractiveness and economic impact* (S. Barrios, M. Mas, E., Navajas, J. Quesada. Institute for Prospective Technological Studies, Joint Research Centre, European Commission, 31 January 2008. MPRA Paper No. 6998, posted 04. February 2008 / 10:08. Retrieved March 3, 2008 from http://mpra.ub.uni-muenchen.de/6998

Mueller, M., Brenden, K., & Pagè, C. (2004). *Reinventing Media Activism: Public Interest Activism in the Making of U.S. Communication-Information Policy 1960-2002.* Retrieved from http://dcc.syr.edu/ford/rma/reinventing.pdf . Last access March 1, 2008.

Narayan, G., & Nerurkar, A. N. (2006). Value-proposition of e-governance services: Bridging rural-urbandigital divide in developing countries. *International Journal of Education and Development using ICT, 2*(3), 33-44.

Nardi, B, Schiano, D., & Gumbrecht (2004, November). *Blogging as social activity, or, would you let 900 million people read your diary?* Paper presented at CSCW 2004, Chicago, Illinois, USA.

Nardi, B., & Harris, J. (2006). Strangers and friends: Collaborative play in *World of Warcraft.* In *Proceedings of CSCW 2006* (pp. 1–10). New York: Association for Computing Machinery.

Nathan, A. J. (2003). Authoritarian resilience. *Journal of Democracy, 14*(1), 6.

National Partnership for Reinventing Government (2000). *Access America: Reengineering through information technology.* Washington.

Naughton, B. (1997). The patterns and logic of China's Economic reform. In the Joint Economic Committee, Congress of the United States (Ed.), *China's Economic Future: Challenges to U.S. Policy* (p. 1). New York and London: M.E. Sharpe.

Needham, C. (2004). The citizen as consumer: e-government in the United Kingdom and the United States. In R.K. Gibson, A. Römmele, & S. Ward, *Electronic Democracy* (pp. 43-69). London and New York: Routledge.

Negroponte, N. (1995). *Being Digital.* London: Coronet Book/Hodder & Stoughton.

Net users angry over slow connections. (2003, March 5). *South China Morning Post.*

Netchaeva, I. (2002). E-government and e-democracy: A comparison of opportunities in the north and south. *Gazette. The International Journal for Communication Studies, 64*(5), 467-477.

NetDailogue (2005). *Clearing house on international internet governance.* Harvard Law School's Berkman Center for Internet and Society (Berkman Center) and Stanford Law School's Center for Internet and Society (CIS).

Newman, E. (1924). *Wagner as man and artist.* New York: Knopf.

Norris, P. (2001). *Digital divide: Civic engagement, information poverty and the Internet worldwide.* Cambridge: Cambridge University Press.

North, D. C. (1990). *Institutions, institutional change and economic performance.* Cambridge: University of Cambridge.

NTIA (National Telecommunications and Information Administration) (2002). *A nation online: How Americans are expanding their use of the Internet.* Retrieved September 23, 2007 from: http://www.ntia.doc.gov

Nugent, J. D. (2001). If e-democracy is the answer, what's the question? *National Civic Review, 90*(3), 221- 233.

Nulens, G., & Van Audenhove, L. (1999). An information society in Africa? An analysis of the information society policy of the World Bank, ITU and ECA. *International Communication Gazette, 61,* 451-471.

Nye, J. S., & Owens, W. A. (1996). America's Information Edge. In *Foreign Affairs, 75*(2), 20-36.

Obeng, K. W. (2003). Ghana pursues justice and development through computer training. *Choices, 12*(4), 20-21.

OECD (1996). *Global information infrastructure and global information society, (GII-GIS). Statement of policy recommendations made by the ICCP Committee.* Retrieved April 10, 2008, from http://www.oecd.org/dataoecd/3/58/1896739.pdf

OECD (1999). *Report on the activities of the working party on indicators for the information society.* Retrieved March 18, 2008, from http://www1.oecd.org/std/na-meet99/Docs/na99_48e.pdf

OECD (2001). *Citizens as partners OECD handbook on information, consultation and public participation in policy-making.*

OECD (2002). *Measuring the information economy.* Retrieved March 20, 2008, from http://www.oecd.org/dataoecd/16/14/1835738.pdf

OECD (2003). *A framework document for information society measurement and analysis.* Paris: Oecd.

OECD (2003). *Policy brief: The e-government imperative: main findings* (OECD Observer, March 2003). Paris: OECD Publications.

Oecd (2005). *E-government as a tool for transformation.* Paris.

Oecd (2005). *Oecd input to the United Nations Working Group on Internet Governance.* Paris.

OECD (Organisation for Economic Co-operation and Development). (2005). *Guide to measuring the information society.* Paris: Oecd.

OECD (Organization for Economic Cooperation and Development) (1986). *Trends in the information economy.* Paris: OECD Publications.

Olesen, A. (2007, October 20). China's Internet controls tightened for politics' sake. *The Associated Press.*

Olsen, J. P. (2002). The many faces of Europeanization. *Arena Working Papers, 1*(2).

Olsen, J. P. (2002). Toward an administrative European space?. *Arena Working Papers, 2*(26).

Onuf, N. (1989). *World of our making. Rules and rule in social theory and international relations.* Columbia: University of Carolina Press.

Osborne, D., & Gaebler, T. (1992). *Reinventing government: How the entrepreneurial spirit is transforming the public sector.* New York: Penguin.

Overeem, A., Witters, J., & Peristeras, V. (2007, January). *An interoperability Framework for Pan-European E-Government Services (PEGS).* Proceedings of the 40th Hawaii International Conference on System Sciences, Maui (Usa): IEEE Press

Overturf, S. F. (1986). *The economic principles of European integration.* New York: Praeger.

Owusu, G. (2005). The role of district capitals in regional development. *International Development Planning Review, 27*(1), 59- 89.

Padovani, C. (2005). WSIS and multi-stakeholdersim. In D. Stauffacher & W. Kleinwachter (Eds.), *The World Summit on the Information Society: Moving fro the Past into the Future* (pp. 147-155). UN ICT Task Force.

Padovani, C. (2005). Civil society organizations beyond WSIS: Roles and potential of a "young" stakeholder. In O. Drossou & H. Jensen (Eds.), *Visions in proc-*

ess II. The World Summit on the Information Society. Geneva2003-Tunis 2005 (pp. 37-45). Berlin: Henrich Boell Foundation.

Padovani, C., & Pavan, E. (2007). Diversity reconsidered in a global multi-stakeholder environment: insights from the online world. In W. Kleinwächer (Ed.), *The power of ideas: Internet governance in a global multistakeholder environment* (pp. 99-109). Berlin: Germany Land of Ideas.

Padovani, C., & Tuzzi, A. (2004). WSIS as a World of Words. Building a common vision of the information society?. In *Continuum. Journal of Media and Society, 18*(3), 360-379.

Padovani, C., & Tuzzi, A. (2006). Communication Governance and the Role Civil Society. Reflections on Participation and the Changing Scope of Political Action. In J. Servaes & N. Carpentier (Eds.), *Towards a Sustainable Information Society Beyond WSIS* (pp. 51-79). Bristol & Portland: Intellect.

Padovani, C., Tuzzi, A., & Nesti, G. (2007). Communication and (e)democracy: assessing European e-democracy discourses. In B. Cammaerts & N. Carpentier (Eds.), *Reclaiming the media. Communication rights and democratic media roles* (pp. 9-30). Bristol & Portland: Intellect.

Palan, R.P., Abbott, J. & Deans, P. (1996). *State strategies in the global political economy.* London: Pinter.

Partnership on Measuring ICT for Development (2005). *Core ICT indicators.* Retrieved February 15, 2008, from www.itu.int/ITU-D/ICT/partnership/material/CoreICTIndicators.pdf

Peddibothla, N. B., & Subramani, M. R. (2007). Contributing to public document repositories: A critical mass perspective. *Organization Studies, 28*(3), 327-348.

Pei, M. (1994). *From reform to revolution: The demise of communism in China and the Soviet Union.* Cambridge, Mass: Harvard University Press.

Péjout, N. (2004). Big brother en Afrique du Sud? Gouvernement électronique et contrôle panoptique sous et après l'apartheid. In P. Guillaume, N. Péjout & A. Wa Kabwe Segatti (Ed.), *L'Afrique du Sud dix ans après: Transition accomplie?* (pp. 79-103). Johannesburg – Paris: Institut Français d'Afrique du Sud – Karthala.

Péjout, N. (2007). *Contrôle et contestation. Sociologie des politiques et modes d'appropriation des technologies de l'information et de la communication (TIC) en Afrique du Sud post-apartheid.* Unpublished doctoral dissertation, Ecole des Hautes Etudes en Sciences Sociales, Paris.

Peters, G. B. (1996). *The future of governing: Four emerging models.* Lawrence: University Press of Kansas.

Peters, G. B., and Pierre, J. (1998). Governance without government? Rethinking public administration. *Journal of Public Administration Research and Theory, 8*(2), 223-243

Piana, D. (2006). *Costruire la democrazia. Ai confini dello spazio pubblico europeo.* Novara: Liviana.

Posel, D. (2000). A mania for measurement: Statistics and statecraft in the transition to apartheid. In S. Dubow (Ed.), *Science and society in Southern Africa* (pp. 116-142). Manchester: Manchester University Press.

Preece, J. (1999). Empathic communities: Balancing emotional and factual communication. *Interacting with Computers, 12*(1), 63-77.

Prensky, M. (2001). *Digital game-based learning.* New York: McGraw-Hill.

Qiu, J. L. (2003). *The Internet in China: Data and issues.* Annenberg Research Seminar on International Communication.

Raboy, M., & Landry, N. (2004). *La communication au coeur de la gouvernance globale.* Enjeux et perspectives de la société civile au Sommet mondial sur la société de l'information. Montréal: Départment de Communication, Université de Montréal. Retrieved from http://www.lrpc.umontreal.ca/smsirapport.pdf. Last access May 31st, 2008.

Radaelli, C. (2000). Policy transfer in The European Union: Institutional isomorphism as a source of legitimacy. *Governance, 13*(1): 25-43.

Radaelli, C. (2000). Whither Europeanization? Concept stretching and substantive change. *European Integration Online Papers, 4*(8).

Radaelli, C. M. (2003). *The open method of coordination: A new governance architecture for the European Union?* Preliminary Report, SWEPS (Swedish Institute for European Policy Studies). Retrieved May 24, 2008 from: http://www.epin.org/pdf/RadaelliSIEPS.pdf

Rahman, A. (2006). *Access to global information. A case of digital divide in Bangladesh.* Northern University Bangladesh: Library and Information Division.

Realini, A. F. (2004). *G2G E-government: the big challenge for Europe.* Unpublished master thesis, University of Zurich, Zurich.

Reding, V. (2005). *Opportunities and challenges of the ubiquitous world and some words on Internet governance 2005.* European Commission.

Reinecke, W. H., & Deng, F. (2000). *Critical choices. The United Nations, networks and the future of global governance.* Ottawa: International Development Research Centre.

Reporters Without Borders, *The 2007 annual report—Dictatorships get to grips with Web 2.0.* Retrieved June 5, 2008, from http://www.rsf.org/rubrique.php3?id_rubrique=675

Reporters Without Borders. (2008, January 4). Retrieved January 12, 2008, from http://www.rsf.org/article.php3?id_article=24946

Reuters, A. (2007). *Europe takes lead in Second Life users,* Reuters Second Life News Center, February 9, Retrieved June 6, 2008, from http://secondlife.reuters.com/stories/2007/02/09/europe-takes-lead-in-second-life-users/.

Rheingold, H. (1993). *The virtual community: Homesteading on the electronic frontier.* Reading, MA: Addison Wesley

Ricci, A. (2000). Measuring information society. Dynamics of European data on usage of information and communication technologies in Europe since 1995. *Telematics and Informatics, 17,* pp. 141-167.

Riekmann Puntscher, S., Mokre, M., & Latzer, M. (2006). *The state of Europe: Transformation of statehood from a European perspective.* IL: Chicago University Press.

Roach, R. (2001). Distance education course to explore African and African-American art. *Black Issues in Higher Education,* April 12, (p. 46).

Roach, R. (2002). African students take MIT course without leaving home campuses, *Black Issues in Higher Education,* May 9, *47.*

Roberts-Witt, S. (2000). Unused fiber serves as springboard for broadband networking in this West African nation. *Internet World,* September 1, (p. 38–39).

Rodotà, S. (2004). *Tecnopolitica. La democrazia e le nuove tecnologie della comunicazione.* Bari-Roma: Laterza.

Rollnick, R. (2002, January 14). Telecommunication Summit: China Concerned at Electronic Threats to Moral Standards. *Earth Times News Service.* Retrieved November 23, 2002, from http://earthtimes.org/jan/telecommunicationchinajan14_02.htm

Rosenau, J. N. (1995). Governance and democracy in a globalizing wold. In D. Archbugi, D. Held, & M. Mohler, (Eds.), *Re-imagining political community: studies in cosmopolitan democracy* (pp. 28-57). Cambridge: Polity Press.

Rosenau, J. N. (2002). Information technologies and the skills, networks, and structures that sustain world affairs. In J. N. Rosenau & J. P. Singh (Eds.), *Information Technologies and Global Politics* (pp. 275-88). Albany: State University of New York Press.

Rymaszewski, M., Au, W. J., Ondrejka, C., Platel, R., Van Gorden, S., Cézanne, J., Cézanne, P., Batstone-Cunningham, B., Krotoski, A., Trollop, C., & Rossignol, J. (2008). *Second Life: The official guide.* Hoboken, NJ: Wiley.

Sadie, J. L. (1950). The political arithmetic of the South African population. *Journal of Racial Affairs, 1*(4), 3-8.

Saltzer, J. H., Reed, D. P., & Clark, D. D. (1981). *End-to-end arguments in system design.* M.I.T. Laboratory for Computer Science.

Sanders, L. (1997). Against deliberation. *Political Theory, 5*(25), 347-377.

Sartori, G. (1970). Concept misformation in comparative politics. *The American Political Science Review,* 64(4), 1033-1053.

Sassen, S. (2005). Electronic markets and activist networks: The weight of social logics in digital formations (pp.54-88). In R. Latham & S. Sassen (Eds.), *The Digital Formations: IT and New Architectures in the Global Realm.* Princeton: Princeton University Press.

Schedler, K. & Proeller, I. (2000). *New public management.* Bern, Stuttgart, Wien: Haupt.

Schelin, S. H. (2003). E-government: An overview. In G.D. Garson (ed) *Public information technology: Policy and management issues,* (p. 120-137). Hershey, PA: IGI Global Publishing.

Schiller, D. (2000). *Digital capitalism.* Cambridge, MA: MIT Press.

Schiller, H. I. (1973). *The mind managers.* Boston: Beacon.

Schiller, H. I. (1985). *Strengths and weaknesses of the new international information empire.* In P. Lee (Ed.), *Communication for All* (pp. 3-23), New York: Orbis.

Schneider, V. (2003, January). The transformation of the state in the digital age. Paper presented at the workshop *"The transformation of statehood from a European perspective",* Austrian Academy of Sciences, Vienna.

Schoenbaum, D. (1966). *Hitler's social revolution.* Garden City, NY: Doubleday.

Schroeder, R., & Bailenson, J. (2008 ,in press). Research uses of multi-user virtual environments. In R. Lee, N. Fielding, & G. Blank (Eds.), *The handbook of Internet research.* London: Sage.

Schulz, C., & Olaya, D. (2005). *Toward an information society measurement instrument for Latin America and the Caribbean: Getting started with census, household and business surveys.* Santiago, Chile: United Nations.

Schware, R. (Ed.) (2005). *E-development: From excitement to effectiveness.* Washington, DC: The World Bank Group.

Senne, D., & Engelbrecht, L. (2007). Home Affairs admits ID inefficiencies. Johannesburg: *ITWeb,* 30 August.

Servaes, J., & Heinderyckx, F. (2002). The 'new' ICTs environment in Europe: Closing or widening the gaps? *Telematics and Informatics, 19,* 91-115.

Sheff, D. (2002). *China dawn: The story of a technology sand business revolution.* New York: Harper Business.

Shoesmith, B., & Hearn, K. (2004). Exploring the roles of elites in managing the Chinese Internet. *Javnost: The Public,* 11(1), 101-14.

Shore, C. (2000). *Building Europe. The cultural politics of European integration.* Routledge: London.

SIBIS (2003). *New e-Europe indicator handbook.* European Commission. Retrieved September 23, 2007 from: http://www.sibis-eu.org/

Siedentopf, H., & Speer, B. (2003). The European administrative space from a German administrative science perspective. *International Review of Administrative Science,* 69(1): 9-28.

Singer, P. A., Salamnca-Buentello, F., & Daar, A. (2005). Harnessing nanotechnology to improve global equity. *Issues in Science and Technology, 4,* 57-64.

Singh, J. P. (2002). Introduction: Information technologies and the changing scope of global power and governance. In J. N. Rosenau & J. P. Singh (Eds.), *Information Technologies and Global Politics* (pp. 1-38). Albany: State University of New York Press.

Slob, G. (1990). *Computerizing apartheid: Export of computer hardware to South Africa.* Amsterdam: Holland Committee on Southern Africa.

Smith, S. (2001). Reflectivist and constructivist approaches to international relations. In J. Baylis and S. Smith (eds.), *The Globalization of World Politics. An introduction for international relations* (pp. 224-51). Cambridge: Cambridge University Press.

Smythe, D. (1994). *Counterclockwise: Perspective on communication.* Ed. Thomas Guback. Boulder, CO: Westview Press.

Sohmen, P. (2001). Taming the dragon: China's efforts to regulate the Internet. *Stanford Journal of East Asian Affairs*, *1*(1), 18.

South Africa Press Agency, 2004. More than half a million domestics registered with unemployment insurance fund (UIF). Pretoria: South Africa Press Agency, 2 June.

Sproull, L., Dutton, W., & Kiesler, S. (2007). Introduction to the Special Issue: On Line Communities. *Organization Studies*, *28*(3), 277-283.

Sraku-Lartey (2006). Developing the professional skills of information managers in the forestry sector in Africa. *IAALD Quarterly Bulletin*, *51*(2), 75-78.

Sraku-Lartey, M. (2003) The role of information in decision making in the forestry sector: Developing a computerized management information system (MIS) for forestry research activities in Ghana. *IAALD Quarterly Bulletin*, *48*(1), 105-108.

Sraku-Lartey, M. (2006). Building capacity for sharing forestry information in Africa. *IAALD Quarterly Bulletin*, *51*(3), 186-190.

Srivastava, S., & Teo, T. S. H. (2007). What facilitates e-government development? A cross-country analysis, Electronic Government. *An International Journal*, *4*(4), 365-378.

Stahl, B. C. (2005). The paradigm of e-commerce in e-government and e-democracy. In W. Huang & K. Siau.(Eds.), *Electronic government strategies.* Hershey, PA: IGI Global Publishing.

Statistics South Africa. (2002). *South African statistics.* Pretoria: Statistics South Africa.

Stehr, N. (1999). *Knowledge societies.* London: Sage.

Stehr, N. (2000). Deciphering information technologies. Modern societies as networks. *European Journal of Social Theory*, *3*(1), 83-94.

Stehr, N. (2001*). The fragility of modern societies. Knowledge and risk in the iInformation age.* London: Sage.

Stephenson, N. (1992). *Snow crash.* New York: Bantam Books.

Stewart, C. M., Gil-Egui, G. Y. T., & Pileggi, M. I. (2006). Framing the digital divide: A comparison of US and EU policy approaches. *New Media & Society*, *8*(5), 731-751.

Street, J. (1997). Remote control? Politics, technology and "electronic democracy". *European Journal of Communication*, *12*(l), 27-42.

Street, J. (2001). Electronic democracy. In N. J., Smelser & P.B. Baltes, *International encyclopedia of the social & behavioral sciences* (pp. 4397-4399). Amsterdam: Elsevier Science.

Strejcek, G., & Michael, T. (2002). Technology push, legislation pull? E-government in the EU. *Decision Support System*, 34: 305-313.

Suchman, L. (2005). Affiliative objects. *Organization*, *12*(3), 379-399.

Tagoe, N., Nyarko E., & Anuwa-Amarh, E. (2005). Financial challenges facing urban SMEs under financial sector liberalization in Ghana. *Journal of Small Business Management*, *43*(3), 331-343.

Tan, Z. (1999). Regulating China's Internet: Convergence toward a coherent regulatory regime. *Telecommunications Policy*, *23*(3), 261-76.

Taylor, J., & Burt, E. (2005). Voluntary organizations as e-democratic actors: political identity, legitimacy and accountability and the need for new research. *Policy and Politics*, *33*(4), 601-616.

Taylor, T. L. (2006). *Play between worlds: Exploring online game culture.* Cambridge, MA: MIT Press.

The Internet Timeline of China, Part II. *China Internet network information center*. Retrieved May 12, 2004, from http://www.cnnic.net.cn/html/Dir/2003/12/12/2001. htm

Tonn, B., & Feldman, D. (1995). Non-spatial government. *Futures, 27,* 11-36.

Trechsel, A. H., Kies, R., Mendez, F. & Schmitter, P. C. (2003). *Evaluation of the use of new technologies in order to facilitate democracy in Europe: E-democratizing the parliaments and parties of Europe (STOA research report).* Strasbourg, European Parliament.

Trechsel, A., Kies, R., Mendez, F., & Schmitter, P. (2003). *Evaluation of the use of new technologies in order to facilitate democracy in Europe, E-democratizing the parliaments and parties of Europe.* European University Institute and University of Genoa.

Triolo, P. S., & Lovelock, P. (1996). Up, up, and away—With strings attached. *The China Business Review, 23*(6), 28.

UN (United Nations). (2003). *Expanding public space for the development of the knowledge society.* Report of the Ad Hoc Expert Group Meeting on Knowledge Systems for Development. 4-5 September 2003. New York: United Nations.

UN General Assembly (2003). *Resolution 57/295.*

UNCTAD (2001). *Electronic commerce and development report 2001.* New York & Geneva: United Nations.

UNCTAD (2002). Trade and Development Report: Globalization, Distribution and Growth, Geneva: United Nations.

UNCTAD (United Nations Conference on Trade and Development). (2003, July 3). *Information society measurements: The case of e-business.* Background Paper by the Unctad Secretariat TD/B/COM.3/EM.19/2.

UNDP (1990). Human Development Report: Concept and Measurement of Human Development, New York, NY: Oxford University Press.

UNICEF (1987), *Adjustment with a Human Face,* Geneva: United Nations.

United Nation Department of Public Information, News and Media Division (2006). *SG/A/1006.PI1717.* Retrieved from http://www.un.org/News/Press/docs/2006/sga1006. doc.htm. Last access March 1st, 2008.

United Nations (2001). General Assembly Resolution 56/183 - 90th plenary meeting. Retrieved November 9, 2007, from http://www.wsis-pct.org/resol-56-183.html

United Nations (2001). *Globalization and the state.* Washington DC.

United Nations (2001). *Human development report.* New York.

United Nations (2005). *Global e-government readiness report 2005. From e-government to e-inclusion.* New York.

United Nations (2005). World public sector report 2005: Unlocking the human potential for public sector performance. New York: United Nations.

United Nations (2008). *Global e-government survey. From e-governmnet to connected governance.* New York.

United Nations General Assembly. *Resolution 56/183.* Retrieved from http://www.un.org/News/Press/docs/2006/ sga1006.doc.htm. Last access March 1, 2008.

US (United States of America) (2001). *Leadership for the new millennium, delivering on digital progress and prosperity, third annual report of the electronic commerce working group.* Department of Commerce. Retrieved September 23, 2007 from: http://www.lib. umich.edu/govdocs/stsci.html

van Ark, B., & Inklaar, R. (2005). *Catching up or getting stuck? Europe's troubles to exploit ICT's productivity potential.* GGDC, University of Groningen, September.

Van Der Berg, R. J. (2004). First ID for !Xhu and !Khwe communities, *Bua News,* Pretoria: Government Communication and Information System – Republic of South Africa, 9 June.

Van Dijk, J. (2006). *The network society.* London: Sage.

Van Tonder, K. (2003). Biometric identifiers and the right to privacy. *De Rebus*, 28 (8), August, Retrieved October 3, 2004, from http://www.derebus.org.za/archives/2003Aug/articles/Biometric.htm

Vastag, B. (2007). Virtual worlds, real science: Epidemiologists, social scientists flock to online world. *Science News*, *172* (October 27), 264-265.

Vecchiatto, P. (2007). Languishing HANIS needs attention. Johannesburg: *ITWeb*, 7 June.

Vecchiatto, P. (2007). SA trials smart ID cards. Johannesburg: *ITWeb*, 2 November.

Venables, A. J. (2001). *Geography and international inequalities: The impact of new technologies*. Retrieved April 12, 2008 from: http://siteresources.worldbank.org

Venturelli, S. (2002). Inventing e-regulation in the US, EU and East Asia: Conflicting social visions of the Information Society. *Telematics and Informatics*, *19*, 69-90.

Vesely, M. (2003). New technology for an old continent. *African Business*, July, 20-21.

Viereck, P. (1941). *Metapolitics: From the romantics to Hitler*. New York: A. A. Knopf

Viesti, G., & Prota, F. (2004). *Le politiche regionali dell'Unione Europea*. Bologna: Il Mulino.

Vinay, V. (1999). What is Free Software?. *Resonance*, 4(4), 1-6.

Vink, M. (2002, November). *What is Europeanization? and other Questions on a New Research Agenda*. Paper presented at Second YEN Research Meeting on Europeanization. Milan: Bocconi University.

Wacquant, L. J. D. (2004). *Punir les pauvres: Le nouveau gouvernement de l'insécurité sociale*, Marseille: Agone.

Wade, R. (1990). *Governing the market: Economic theory and the role of government in East Asian industrialization*. Princeton, NJ: Princeton University Press.

Wade, R. (1996). Japan, the World Bank, and the art of paradigm maintenance: The East Asian miracle in political perspective. *New Left Review*, I/217, 3-36.

Wade, R. (2002). Bridging the digital divide: New route to development or new form of dependency?. *Global Governance*, *8*(4), 443-466.

Wade, R. (2003). What strategies are viable for developing countries today? The World Trade Organization and the shrinking of 'development space'. *Review of International Political Economy*, *10*(4), 621-644.

Wade, R. H. (2005). Failing states and cumulative causation in the world system. *International Political Science Review*, *26*(1), 17-36.

Wagner, R. (1893). *The art-work of the future*. In *Richard Wagner's prose works* (vol. 1, pp. 69-213). London: K. Paul, Trench, Trübner.

Wajcman, J. (2002). Addressing technological change: The challenge to social theory. *Current Sociology*, *50*(3), 347–363.

Wajcman, J. (2006). New connections: Social studies of science and technology and studies of work. *Work, Employment and Society*, *20*(4), 773–786.

Wallerstein, I. (1991). *Unthinking social science: The limits of nineteenth century paradigms*. Cambridge: Polity.

Wallerstein, I. (2005). After developmentalism and globalization, what? *Social Forces*, *83*(3), 1263-1278.

Wallerstein, I. (2006). *European universalism: The rhetoric of power*. New York: The New Press.

Walton, G. (2001). *China's golden shield: Corporations and the development of surveillance technology in the People's Republic of China*. Retrieved June 2, 2008, from http://www.ichrdd.ca/english/commdoc/publications/globalization/goldenShieldEng.html

Walton, G. (2003, March 10). *Great Wall, Small World*. Congressional-Executive Commission on China (CECC). CECC Open Forum. Retrieved May 26, 2008, from http://www.cecc.gov/pages/roundtables/031003/walton.php

Warschauer, M. (2003). Demystifying the digital divide. *Scientific American, 289*(2), 42-47.

WB (World Bank) (2002). *The e-government handbook for developing countries.* Infodev, Centre for Democracy and Technology, November. Retrieved April 12, 2008 from: http://www.cdt.org/egov/handbook/2002-11-14egovhandbook.pdf

Weber, M. (1978). *Economy and society: An outline of interpretative sociology*, Berkeley: University of California Press.

Weber, S. (2005). The Political Economy of Open Source Software and Why it Matters (pp. 178-212). In Latham, R., & Sassen, S. (2005b). *The Digital Formations: IT and New Architectures in the Global Realm.* Princeton: Princeton University Press.

Webster, T. (2002). *Theories of the information society 2nd Edition.* London: Routledge.

Wellman, B. (2001). Computer networks as social networks. *Science, 293*(14), 2031-2034.

Wellman, B., & Giulia, M. (1999). Net Surfer do not ride alone. Virtual communities as communities. In Kollock and Smith, *Communities and cyberspace.* New York: Routledge.

Wendt, A. (1992) Anarchy is what states make of it: the Social construction of power politics. *International Organizations* 46: 391-425

Werlin, H. H. (1998.) *The mysteries of development: Studies using political elasticity theory.* Lanham, MD: University Press of America.

Werlin, H. H. (2003). Poor nations, rich nations: A theory of governance. *Public Administration Review, 63*(3), 329-342.

Werry, C., & Mowbray, M. (2001). *Online communities: Commerce community action, and the virtual university.* Upper Saddle River, N.J.: Prentice Hall.

WGIG (2005). *WGIG Final Report.* Retrieved from http://www.wgig.org/WGIG-Report.html. Last access March 1, 2008.

White House (1997). *A framework for global electronic commerce.* WashingtonDC.

Williams, D., Ducheneaut, N., Xiong, L., Zhang, Y., Yee, N., & Nickell, N. (2006). From tree house to barracks: The social life of guilds in *World of Warcraft. Games and Culture, 1*, 338-361;

Wimmer, M. A. (2001, October-November) *European development toward online one-stop government: The "e-gov" project.* Proceedings of the ICEC2001 Conference, Vienna.

Wimmer, M. A. (2002). A European perspective toward online one-stop government: The eGOV project. *Electronic Commerce Research and Applications, 1*: 92-103.

Wimmer, M., Traunmüller, R., & Lenk, K. (2001). Electronic business invading the public sector: Considerations on change and design. *Proceedings of the 34th Hawaii International Conference on System Sciences.* Maui (Usa): IEEE Press.

Wittel, A. (2001). Towards a network sociality. *Theory, Culture and Society, 18*(6), 51-76.

Woo-Cumings, M. (Ed.) (1999). *The developmental state.* Ithaca, NY: Cornell University Press.

World Bank (1980). *World development report, 1980.* New York: Oxford University Press.

World Bank (1999). *World development report: Knowledge for development.* New York: Oxford University Press.

World Bank (2002). *The handbook of e-government. The information for development program.* New York: infoDev

Wright, B. (2004). Telecoms around the continent. *African Business*, May, 16-17.

WSIS (2003). *Plan of action.* Retrieved April 16, 2008, from http://www.itu.int/dms_pub/itus/md/03/wsis/doc/S03-WSIS-DOC-0005!!MSW-E.doc

WSIS (2005). *Tunis agenda for the information society.* WSIS-05/TUNIS/DOC/6(Rev.1)-E

WSIS (World Summit on the Information Society). (2005). *Tunis agenda for the information society.* Retrieved April 18, 2008, from http://www.itu.int/wsis/docs2/tunis/off/6rev1.html

Wu, Guoguang. (2001). One head, many mouths: Diversifying press structures in reform China. In C. C. Lee (Ed.), *Power, money, and media: Communication patters and bureaucratic control in cultural China.* Evanston, Ill.: Northwestern University Press.

Xinhua News Agency. (2008, January 17). Retrieved January 20, 2008, from http://news.xinhuanet.com/newscenter/2008-01/17/content_7439151.htm

Yang, D. L. (2001). The great net of China. *Harvard International Review, 22*(4), 65.

Yildiz, M (2007) E-government research: Reviewing the literature, limitations, and ways forward. *Government Information Quarterly, 24*(3), 646-665.

Zachary, P. G. (2002) Ghana's digital dilemma, *Technological Review, July/August,* (pp. 66-73).

Zachary, P. G. (2003). A program for Africa's computer people. *Issues in Science & Technology,* Spring, *79.*

Zelnick, N. (2000). Colonialism? Not again. *Internet World, 6*(18), 15.

Zhang, J. (2003). Network convergence and bureaucratic turf wars. In C. R. Hughes & G. Wacker (Eds.), *China and the Internet: Politics of the digital leap forward* (p. 85). London and New York: Routledge Curzon.

Zhang, J., & Woesler, M. (Eds.). (2002). *China's digital dream: The impact of the Internet on Chinese society.* (pp. 35-67). Bochum: The University Press Bochum.

Zhao, Y. (1998). *Media, Market, and Democracy in China: Between the Party Line and the Bottom Line.* Urbana and Chicago: University of Illinois Press.

Zhengfu Shang-wang Gongcheng: Huigu yu Zhanwang (Government Online Project: Reviews and Prospects). Retrieved March 6, 2001, from http://www.gov.cn/govonlinereview/6future/01.htm

Zhou, R. (2002, December 4). Internet cafes percolating despite legal clampdown. *China Daily.* Retrieved January 16, 2003, from http://www3.chinadaily.com.cn/en/doc/2002-12/04/content_146503.htm

Zhou, Y. (2006). *Historicizing online politics: Telegraphy, the Internet, and political participation in China.* Stanford: Stanford University Press.

Zittrain, J. (May 2006). The generative Internet. *Harvard Law Review, 119,* 1974-2040.

Zittrain, J., & Edelman, B. (2003). *Internet filtering in China.* Harvard Law School Public Law. Research Paper, No. 62. (pp. 70-7). Retrieved January 24, 2008, from http://unpan1.un.org/intradoc/groups/public/documents/apcity/unpan011043.pdf

Zittrain, J., & Edelman, B. (2005, April 14). *Internet filtering in China in 2004-2005: A country study.* Retrieved June 6, 2008, from http://opennet.net/sites/opennet.net/files/ONI_China_Country_Study.pdf

About the Contributors

Francesco Amoretti is professor of Political Communication, and of E-democracy and E-Government Policies, University of Salerno, Graduate Degree Course in Communication Science. Since 1999 he is Member of the Directive Committee of the Political Communication Review. He has published journal articles in several areas, including social policies, administrative reforms, and mass media and political systems. Currently his interests focus broadly on new technologies and politics – e-democracy and e-government - communication policy, European public space, and cyberspace. Recent publications are in the Encyclopedia of Digital Government, Encyclopedia of Information Science and Technology, and in a issue of Review of Policy Research.

* * *

Clementina Casula is researcher in Economic Sociology at the Department of Economic and Social Research (DRES) at the Faculty of Education, University of Cagliari (Italy). She has published works on territorial cooperation, public administration, women and ICT. Her most recent research interests focus on the analysis of EU policies in the field of regional development, labour market, governance, information society, equal opportunities.

Chin-fu Hung holds a PhD in Politics and International Studies from the University of Warwick in the United Kingdom. He is currently Assistant professor in the Department of Political Science and Graduate Institute of Political Economy at the National Cheng Kung University in Tainan, Taiwan. His research interests include the political and economic transition of China, the impact of the Internet upon democratization, especially in authoritarian regimes, and the political economy of development in East Asia. He has written a number of English papers on the topics of cyber participation in democratic and (semi-)authoritarian states, including China, Taiwan, Singapore, and Malaysia. He has also published several articles in the journal of Issues & Studies on the impact of new information and communication technologies upon China's `virtual' politics (2003), state-business relations (2005), and the awareness of citizens' rights in the information age (2006). In addition, he has also written book chapters on the Cross-Taiwan's Strait relations in the Internet age (appeared in the book of Cyber China, edited by Françoise Mengin, 2004), as well as the journal articles on the subject of the politics of economic integration in East Asia (2007) and the changing concept of the Chinese governance in the age of the Internet and globalization (2008).

Enrico Gargiulio is PhD in sociology and social research at the Department of Sociology of the University of Naples "Federico II". He has published works on citizenship and on the global processes of social, political and economic exclusion. His most recent research interests focus on the new forms - sovranational and transnational - of citizenship, the relations between citizenship and migratory dynamics, the theoretical and material links between the construction of the image of the modern citizen and the construction of its mirror image, the non-citizen, the study of transnational élites and the critical analysis of the process of individualization

Diego Giannone is a PhD in "sociology and social research". He defended his doctoral dissertation, titled "*Modelli di misurazione e diffusione della democrazia. L'Unione Europea nel contesto internazionale*" *[Models of measurement and spreading democracy. The European Union in the internationl context]*, on April 8, 2008 at the Department of Sociology and Political Science of the University of Salerno. Currently he is working with the chair of Contemporary History of Europe. His major themes and research interests include: the process of democratization, measurement of democracy, the process of European integration, new technologies and new rights. Among his publications: "*Misurare la democrazia: strumenti vecchi, problemi nuovi*" *["Measuring Democracy: old tools, new problems"]*, in Associazione Italiana di Sociologia (ed.), Giovani sociologi 2007, Naples: Scriptaweb.

Paolo Landri is a researcher at the Institute of Research on Population and Social Policies at the National Research Council of Italy (CNR-IRPPS). His main research interests concern educational organizations and policies. Currently, he is studying the relationship between knowledge, learning and practice in organizations. He teaches qualitative methods at the Faculty of Sociology of the University of Naples "Federico II". He has published "*The pragmatic of passions. A sociology of attachment to mathematics*" (Special Issue – The passion of knowing and learning, 2007, Organization), and recently "*Challenges of educating European Managers of Lifelong Learning*" (with Armstrong, S.J., Ponzini, G. & Thursfield, D.) in Wankel, C. & DeFilippi, R. (eds), "*University and Corporate Innovations in Lifetime Learning*", *Research in Management Education and Development, vol. 6, Information Age Publishing, 2008.*

Fortunato Musella has got PhD in political science at the University of Florence, with a dissertation thesis dedicated to the Italian regional governments, and he is currently researcher at the University of Naples. His main research interests embrace (a) the study of government, (b) elections and personalization of vote, (c) new technologies and politics. His recent publications include the volume "*Governo monocratico. La svolta presidenziale delle regioni italiane*" (Bologna, il Mulino, 2008), as well as several articles and book's chapters appearing in *Quaderni di Scienza Politica, Quaderni dell'Osservatorio Elettorale, Polis, Comunicazione Politica*. Recently he has written with Francesco Amoretti some entries for the *Encyclopedia of Information Science and Technology* edited by Mehdi Khosrow-Pour (Hershey, IGI Global publications, 2008).

Joseph Ofori-Dankwa holds a PhD from Michigan State University and is currently a professor of management at Saginaw Valley State University. His research interests center on identifying firm performance indicators in developing economies, and also in developing integrative research models using the "diversimilarity" theoretical approach, a term which he first utilized in 1996.

Connie Ofori-Dankwa is a BBA student at the Ross School of Business at the University of Michigan, and an associate editor of the University of Michigan's academic journal, The Michigan Journal of Business. Her areas of interest include management and international business.

Claudia Padovani is assistant professor of political science and international relations at the Department of Historical and Political Studies at the University of Padova, Italy. She teaches international communication as well as institutions and governance of communication while conducting research in the fields of the global governance of the information and knowledge society. She has written extensively about the UN promoted World summit on the Information Society process, with a specific interest in the multi-stakeholder approach, the role of civil-society actors and the historical legacies of international debates concerning communication imbalances. She has organized a number of WSIS-related academic symposiums and guest-edited a series of special issues of academic journals focused on discussing WSIS/IGF within the broader perspective of communication issues in the 21st century.

Elena Pavan is a PhD candidate at the Department of Sociology and Social Research, University of Trento, Italy. Her research focuses on social and issue networks in the context of Internet governance debates, with a particular focus on the UN promoted Internet Governance Forum and mobilization networks dealing with media reform and communication rights. She has published and presented research related to those issues in various venues.

Nicolas Pejout is associate researcher at the Centre d'Etude d'Afrique Noire (CEAN, Bordeaux, France). His current research focuses on the development of the information society in developing countries, particularly in Africa. He is investigating three main dynamics: the emergence of e-government and e-politics; the development of an ICT-based "new economy"; the expansion of online social networks. Looking at public policies and everyday usages, he is questioning ICTs as tools that facilitate both, on the one hand, domination, control and the reproduction of power structures and, on the other hand, social change in favour of dominated individuals and groups. He received his PhD in development socio-economics from the Ecole des Hautes Etudes en Sciences Sociales (EHESS, Paris, France) in 2007. His dissertation is a study of the South African information society for the last 50 years.

Mauro Santaniello is PhD in communications sciences at the University of Salerno. His research interests focuse on Internet governance and e-democracy policies. Among his publications the review article *"Progetti on-line di Democrazia Elettronica"* in *Comunicazione Politica* (2006) and, with Francesco Amoretti, the entry *"Community of Production: Dual reality of Social Software"* in the *Encyclopedia of Multimedia and Networking*, Second Edition (Idea Group Inc., 2008).

Oreste Ventrone is professor of Sociology at the University of Naples Federico II, Graduate Degree Course in Sociology. His main teaching and research focus on: social theory and methodology; historical sociology; global political economy; world-systems analysis, with a special focus on global processes and institutions. Recent works include *Globalizzazione: Breve storia di un'ideologia, Milano,* Franco Angeli, 2004; (Ed.) S Napoli, Gesco edizioni, 2007.

William Sims Bainbridge is the author of 14 books, 4 textbook-software packages, and about 200 shorter publications in the social science of technology, information science, and culture. *Goals in Space* was a questionnaire study of motivations for space exploration, and *Dimensions of Science Fiction* explored popular conceptions of the future in space. In 2006 he published *God from the Machine, applying artificial intelligence techniques to understand religious cognition,* and he has just published *Across the Secular Abyss and Nanoconvergence* about the tensions between religion, cognitive science, social science, and emerging technologies. Among recent projects are editing *The Encyclopedia of Human Computer Interaction* (2004) and co-editing *Nanotechnology: Societal Implications - Improving Benefits for Humanity* (2006) and *Managing Nano-Bio-Info-Cogno Innovations: convergine Technologies in Society* (2006). He represented the social sciences on five advanced technology initiatives: High Performance Computing and Communications, Knowledge and Distributed Intelligence, Digital Libraries, Information Technology Research, and Nanotechnology, and he represented computer and information science on the Nanotechnology and Human and Social Dynamics initiatives.

Index